Synthese Library

Studies in Epistemology, Logic, Methodology, and Philosophy of Science

Volume 433

The aim of *Synthese Library* is to provide a forum for the best current work in the methodology and philosophy of science and in epistemology. A wide variety of different approaches have traditionally been represented in the Library, and every effort is made to maintain this variety, not for its own sake, but because we believe that there are many fruitful and illuminating approaches to the philosophy of science and related disciplines.

Special attention is paid to methodological studies which illustrate the interplay of empirical and philosophical viewpoints and to contributions to the formal (logical, set-theoretical, mathematical, information-theoretical, decision-theoretical, etc.) methodology of empirical sciences. Likewise, the applications of logical methods to epistemology as well as philosophically and methodologically relevant studies in logic are strongly encouraged. The emphasis on logic will be tempered by interest in the psychological, historical, and sociological aspects of science.

Besides monographs *Synthese Library* publishes thematically unified anthologies and edited volumes with a well-defined topical focus inside the aim and scope of the book series. The contributions in the volumes are expected to be focused and structurally organized in accordance with the central theme(s), and should be tied together by an extensive editorial introduction or set of introductions if the volume is divided into parts. An extensive bibliography and index are mandatory.

More information about this series at http://www.springer.com/series/6607

Anton Killin • Sean Allen-Hermanson
Editors

Explorations in Archaeology and Philosophy

Springer

Editors
Anton Killin
Department of Philosophy
Mount Allison University
Sackville, NB, Canada

School of Philosophy and Centre of
Excellence for the Dynamics of Language
Australian National University
Acton, ACT, Australia

Sean Allen-Hermanson
Department of Philosophy
Florida International University
Miami, FL, USA

Chapter 3, "Mortar and Pestle or Cooking Vessel? When Archaeology Makes Progress Through Failed Analogies", authored by Dr. Rune Nyrup, is published Open Access with funding of the Wellcome Trust.

This work was supported by the Wellcome Trust [grant no. 213660/Z/18/Z] and the Leverhulme Trust through the Leverhulme Centre for the Future of Intelligence [RC-2015-067].

ISSN 0166-6991   ISSN 2542-8292   (electronic)
Synthese Library
ISBN 978-3-030-61054-8   ISBN 978-3-030-61052-4   (eBook)
https://doi.org/10.1007/978-3-030-61052-4

This Springer imprint is published by the registered company Springer Nature Switzerland AG
The registered company address is: Gewerbestrasse 11, 6330 Cham, Switzerland

Contents

About the Editors

Anton Killin is a postdoctoral fellow in the School of Philosophy and Centre of Excellence for the Dynamics of Language at the Australian National University, and the Department of Philosophy at Mount Allison University in Sackville, NB, Canada. Anton's primary areas of research include philosophy of the sciences, philosophy of the arts (especially music), and evolutionary theory. Anton has published on the evolution of music and language and, with Adrian Currie, on cognitive archaeology.

Sean Allen-Hermanson is professor of philosophy at Florida International University in Miami, FL, USA. Sean's primary areas of research include philosophy of mind and cognitive science. Sean has published on extended cognition, cultural evolution, comparative psychology, and implicit bias.

Contributors

Sean Allen-Hermanson Department of Philosophy, Florida International University, Miami, FL, USA

Carl Brusse Department of Philosophy and Charles Perkins Centre, University of Sydney, Sydney, NSW, Australia

School of Philosophy, Australian National University, Acton, ACT, Australia

Murray Clarke Department of Philosophy, Concordia University, Montreal, QC, Canada

Fabio Di Vincenzo Dipartimento di Biologia Ambientale, Sapienza Università di Roma, Rome, Italy

Caleb Everett Department of Anthropology, University of Miami, Miami, FL, USA

Marilynn Johnson Department of Philosophy, University of San Diego, San Diego, CA, USA

Anton Killin School of Philosophy and Centre of Excellence for the Dynamics of Language, Australian National University, Acton, ACT, Australia

Department of Philosophy, Mount Allison University, Sackville, NB, Canada

Mark Nielsen School of Psychology, University of Queensland, Brisbane, QLD, Australia

Rune Nyrup Department of History and Philosophy of Science and Leverhulme Centre for the Future of Intelligence, University of Cambridge, Cambridge, UK

Ronald J. Planer School of Philosophy, Australian National University, Acton, ACT, Australia

Artur Ribeiro Institute of Prehistoric and Protohistoric Archaeology, University of Kiel, Kiel, Germany

Bruce Routledge Department of Archaeology, Classics and Egyptology, University of Liverpool, Liverpool, UK

Elizabeth Scarbrough Department of Philosophy, Florida International University, Miami, FL, USA

Armin W. Schulz Department of Philosophy, University of Kansas, Lawrence, KS, USA

Ceri Shipton Centre of Excellence for Australian Biodiversity and Heritage, College of Asia and the Pacific, Australian National University, Acton, ACT, Australia

Institute of Archaeology, University College London, London, UK

Derek D. Turner Department of Philosophy, Connecticut College, New London, CT, USA

Michelle I. Turner Crow Canyon Archaeological Center, Cortez, CO, USA

Chapter 1
The Twain Shall Meet: Themes at the Intersection of Archaeology and Philosophy

Anton Killin and Sean Allen-Hermanson

Abstract *Explorations in Archaeology and Philosophy* grew out of an interdisciplinary conference on the Upper Palaeolithic, "Digging Deeper: Archaeological and Philosophical Perspectives", held on Miami Beach, Florida, in December 2017. The previous decade had seen increasing numbers of publications on topics of interest to both philosophers and archaeologists, so the time was ripe for a conference which served to generate constructive dialogue between researchers from both disciplines. Themes discussed included art, music, the mind, symbols, mortuary practices, and archaeological methodology. This volume draws from several contributors to that conference as well as other leading researchers at the forefront of the intersection of archaeology and philosophy. Its scope is expanded from the Upper Palaeolithic to archaeological research in general—adopting a broad conception inviting many perspectives from philosophy and interdisciplinary evolutionary studies to the table—and is organized by topics into four sections: Theory and Inference; Interdisciplinary Connections; Cognition, Language and Normativity; Ethical Issues.

Keywords Archaeology · Philosophy · Theory and inference · Interdisciplinary evolutionary studies

A. Killin (✉)
Department of Philosophy, Mount Allison University, Sackville, NB, Canada

School of Philosophy and Centre of Excellence for the Dynamics of Language, Australian National University, Acton, ACT, Australia
e-mail: anton.killin@anu.edu.au

S. Allen-Hermanson (✉)
Department of Philosophy, Florida International University, Miami, FL, USA
e-mail: hermanso@fiu.edu

© Springer Nature Switzerland AG 2021
A. Killin, S. Allen-Hermanson (eds.), *Explorations in Archaeology and Philosophy*, Synthese Library 433, https://doi.org/10.1007/978-3-030-61052-4_1

This volume showcases the heterogeneity of topics discussed where philosophy and archaeology overlap, digging into a sample of themes of interest to researchers from both disciplines as well as interdisciplinary evolutionary, prehistorical, and cognitive research more generally. Our aim is to flag this diversity by presenting the reader with a wide array of topics, subdisciplines, and perspectives, rather than a more streamlined though narrower examination of, say, philosophy of archaeology conceived as a branch of analytic philosophy of science (though, this will feature, of course!). Some contributors home in on direct archaeological issues; for others, the archaeological record provides data for testing ideas, requiring explanation, or simply producing inspiration, while philosophical or cognitive matters are emphasized. All that said, this volume is far from exhaustive and only represents a slither of the fascinating interdisciplinary research it aims to put on display. As a result, several important areas of research are not directly represented, such as Indigenous and community archaeology, and this is suggestive of future work.

We begin with three chapters collected under the banner of Theory and Inference. The opener, Derek Turner and Michelle Turner's "'I'm Not Saying it was Aliens': An Archaeological and Philosophical Analysis of a Conspiracy Theory", puts the popular 'ancient aliens' narrative under the microscope. Their analysis not only aims to show why this conspiracy theory doesn't work, it highlights the epistemically problematic and ideological heavy-lifting that this popular theory seems to have achieved.

Despite some skepticism over its merits, the method of analogy is a central piece in the archaeologists' theoretical toolkit. Responding to that skepticism, Rune Nyrup's case study of the ancient Roman mortarium in "Mortar and Pestle or Cooking Vessel? When Archaeology Makes Progress through Failed Analogies" shows that even when analogical reasoning misses its mark, the method can provide epistemic goods. Nyrup argues that since one aim of archaeological inquiry is to provide a clearer understanding of the difficulties involved in interpreting the past, various inferential strategies involving analogies can shed real light.

In "Scaffolding and Concept-Metaphors: Building Archaeological Knowledge in Practice", Bruce Routledge traces the use of the concepts of scaffolding and concept-metaphor in archaeological theorizing and defends the role they play in the analysis of knowledge formation. He draws attention not only to the more recent work of the likes of Alison Wylie and Adrian Currie, but also the classic work of V. Gordon Childe and his functional-economic analysis of the Three-Age System (Stone Age, Bronze Age, Iron Age) as a once useful scaffold which is (arguably) ultimately redundant.

The next three chapters emphasize wider interdisciplinary connections. Armin Schulz's "Human Curiosity Then and Now: The Anthropology, Archaeology, and Psychology of Patent Protections" probes an interdisciplinary sample of data for evidence of humanity's deep sense of inquisitiveness and innovation, a sense which Schulz argues is fueled not primarily by the prospect of external material gains. For Schulz, humans are typically internally motivated to seek novel information. This perspective is contrasted with economics-based hypotheses of human innovative activity. It is brought to bear on theorizing about contemporary patent regimes, aiming to undermine unreflective confidence in those regimes.

Humans are also musical creatures, as Schulz considers in one of his examples. The use of bird bone and mammoth ivory in flute construction dates back, as far as we know, to around 40,000 years ago. In "Music Archaeology, Signaling Theory, Social Differentiation", Anton Killin situates discussion about the significance of mammoth ivory as a raw material for flute production in the context of the archaeological application of biological signaling theory more generally. Whereas for Schulz, ancient musical instruments are one result of curious experimentation in humanity's past, Killin's discussion suggests a compatible, though distinct, conclusion: that mammoth ivory flutes are part of a category of artifacts signaling social differentiation in Upper Palaeolithic human life.

Carl Brusse's "The Archaeology and Philosophy of Health: Navigating the New Normal Problem" concludes this section with a discussion of assessing the health of past human populations via the material record. Brusse outlines the inadequacies for this job of one prominent naturalistic theory of health, Christopher Boorse's biostatistical health concept. Brusse goes on to explore the prospects of a biological functional theory, and although he suggests that this account's merits place it ahead of Boorse's, he also says that this approach too has its limits.

Next up is the Cognition, Language and Normativity section. Marilynn Johnson and Caleb Everett, in their "Embodied and Extended Numerical Cognition", explore the nature and origins of human numeral cognition through the philosophical framework of theories of embodied and extended mind. According to these theories, one's mind encompasses more than 'what's between the ears', to include mindedness resulting from the coupling of interior neural resources to external cognitive supports, including bodily movements as well as tools and other features of the environment. These authors provide a cross-linguistic analysis of numeracy, and they probe the archaeological record for the earliest evidence currently known of the numerical abilities of ancient hominins.

Captivated by the incredible story of human origins and dispersal, Murray Clarke sets out to explain how language, global migration, and complex artistic innovation and technological achievement might have been possible in "Late Pleistocene Dual Process Minds". Integrating a Two-Systems theory of mind (a view of the human mind as consisting of a fast, reflective, automatic thinking system and a slow, deliberate, conscious thinking system) with insights from Leda Cosmides' and John Tooby's Evolutionary Psychology research program, Clarke argues that modular mental adaptations in concert with upgraded domain-general reasoning capacities provided our ancestors the cognitive wherewithal for global expansion and complex cultural evolution, underwriting the great story the archaeological record details. Clarke contrasts his view against less nativist, gene-culture coevolutionary views as well as the anti-nativist, cultural evolutionary psychology picture of Cecilia Heyes.

Like Clarke, Ronald Planer adopts the Two-Systems theory of mind in his "Theory of Mind, System-2 Thinking, and the Origins of Language". Planer aims to provide an explanation of how 'mindreading' (the ability to attribute mental states to other agents) developed in the hominin line compared to its great ape baseline.

Planer's hypothesis is that mindreading was not upgraded due to genetic selection acting on a domain-specific module shared with other apes, but rather as a result of upgrades in working memory and domain-general reasoning capacities. Planer probes the developmental and comparative literature for supporting evidence, and he distinguishes his project from that of other cognitive archaeologists interested in working memory such as Thomas Wynn and Frederick Coolidge.

In "The Acheulean Origins of Normativity", Ceri Shipton, Mark Nielsen, and Fabio Di Vincenzo assess the stone tool record for evidence of overimitation—the human bias for replicating even causally superfluous actions when reenacting the action sequence of a demonstrator—which these authors link with the origins of normativity. They argue that such a capacity would have been adaptive in a wide range of contexts including large-group living and large-game hunting.

We then turn to ethical issues. Artur Ribeiro's "Social Archaeology as the Study of Ethical Life: Agency, Intentionality, and Responsibility" advances a conception of agency based on ethical normativism; a concept-metaphor of agency as an historical, social institutional product, rather than a property of individual humans or groups of humans. Ribeiro proposes a social archaeological research program examining how, throughout history, societies and their institutions restrict or endow individual freedoms and thus influence the moral responsibilities of individuals in practice.

Finally, in "Are Archaeological Parks the New Amusement Parks? UNESCO World Heritage Status and Tourism", Elizabeth Scarbrough considers the ethical impact of the phenomena of 'UNESCO-cide' and 'Disneyfication' which plague cultural heritage tourism, that is, overtourism and its resulting harms to the site and local populations, and the sanitizing of the cultural experience of the site. She provides two cases studies—Angkor Archaeological Complex, Cambodia, and George Town, Malaysia—drawing attention to features of these two phenomena at these sites, and she evaluates potential strategies for alleviating the problems. Her conclusion is sobering; the kind of tourism which has financially supported the preservation of much cultural heritage may be the very thing threatening its continued preservation.

The breadth of topics and depth with which our authors have investigated them greatly excites us and, we hope, showcases a thriving interdisciplinary intersection. For this, we're extremely grateful to our authors for their contributions. A generous donation to Florida International University from archaeology and philosophy enthusiast Allan Wesler enabled a year-long research project (hosted by FIU's philosophy department) and an interdisciplinary conference; this volume is one output of that project. We thank Otávio Bueno for his interest and support, an anonymous referee for their valuable and constructive feedback, and, in addition to our authors, a host of archaeologists and philosophers for very fruitful discussions, including Fred Adams, Adrian Currie, John Darnell, Colleen Manassa Darnell, Gwyn Davies, Tony Dennis, Francesco d'Errico, Edouard Machery, Jesse Prinz, and Mary C. Stiner.

Part I
Theory and Inference

Chapter 2
"I'm Not Saying It Was Aliens": An Archaeological and Philosophical Analysis of a Conspiracy Theory

Derek D. Turner and Michelle I. Turner

Abstract This chapter draws upon the archaeological and philosophical literature to offer an analysis and diagnosis of the popular 'ancient aliens' theory. First, we argue that ancient aliens theory is a form of conspiracy theory. Second, we argue that it differs from other familiar conspiracy theories because it does distinctive (and distinctively problematic) ideological work. Third, we argue that ancient aliens theory is a form of *non-contextualized inquiry* that sacrifices the very thing that makes archaeological research successful, and does so for the sake of popular accessibility. Rather than merely dismissing ancient aliens as 'pseudoarchaeology' on demarcationist grounds, we offer a more complicated account of how the theory works, and what ideological work it does.

Keywords Aliens · Archaeology · Conspiracy theory · Demarcation · Pseudoarchaeology

2.1 Introduction

According to what we will call the 'ancient aliens' narrative, extraterrestrials visited Earth long ago and interacted with people in ways that help explain the material remains studied by archaeologists today, as well as various myths and scriptural

D. D. Turner (✉)
Department of Philosophy, Connecticut College, New London, CT, USA
e-mail: dtur@conncoll.edu

M. I. Turner
Crow Canyon Archaeological Center, Cortez, CO, USA

© Springer Nature Switzerland AG 2021
A. Killin, S. Allen-Hermanson (eds.), *Explorations in Archaeology and Philosophy*, Synthese Library 433, https://doi.org/10.1007/978-3-030-61052-4_2

texts.[1] Popular sources such as Erich von Däniken's 1970 book, *Chariots of the Gods?* and The History Channel's *Ancient Aliens* series have helped to establish the ancient aliens theory in the broader culture, much to the chagrin of professional archaeologists.[2] Indeed, purveyors of the ancient aliens narrative often portray professional archaeologists as closed-minded and turf-protective, unwilling or unable to acknowledge the evidence of ancient aliens that is staring them in the face. Sometimes they portray archaeologists as conspiring together to keep the public in the dark. Some professional archaeologists, in turn, have taken great care to explain why the evidence simply does not support the ancient aliens narrative (see especially Feder 2020). There has also been considerable debate among professional archaeologists with respect to whether and how to engage with ancient aliens theory and other forms of 'alternative archaeology' (Anderson 2019; Fagan and Feder 2006; Feder 1984, 2016; Feder et al. 2016; Holtorf 2005; Derricourt 2012; Holly 2015; Card and Anderson 2016; Wilson 2012).

Scholars may scoff, but ancient aliens theory is a significant cultural phenomenon, one that overlaps with familiar science fiction narratives while tapping into broader interest in UFOs. In January 2020, the *Ancient Aliens* series premiered its 15th season. Its website lists 177 episodes (History Channel 2020). Its viewership, though waning after so many seasons, remains sizeable; the September 6, 2019 episode had over a million live viewers according to Nielsen ratings (Rejent 2019). Ten thousand people paid to attend the 2018 AlienCon conference to meet personalities from the show (Kurutz 2018).[3] For many viewers, it is likely that this show or similar sources are the first place they heard about a particular archaeological site, and perhaps their only source of information about it. As the popular "I'm not saying it was aliens…" meme implies, the show functions less by providing a coherent narrative of the aliens' activities and motives than by sowing doubt about whether mainstream explanations are really sufficient to explain human accomplishments in the past. Viewers might say they watch it for its entertainment value, but after 14 years of questioning mainstream archaeology, the show has built a following of true believers. It is part of a general trend. Chapman University's (2018) Survey of American Fears has documented a rising wave of belief in ancient aliens. In 2018, 41.4% of participants agreed with the statement "aliens have visited Earth in our ancient past," up from 27% in 2016. Respondents who agree with this statement might not necessarily think that aliens interacted with or shared technology with ancient people, and 2016–18 is a fairly narrow time interval; nevertheless, the

---

[1] We will sometimes refer to the 'ancient aliens narrative', with the idea that it is the sort of narrative explanation that is quite common in historical science. At times, we'll refer to 'ancient aliens theory', though we do not mean to use 'theory' as an honorific term. Calling something a theory need not imply that it is scientific or well-supported.

[2] For some discussion of the history of ancient aliens theory, see Grünschloss (2007), and for more on its place in the popular media, see Parker (2016).

[3] The AlienCon conference reminds one of Koertge's (2013) contrast between different kinds of social and institutional set-ups that characterize the scientific community vs. other communities that might look like science but aren't.

results do at least suggest that some version(s) of the ancient aliens narrative could be growing in popularity. Archaeologists who work with the public routinely field questions about ancient aliens. And it is far from a harmless idea; watchdog groups have noted an overlap between people who espouse ancient alien narratives and those who espouse white supremacy (Zaitchik 2018).

In this chapter, we offer a diagnosis of ancient alien narratives that draws upon insights from both the archaeological and the philosophical literature. Philosophers who write about pseudoscience have not engaged much with the archaeological literature or with issues specific to archaeology (see, e.g., the papers collected in Pigliucci and Boudry 2013, which scarcely mention archaeology). And philosophers who have written on the epistemology of historical reconstruction have not been sensitive to the distinctive ideological functions that certain historical narratives can play in archaeological contexts (see, e.g., Cleland 2002; Turner 2007; or Currie 2018). On the other hand, archaeologists such as Derricourt (2012) who are interested in the appropriateness of the term 'pseudoarchaeology' do not engage with the extensive philosophical literature on the demarcation problem, especially as it pertains to other popular challenges to historical science, such as creationism and intelligent design theory (e.g., Kitcher 1983; Ruse 1988; Pigliucci and Boudry 2013). And although some archaeologists have observed that ancient aliens theory is a bit like a conspiracy theory, they seem unaware of relevant philosophical discussion of the structure of conspiracy theories (Keeley 1999, 2003; Basham 2003).[4]

We offer an analysis of ancient alien narratives that combines ideas from both disciplines. From the direction of philosophy, we draw upon Brian Keeley's (1999) analysis of conspiracy theories, and we argue that the ancient aliens theory is a good example of a conspiracy theory. Consider Keeley's first pass at a definition of conspiracy theory:

> What is a conspiracy theory? A conspiracy theory is a proposed explanation of some historical event (or events) in terms of the significant causal agency of a relatively small group of persons—the conspirators—acting in secret (Keeley 1999, p. 116).

The ancient aliens theory fits this description quite well, with the extra-terrestrial visitors playing the role of conspirators who are carrying out a specific plan with respect to Earth and/or humanity. Although the aliens themselves are the primary conspirators, proponents of the ancient aliens theory often also cast professional archaeologists as (possibly unwitting) abettors of the conspiracy. There are some differences between the ancient aliens theory (in its various versions) and other paradigmatic conspiracy theories, but we argue in Sect. 2.2 that the ancient aliens theory is indeed best understood as a conspiracy theory.

---

[4] In this chapter, we draw primarily on the philosophical discussion of conspiracy theory, especially Keeley (1999). However, researchers in other disciplines have had much to say on the topic. See Hofstadter (1964) for one classic historical discussion of conspiracy theory in the American context. From a political science perspective, Barkun (2013) traces the recent explosion of conspiracy theory in American cultural and political life. And see Harambam and Aupers (2015) for interesting sociological research on conspiracy theory.

Keeley argues that conspiracy theories are popular "because they exhibit several well-known explanatory virtues" (Keeley 1999, p. 119). Many philosophers of science think of explanatory unification as a theoretical virtue, and it is easy to point to examples of scientific theories that did this well: famously, Darwin's theory provided a unified explanation of adaptation, homologous traits, and the biogeographical distribution of populations. Historical conspiracy theories, Keeley observes, also have great unifying power. Proponents of ancient alien narratives have used them to explain everything from the Nazca lines, to the Biblical story of Ezekiel and the wheel, and even the extinction of the dinosaurs.

While we agree that some of the appeal of the ancient aliens theory is a matter of its unifying explanatory power, we do not think that this is the whole story. Here archaeologists have had much more to say about the ideological function(s) of ancient aliens, and we explore some of those ideas in Sect. 2.3. Where philosophers of science have focused a great deal on the nature and testability of explanatory narratives, archaeologists have been far more sensitive to the ways in which historical narratives often get conscripted to do political and ideological work.

In short, our analysis of ancient aliens theory focuses on two questions:

1. How does the ancient aliens theory work?
2. What work does the theory do?

Our answer to the first question is that it works like a conspiracy theory. Our answer to the second question is that in addition to the explanatory work that one might expect of a conspiracy theory, ancient aliens narratives do a great deal of ideological heavy lifting—for example, they diminish the cultural achievements of non-European societies, and at the same time they offer reinterpretations of religious narratives that seem to retain much of the religious content while jettisoning the supernatural metaphysical commitments. The gods, it is alleged, are extra-terrestrial but still 'this-worldly' beings. Our central claims are that these two questions need to be addressed in tandem, and that the best answers draw upon insights from both archaeology and philosophy of science.

In Sect. 2.4, we build on this analysis and develop a fuller diagnosis of ancient aliens theory. There we draw upon recent work in the philosophy of historical and archaeological research (Chapman and Wylie 2016; Currie 2018). One theme of that work is that successful archaeological investigation has everything to do with epistemic context. Background knowledge of sites, regions, and methods is paramount. Archaeologists make progress by leveraging things that they already know. And this is precisely what ancient aliens enthusiasts refuse to do. Ancient aliens theory exemplifies what we will call *non-contextualized inquiry*. This very defect, however, helps explain the popularity of ancient aliens, because the refusal to engage with or to leverage epistemic context also makes ancient aliens thinking accessible to non-specialists.

2.2 How the Ancient Aliens Theory Works

According to Keeley (1999), a conspiracy theory offers a potential explanation of some historical event(s) in terms of the secret activities of a group of conspirators. The conspirators, he observes, need not be all-powerful. And the group of conspirators is typically fairly small. In order for there to be a conspiracy at all, there must be many of us who are not in on it. Keeley then seeks to characterize what he calls UCTs or unwarranted conspiracy theories. Noting that some conspiracy theories (like the Watergate conspiracy) turn out to be both warranted and true, he argues that there is nevertheless a pattern of features that co-occur in UCTs. These include:

> (1) A UCT is an explanation that runs counter to some received, official, or 'obvious' account.
> (2) The true intentions behind the conspiracy are invariably nefarious.
> (3) UCTs typically seek to tie together seemingly unrelated events.
> (4) [T]he truths behind the events explained by conspiracy theories are typically well-guarded secrets, even if the ultimate perpetrators are sometimes well-known public figures.
> (5) The chief tool of the conspiracy theorist is what I shall call *errant data* (Keeley 1999, pp. 116–117).

Ancient aliens theory very obviously has features (1) and (3). Proponents of ancient aliens, such as the 'ancient astronaut theorists' who appear as talking heads in The History Channel's series, often present it as an alternative to the official accounts of 'mainstream' archaeology. What's more, the structure of each episode is designed to highlight the way in which the ancient aliens theory offers a surprisingly unified account of seemingly unrelated phenomena, such as archaeological sites on different continents, or religious narratives from different cultural traditions.

The way in which ancient aliens theory ties together seemingly unrelated events deserves a bit more careful scrutiny. Some philosophers of science have pointed out that historical scientists often give *common cause explanations* of disparate historical traces (Cleland 2002, 2011). Indeed, Carol Cleland argues that 'prototypical historical science' involves formulating and testing common cause explanations of trace patterns. The thought is that where two things, A and B, are correlated, A could be the cause of B, or B could be the cause of A. A third possibility is that there is a distinct prior common cause, C, that is responsible for both. To revert to one of Cleland's stock examples, the asteroid collision at the end of the Cretaceous period, ~66 million years ago, was the common cause of the iridium spike and shocked quartz spherules observable in the clay layer formed at the very end of the Cretaceous, as well as the Chicxulub crater. One need not accept Cleland's account of prototypical historical science in order to appreciate that common cause explanations are very common.

Thus, insofar as it offers a common cause explanation of disparate phenomena, the ancient aliens theory works in pretty much the same way that many other good historical theories work. If there is a difference, it is that the ancient aliens theory looks like an overexuberant common cause explanation. An asteroid colliding with the earth at a certain moment in the past is only capable of generating certain sorts of biological and geological effects. There are many things that asteroids can't do. However, when the causal agents are secretive extraterrestrials with unimaginably advanced technology, the range of phenomena attributable to them is far less constrained.

One feature of UCTs that the ancient aliens theory does not necessarily share is (2) the suggestion that the conspirators have nefarious intentions. Indeed, one of the most interesting features of ancient aliens theory is that the aliens' intentions vis-à-vis humanity are somewhat inscrutable. One can find all sorts of shifting, ambiguous narratives about the aliens' 'plan': on the more nefarious end of the spectrum, one can find the suggestion that aliens have been running secretive extraction operations on Earth, and that they are particularly interested in precious metals such as gold. Out on the nefarious extreme, one can find suggestions that ancient aliens bred humans to work as miners in these operations. But some speculations about the aliens' intentions skew in a much more benign direction, with some saying that aliens shared technology with (some) people in the past, with the aim of guiding and cultivating human civilization. The possibility of nefarious intent does loom over ancient aliens theory, but it could well be, in this case, that the lack of clarity about the aliens' intentions is part of the appeal. In many other more mundane conspiracies, whether imagined or real, the conspirators' goals involve acquiring and maintaining power, and that's what makes them a bit nefarious. Most versions of the ancient aliens theory, though, treat the aliens as so far superior to us, technologically, that acquiring and maintaining power is not really an issue.

Keeley observes that in most UCTs the truth is 'a well-guarded secret', and that is also the case with ancient aliens. But one especially interesting feature of ancient aliens theory is just how it says the secret is guarded. In earlier times, the aliens were not exactly hiding from anyone. In fact, ancient aliens theory assumes that at certain times in the past, the aliens were not being terribly secretive at all. Consider how the theory explains certain scriptural passages. In the first chapter of Ezekiel, the prophet describes a vision involving four strange glowing humanoid creatures in the air. And then there is the famous wheel:

> Now as I beheld the living creatures, behold one wheel upon the earth by the living creatures, with his four faces. The appearance of the wheels and their work was like unto the color of a beryl: and they four had one likeness: and their appearance and their work was as it were a wheel in the middle of a wheel (Ezekiel 1: 15–16).

Von Däniken (1970) treats this as a veridical report of an encounter with an alien spacecraft, but one where the observer has little idea what they are actually seeing, and so describes both the craft and the aliens in metaphorical terms. The ancient aliens theory 'take' on passages such as this one clearly implies that aliens did not mind much if people saw them, at least on occasion. And yet the whole theory nevertheless treats alien activity on Earth as the great secret that is key to understanding history. What 'guards' the secret is an interesting combination of factors:

i. The aliens only selectively appeared to some people who were 'in the know'.
ii. People in the past misinterpreted what they were experiencing, and ended up describing the aliens in religious or mythological terms.

iii. The only traces left by the aliens are indirect ones, via their impact on human art, architecture, and material remains, as well as myth and religious scripture.

iv. The aliens, for whatever reason, are not sharing information with us, today, about their previous activities on Earth. The aliens presumably know what they did, but they are keeping that a secret.

v. Professional archaeologists are working in a coordinated way to discredit ancient aliens theory and brand it as illegitimate and unscientific.

One thing, perhaps, that distinguishes the ancient aliens narrative from other conspiracy theories is that it asserts a kind of double conspiracy. On the one hand, believers in ancient aliens must think that the aliens themselves are withholding information. They aren't telling us when they visited Earth in the past, or what they did, or why. On the other hand, there is the academic establishment: professional archaeologists are using their institutional authority to try to suppress the ancient aliens narrative. Believers in ancient aliens interpret this as a coordinated effort by professional insiders to thwart investigation.

Finally, ancient aliens theory is, in a sense, all about what Keeley calls *errant data*. According to Keeley, errant data come in two varieties. First, there are data that the going, mainstream explanation just fails to account for. Second, there are data which, if true, would contradict the mainstream account. Some of the data that proponents of ancient aliens theory cite are errant data in these two senses. To take the second kind of errant data first, there is at least one episode of the History Channel's *Ancient Aliens* series devoted to crystal skulls. These crystal skulls are errant data in the second sense: if they really were as old as claimed, and if they really had some of the properties alleged, then they might indeed cause problems for the mainstream account. The stories about them are not true, however (Feder 2020, pp. 291–292).

The first sort of errant data is potentially more important. We think the issue is not so much that there are data that mainstream archaeological narratives fail to account for, but rather that actual archaeological explanations are often messy, complex, contested, and fraught with historical contingency. This is very much what one should expect in historical reconstruction. For example, Currie (2014) argues that progress in historical science often involves a shift away from simpler explanatory models to more complex explanations. One challenge that archaeologists face, for example, is explaining the appearance of monumental architecture at a particular site at a particular moment. Why did people at this place and time begin building temples or royal tombs? Of course, archaeologists have much to say about these sorts of questions, but their explanatory narratives are messy and complicated, and vary regionally. They might cite local environmental conditions, or migration, or cultural traditions, or gradual development of knowledge, or technological innovations, or any number of other contributing factors. Because such questions have no simple answers, it can look—especially to someone *expecting* a simple story—like mainstream archaeology has not fully accounted for the data. And of course the ancient aliens theory does offer a simple sounding story: monumental architecture arose because the aliens started coaching people, showing them how to live, and sharing technological secrets. For someone expecting explanatory simplicity, it might look like the ancient aliens theory is accounting for errant data, in Keeley's first sense.

We think there is even a third kind of errant data: data that show up as errant only given certain background assumptions that professional archaeologists rightly reject. Consider the background assumption that only some people (say, white Europeans) are capable of developing sophisticated technology, solving difficult engineering problems, doing astronomy, and creating great art and architecture. Against that racist background assumption, an awful lot of archaeological sites in non-European parts of the world will look like errant data that mainstream archaeology has failed sufficiently to account for. The racist background assumption need not be stated explicitly. In practice, though, ancient aliens theorists often treat the material remains of non-white, non-European cultures as errant data, which they then account for by supposing that aliens imparted their wisdom to Indigenous communities.

Overall, then, ancient aliens theory fits Keeley's description of conspiracy theories very well. The only exception is that unlike many other conspiracy theories, ancient aliens theory is flexible with respect to the intentions of the alien conspirators.

Keeley argues that the popularity of conspiracy theories reflects many people's yearning for an orderly, manageable, world. As he puts it:

> Conspiracy theorists are, I submit, some of the last believers in an ordered universe. By supposing that current events are under the control of nefarious agents, conspiracy theories entail that such events are *capable of being controlled* (Keeley 1999, p. 123).

It can be difficult to accept the thought that things just happen, and not necessarily for any particular reason. And it's hard to accept the fact that historical events are highly contingent, with many contributing causes that are difficult to disentangle (Gould 1989; Beatty 1995). If ancient aliens theorists are committed to anything, it is that important events and trends in human history have been under the control of others, even if the intentions of the controlling conspirators are a little unclear. Keeley goes on to suggest that conspiracy theories represent a secularized version of the same sorts of tendencies that, historically, have led many people to believe in divine providence. We think that Keeley is basically right about this, but that this observation is just the first step toward understanding what's really behind ancient aliens theory. Even if conspiracy theories have this feature in common, different conspiracy theories may yet do different kinds of ideological work. In the next section, we augment Keeley's account with some observations drawn from the archaeological literature. We argue that ancient aliens theory does some very distinctive ideological work that sets it apart from other familiar conspiracy theories.

2.3 The Ideological Work That Ancient Aliens Theory Does

In the late nineteenth and early twentieth century, many archaeologists were themselves in the business of seeking grand theories that explained technological advances across vast spaces. Lacking good dating technologies or scientific methods to study material remains, they explained cultural change largely in terms of cultural diffusion. Pottery could not have been invented more than once, it was thought, so after it was developed in one place, it must have spread out from there as cultural groups adopted their neighbors' marvelous new technology. Some extreme versions of these theories essentially proposed a single origin point for the entirety of human accomplishment and are now seen as 'hyperdiffusionist'. Grafton Elliot Smith, an anatomist and early Egyptologist, theorized that agriculture, pottery, clothing, architecture, and politics all diffused out of Ancient Egypt to both the Old World and the New World. Ideas like his became highly influential; University of London cultural anthropologist W.J. Perry made similar claims in the 1920s (Storey and Jones 2011; Trigger 2006, pp. 218–221). As archaeologists professionalized, developed new dating methods that undermined the evidence for hyperdiffusionist theories, and increasingly saw themselves as empirical scientists, they came to recognize that cultures change and develop in many ways other than diffusion. Many archaeologists today embrace a particularism that focuses on specific social groups or sites rather than the more sweeping narratives of diffusionism, and transoceanic hyperdiffusionism is largely discredited (Wilson 2012).

But non-archaeologists had already seized on hyperdiffusionist theories. Most notably, a lawyer and congressman named Ignatius Donnelly wrote the 1882 blockbuster book *Atlantis: The Antediluvian World*. He argued that that lost continent's ancient, advanced civilization was the source of all agriculture and all advanced knowledge worldwide. The appeal of theories that explain so much of human history is clear, and as archaeologists have stopped offering them, others have stepped up to provide them (Wilson 2012). Graham Hancock's 2019 book, *America Before: The Key to Earth's Lost Civilization*, was a New York Times bestseller in the summer of 2019. It argues for an Atlantis-like ancient civilization (but based in Antarctica) that brought sophisticated knowledge to Native Americans (Hoopes 2019).

The tradition of hyperdiffusionist thinking in archaeology has a close relationship to one feature of conspiracy theories that we noted in Sect. 2.2. Conspiracy theories typically offer common cause explanations of disparate evidential traces. In doing so, they seek explanatory unification by means of an appealingly simple story about the activities of the conspirators. Diffusionist explanations in archaeology are also good examples of common cause explanations: the appearance of a certain cultural trait—say, a certain architectural style—at two distant sites can be explained by supposing that there was a common source in the past. There is nothing wrong with diffusionist explanations *per se*, but hyperdiffusionists and conspiracy theorists alike carry this mode of explanation to implausible extremes.

The idea that it was specifically aliens who brought knowledge or technology to humans has fairly clear roots in the space race and the Cold War. Aliens first appeared in popular culture in the late 1800s, with early stories like H.G. Wells' book *The War of the Worlds*. UFO sightings became common beginning with the Roswell incident of 1947 (Sagan 1996), and aliens really came into their own in the 1960s. Carl Sagan (1996, p. 101) traces the first alien abduction stories to 1961. Certainly the space race, Sputnik, and the Cold War fear of attack and invasion from above were part of the brew in which ancient aliens theory appeared. Secretive U.S. Cold War government projects involving missiles and weather balloons were the likely sources of both UFO sightings and a growing suspicion that the government was not sharing everything it knew (Sagan 1996). But ancient aliens theory can largely be traced to one man, Erich von Däniken, whose book *Chariots of the Gods?: Unsolved Mysteries of the Past* (written in 1968 and translated to English in 1970) set forth much of the basis for ancient aliens theory. The television show and other recent sources all draw broadly on von Däniken's thinking. However, scholars have recently noted a new strain of ancient alien thinking in which the aliens did not actually come to Earth but instead are "extra-dimensional, trans-dimensional, hyper-dimensional paranormal, shape-shifting entities" (Feder 2020, p. 211, citing Colavito 2005). In von Däniken's version, the aliens were physical, naturalistic beings with fancy technology; in some more recent versions, the aliens are made to seem a lot more like traditional spiritual or supernatural beings.

One important strand of ancient aliens theory, of course, is that the feats that prehistoric people performed were too difficult for 'primitive' humans to have pulled off on their own. Ancient aliens theory functions mostly by asking questions that seem compelling to a non-archaeologist audience. How could humans possibly have moved such huge blocks of stone, how could they have such a clear understanding of astronomy, why would people build pyramids all over the world? Meanwhile they ignore extensive archaeological evidence of how it was all done and of the gradual cultural processes by which humans developed their knowledge and technologies and learned to achieve these marvels of engineering.

Importantly, the monuments that require explanation in this narrative are not the medieval churches of Europe or the temples of the Greek city states. The ability of those people to build their structures is not in serious doubt. Instead, the question is how ancient Egyptians could build pyramids, how the Nazca of South America could create their massive geoglyphs, how the Ancestral Puebloans of Chaco Canyon could understand astronomy so well. Where European structures are called into question, they are very ancient ones such as Stonehenge or the Neolithic mounds of Ireland. But in spite of some puzzling unevenness—why, for example, did people need extraterrestrial help building Stonehenge but not gothic cathedrals?—there is a clear white supremacist angle to the theory. For ancient aliens theorists, European innovation and skill generally require no particular explanation, but non-European prehistoric people could not possibly have accomplished what they did on their own (Bond 2018; Feder 2020, p. 216).

Moreover, the idea of alien intervention must be viewed in the context of a long history of denying the abilities of non-Europeans. In North America, for example, we have the Moundbuilder myth, a long tradition of European and American settlers proposing alternate explanations for what we now know were the products of Native American knowledge and skill. Much of the North American continent was dotted with enormous and elaborate mounds and earthworks built by Native Americans, often with exquisitely worked artifacts associated with them. Hernando de Soto and other Spanish and French explorers observed Native American communities building and using such mounds. But those communities and others like them were decimated by the diseases brought by the explorers, and by the time that eighteenth and nineteenth-century settlers began moving into these areas, Native American communities had stopped building mounds. Many had also already been pushed out of their ancestral lands. Some settlers were content to merely plunder the mounds for artifacts, but others sought to understand their origins, theorizing that they were built by a superior, now absent, race (Colavito 2020; Feder 2020, pp. 138–167; Silverberg 1968).

Discovering the true identity of the Moundbuilders was a major preoccupation of nineteenth century anthropologists and amateur excavators. Sometimes the 'Moundbuilders' were ancient lost tribes of Israel, or Scandinavians, or Welshmen, or Phoenecians, or even Biblical giants. They were not, however, the local Native people, who were widely considered too 'primitive' to have built such structures. Moundbuilder theories were sometimes bolstered by a single artifact, later revealed as a hoax, with Hebrew or Phoenician script. An important element of the Moundbuilder myth was the inevitable conclusion that this ancient, superior race had actually been displaced by Native Americans, who were therefore recent interlopers on their land. So the Moundbuilder myth not only served to separate Indigenous people from their ancestors' monuments and land but also offered balm on the conscience of the new settlers. The Native Americans they encountered, it seemed, had only been there a little while longer than themselves and had driven away other, earlier residents (Feder 2020, p. 159; Silverberg 1968).

Ancient aliens theory functions in essentially the same way as the Moundbuilder myth. It questions the abilities of earlier people, especially the non-European ones, and replaces their accomplishments and knowledge with mysterious advanced technology from outer space. Aliens in pop culture have long been a metaphor for thinking through encounters with the Other—consider Star Trek and its prime directive, meant to prevent just the kind of ill-fated colonial encounters that took place here on Earth. Or Stephen Hawking's explicit prediction that "If aliens visit us, the outcome would be much as when Columbus landed in America, which didn't turn out well for the [N]ative Americans" (Grier 2010). Ancient aliens theory uses the same metaphor to offer a kind of comfort—Europeans were not the first worldwide colonialists after all. And by extension, the monuments of the past really don't belong to Indigenous people any more than they do to the rest of us, since they could only have been built with extraterrestrial assistance. It also reframes colonialists as bringing 'the gift of civilization' (Bond 2018). The Southern Poverty Law Center has warned that ancient aliens theory is widely discussed on white supremacist forums, where it often overlaps with racist and antisemitic conspiracy theories (Zaitchik 2018).

Ancient aliens theory also clearly has a religious undertone. Von Däniken (1970) views all of humanity's deities as representations of ancient alien visitors who our primitive ancestors could not explain except in religious terms. He explicitly explains Biblical miracles such as Ezekiel's wheel as the work of aliens, and the TV series follows his lead, proposing an alien nuclear reactor that produced manna from heaven, for example. But even when not concerned specifically with Biblical stories, there is often a sense in watching *Ancient Aliens* that the aliens are essentially angels, come to help humanity, or maybe demons come to exploit and manipulate us. As noted above, the ambivalence about the aliens' intentions may be part of the appeal. Are benevolent ancient aliens easier to believe in than angels? Ancient aliens theories seem, for some people, to be stepping into the place of a religious belief that is somehow less tenable. Like a religion, it also offers a sense of being part of something bigger than one's self, of being in on secrets, and it provides a community of fellow believers who find each other on internet discussions and even at conventions. The alleged alien intervention in human history, combined with their apparent absence today, also seems like a transposition of distinctively Christian themes. Just as some Christians anticipate a second coming, believers in ancient aliens might wonder what will happen when the aliens return.

In short, viewing ancient aliens theorists as just another set of crackpots fails to recognize the intellectual history and cultural milieu in which their views have arisen, as well as the ideological work that this theory does for people in a modern world. It supports a resurgent white supremacy, soothes religious anxieties and colonialist guilt, plays into a growing suspicion of experts and intellectuals and, of course, offers a much more interesting and accessible worldview than do archaeologists with their nuanced and particularist explanations.

2.4 Why Ancient Aliens Theory Doesn't Work

So far, we've argued (in Sect. 2.2) that ancient aliens theory is a conspiracy theory, *sensu* Keeley (1999). That is our answer to the question about how ancient aliens theory works. In Sect. 2.3, we argued that ancient aliens theory differs from other conspiracy theories because it does distinctive (and distinctively problematic) kinds of ideological work. But there is more to say about why exactly it is wrong to believe in ancient aliens theory, and why ancient aliens theory goes off the rails when it comes to historical scientific and archaeological investigation. In further developing this diagnosis of ancient aliens theory, we shall avoid the quick move of dismissing it as 'pseudoscience' or 'pseudoarchaeology' (see also Derricourt 2012). We're skeptical that the traditional Popperian project of drawing a sharp line of demarcation between science and pseudoscience is a very useful one (Popper 1959). For one thing, it's hard to find a principled way of drawing the boundary that gets the right results: traditional demarcationist proposals have tended to either include or exclude too much (Laudan 1988; although see also Pigliucci 2013; Mahner 2013; Nickles 2013; and Baudry 2013 for a variety of efforts to update the demarcationist project in ways that meet Laudan's challenge). However, one does not need an account of the essence of science, or of archaeology, in order to point to some features of archaeological investigation that contribute to its success. If it were to turn out that the appeal of ancient aliens theory has to do (perversely) with the rejection of one of the things that makes archaeology so successful, that would be a problem.

To help us get our bearings, it may help to begin with Keeley's (1999) diagnosis of UCTs, or unwarranted conspiracy theories. One might be tempted to dismiss conspiracy theories in general on the grounds that they are unfalsifiable. But Keeley argues that we should resist this temptation. Such an approach could lead us to reject true theories. Here it is important to bear in mind that some conspiracy theories turn out to be correct (e.g., Watergate).

> Richard Nixon and [Oliver] North actively sought to divert investigations into their respective activities and both could call upon significant resources to maintain their conspiracies. They saw to it that investigators were thwarted in many of their early attempts to uncover what they accurately suspected was occurring. Strictly hewing to the dogma of falsifiability in these cases would have led to a rejection of conspiracy theories at too early a point in the investigations, and may have left the conspiracies undiscovered (Keeley 1999, p. 121).

Unfalsifiability is something that the unwarranted conspiracy theories (the UCTs) will generally have in common with the true ones. Thus, it might be a mistake to reject conspiracy theories *tout court* on grounds of unfalsifiability. Keeley thinks that falsifiability is an unreasonable standard to apply in cases where the theory in question asserts that powerful agents are working to obstruct investigation.

Keeley's own diagnosis of unwarranted conspiracy theories focuses on the intellectual consequences of sticking by one's favorite conspiracy theory in the face of the accumulation of apparently contradictory evidence:

> It is this pervasive skepticism of people and public institutions entailed by some mature conspiracy theories which ultimately provides us with the grounds with which to identify them as unwarranted. It is not their lack of falsifiability per se, but the increasing amount of skepticism required to maintain faith in a conspiracy theory as time passes and the conspiracy is not uncovered in a convincing fashion. As this skepticism grows to include more and more people and institutions, the less plausible any conspiracy becomes (Keeley 1999, p. 123).

Keeley's point here is just that over time, the intellectual costs of upholding a conspiracy theory can become exceedingly—perhaps ridiculously—high. To take a contemporary example, someone who believes that the Sandy Hook massacre in Newtown, Connecticut, did not really happen, will be committed to a kind of blanket distrust of huge numbers of ordinary people (including victims' family members and many others who were impacted), institutions (including virtually all respectable news organizations), and government officials. Keeley argues, quite plausibly, that these high intellectual costs help explain why so many conspiracy theories are unwarranted and irrational.

Keeley's diagnosis of unwarranted conspiracy theories certainly gets things right in a wide variety of cases.[5] It's not totally clear, though, whether belief in ancient aliens necessarily leads to the "pervasive skepticism of people and public institutions" that Keeley is so concerned about. The mistrust that goes with enthusiasm for ancient aliens is more localized. A believer in ancient aliens has to be skeptical of academic archaeologists and academic institutions, but that localized skepticism could turn out to be fairly easy for many people to maintain. Indeed, many creationists and climate change deniers seem to have little trouble maintaining localized skepticism about academic expertise. Without necessarily disagreeing with Keeley's diagnosis of many unwarranted conspiracy theories, we think more needs to be said about how and why ancient aliens theory goes off the rails.

In recent years, philosophers of science have had much to say about how historical researchers make progress when confronted with incomplete evidence and/or faint and difficult-to-interpret evidential signals (see especially Chapman and Wylie 2016; Currie 2018). One common theme of this work is the paramount importance of contextual background knowledge. When archaeologists are investigating a particular question about a particular site, they do so in a rich informational context that includes established investigative practices and lots of relevant background knowledge, as well as multiple lines of evidence. This is a way of leveraging previous epistemic successes. Understanding the imagery on a Mayan ruler's tomb does not involve just looking at it, as von Däniken did with Lord Pakal's sarcophagus lid before declaring it to be an image of an ancient astronaut. Instead it requires understanding context such as the identity of the dead king and his place in Mayan history as revealed by past research on Mayan texts. Understanding how Mayans adorned their bodies might be relevant to understanding the object near Pakal's nose. Also highly relevant are Mayan cosmology, symbolism and artistic conventions as understood from texts, inscriptions, thousands of objects of art, and architecture at many other sites. One might want to consider the social and political nature of the city where Pakal ruled, as understood from texts and excavations, as well as the layout and social meanings of the pyramid in which the king was buried. Studies of the bones found inside the tomb and the tools and materials used to make the sarcophagus might also be relevant (Feder 2020, pp. 231–235). These are good examples of what Adrian Currie (2018) calls investigative scaffolding. When archaeologists seek to interpret something like Lord Pakal's sarcophagus lid, they rely heavily on a supporting structure of previous research, even if very little of that previous work actually focused on the artifact in question.

One fascinating feature of ancient aliens theory is that it operates in a way that is almost entirely disconnected from relevant contextual archaeological information. This is especially evident in the History Channel series, where instead of bringing to bear all that's known about a particular site, the show often focuses on only one artifact or building at a site that fits the narrative, then jumps from site to site around

---

[5] See, however, Basham (2003) for an interesting critical response to Keeley. Basham is not convinced that Keeley's account explains why we should not believe in a 'global malevolent conspiracy'.

the world. Rather than reading past excavation reports or analyzing thousands of potsherds or understanding how one site fits into an entire cultural context, ancient aliens theorists presume that they need only the evidence of their own eyes to interpret what a winged artifact represents or what an image of a figure with antennae means. We might refer to this as *non-contextualized inquiry*. Non-contextualized inquiry means investigating an artifact or a site more or less from scratch, with minimal reliance on what's already known. This is one of the most serious defects of ancient aliens theory. It's a defect because contextualization—scaffolding, reliance upon background knowledge—contributes so heavily to success in archaeological and other historical scientific research.

Of course, the very thing that makes non-contextualized inquiry defective also explains its popular appeal. The non-contextualized nature of ancient aliens theory makes it fully accessible to non-specialists. No one needs any background knowledge about particular sites or regions, or any expertise in archaeological research methods, or any grasp of the history of archaeological theory, in order to be fully up to speed with 'research' focusing on ancient aliens. Ancient aliens theory sacrifices the very thing that's crucial for successful historical investigation, and it does so for the sake of accessibility and popular appeal.

Note that our argument here is not a demarcationist one, at least not in the narrow sense. There is an important difference between the following two claims:

(i) Leveraging background knowledge is a necessary condition for research activity to count as real archaeology.
(ii) Leveraging background knowledge is one of the things that professional archaeologists do that contributes to their investigative success.

Claim (i) is a narrower demarcationist claim, and the sort of claim that Laudan (1988) was concerned about. Our suggestion is that claim (ii) is the one that matters, and that it is possible to diagnose ancient aliens theory without defending an essentialist picture of archaeology. (For further anti-essentialist reflections about historical science, see Currie and Turner 2016.) Our diagnostic proposal, rather, is that there is a feature of (most) professional archaeology that contributes to its investigative success, but which ancient aliens theory lacks, and that is the leveraging of epistemic context.

2.5 Conclusion

Analysis of ancient aliens theory represents an opportunity for mutual engagement between archaeologists and philosophers. In Sect. 2.2 above, we argued that archaeologists interested in understanding how ancient aliens theory works would do well to draw upon the philosophical literature on conspiracy theory (especially Keeley 1999). But in Sect. 2.3, we argued that the case of ancient aliens theory also highlights the need to augment Keeley's account. That is because ancient aliens theory has some features that set it apart from other conspiracy theories: it draws on a longer tradition of diffusionist thinking in archaeology, and it does some distinctively problematic ideological work. Ancient aliens theory is a conspiracy theory that works subtly in the service of colonialism, white supremacy, and religious demythologizing. Finally, in Sect. 2.4, drawing upon recent work in the philosophy of historical science, we've offered a fresh diagnosis of where ancient aliens theory goes off the rails as a form of historical investigation. We've resisted the temptation to dismiss it out of hand as 'pseudoarchaeology'. Instead, we've argued that it trades one thing for another: it purchases accessibility and popular appeal at the price of the very thing that makes historical reconstruction successful—namely, the reliance on a rich epistemic context.

In this chapter, we've sought to offer an analysis and diagnosis of ancient aliens theory from the perspectives of professional archaeology and philosophy. However, there is a lingering worry that no such account—neither Keeley's, nor ours—could ever get any traction against a conspiracy theory. The problem is that virtually any critique of a conspiracy theory can be interpreted, from the perspective of someone in the grip of that theory, as part of the conspiracy. But it is nevertheless worthwhile to try to understand, from an external perspective, why ancient aliens theory is so gripping, and also why it is problematic that so many people are in the grip of it.

References

Anderson DS (2019) 'I don't believe, I know': the faith of modern pseudoarchaeology. SAA Archaeol Rec 19(5):31–34
Barkun M (2013) A culture of conspiracy: apocalyptic visions in contemporary America, 2nd edn. University of California Press, Berkeley
Basham L (2003) Malevolent global conspiracy. J Soc Philos 34(1):91–103
Baudry M (2013) Loki's wager and Laudan's error: on genuine and territorial demarcation. In: Pigliucci M, Boudry M (eds) Philosophy of pseudoscience: reconsidering the demarcation problem. University of Chicago Press, Chicago, pp 79–98
Beatty J (1995) The evolutionary contingency thesis. In: Wolters G, Lennox JG (eds) Concepts, theories, and rationality in the biological sciences. University of Pittsburgh Press, Pittsburgh, pp 45–81
Bond SE (2018) Pseudoarchaeology and the racism behind ancient aliens. Hyperallergic. November 13, 2018. https://hyperallergic.com/470795/pseudoarchaeology-and-the-racism-behind-ancient-aliens/. Accessed 8 Jan 2020
Card J, Anderson D (2016) Alternatives and pseudosciences: a history of archaeological engagement with extraordinary claims. In: Card JJ, Anderson DS (eds) Lost city, found pyramid: understanding alternative archaeologies and pseudoscientific practices. University of Alabama Press, Tuscaloosa, pp 1–18
Chapman R, Wylie A (2016) Evidential reasoning in archaeology. Bloomsbury, London
Chapman University (2018) Paranormal America: Chapman University survey of American fears. https://blogs.chapman.edu/wilkinson/2018/10/16/paranormal-america-2018/. Accessed 6 Jan 2020
Cleland C (2002) Methodological and epistemic differences between historical science and experimental science. Philos Sci 69:474–496
Cleland C (2011) Prediction and explanation in historical natural science. Br J Philos Sci 62(3):551–582

Colavito J (2005) The cult of alien gods: H.P. Lovecraft and extraterrestrial pop culture. Prometheus, Amherst, NY.

Colavito J (2020) The mound builder myth: fake history and the hunt for a lost white race. University of Oklahoma Press, Norman

Currie A (2014) Narratives, mechanisms, and progress in historical science. Synthese 191(6):1163–1183

Currie A (2018) Rock, bone, and ruin: an optimist's guide to the historical sciences. MIT Press, Cambridge, MA

Currie A, Turner D (2016) Scientific knowledge of the deep past. Stud Hist Philos Sci 55:43–46

Derricourt R (2012) Pseudoarchaeology: the concept and its limitations. Antiquity 86(332):524–531

Fagan GG, Feder KL (2006) Crusading against straw men: an alternative view of alternative archaeologies: response to Holtorf (2005). World Archaeol 38(4):718–729

Feder KL (1984) Irrationality and popular archaeology. Am Antiq 49(3):525–541

Feder KL (2016) Answering pseudoarchaeology. In: Card JJ, Anderson DS (eds) Lost city, found pyramid: understanding alternative archaeologies and pseudoscientific practices. University of Alabama Press, Tuscaloosa, pp 199–210

Feder KL (2020) Frauds, myths, and mysteries: science and pseudoscience in archaeology, 10th edn. Oxford University Press, New York

Feder KL, Barnhart T, Bolnick DA, Lepper BT (2016) Lessons learned from lost civilizations. In: Card JJ, Anderson DS (eds) Lost city, found pyramid: understanding alternative archaeologies and pseudoscientific practices. University of Alabama Press, Tuscaloosa, pp 167–184

Gould SJ (1989) Wonderful life: the burgess shale and the nature of history. WW Norton, New York

Grier P (2010) Stephen Hawking aliens warning: should we hide? https://www.csmonitor.com/Science/2010/0426/Stephen-Hawking-aliens-warning-Should-we-hide/. Accessed 7 Jan 2020

Grünschloss A (2007) 'Ancient Astronaut' narrations. Fabula 48:205–228

Harambam J, Aupers S (2015) Contesting epistemic authority: conspiracy theories on the boundaries of science. Public Understanding Sci 24(4):466–480

History Channel (2020) Ancient aliens. https://www.history.com/shows/ancient-aliens/. Accessed 6 Jan 2020

Hofstadter R (1964) The paranoid style in American politics. Harpers Magazine 1964:77–86

Holly DH (2015) Talking to the guy on the airplane. Am Antiq 80(3):615–629

Holtorf C (2005) Beyond crusades: how (not) to engage with alternative archaeologies. World Archaeol 37(4):544–551

Hoopes JW (2019) Introduction: pseudoarchaeology, scholarship, and popular interests in the past in the present. SAA Archaeol Rec 19(5):8–9

Keeley B (1999) Of conspiracy theories. J Philos 96(3):109–126

Keeley B (2003) Nobody expects the Spanish Inquisition! more thoughts on conspiracy theory. J Soc Philos 34(1):104–110

Kitcher P (1983) Abusing science: the case against creationism. MIT Press, Cambridge, MA

Koertge N (2013) Belief buddies vs. critical communities: the social organization of pseudoscience. In: Pigliucci M, Baudry M (eds) Philosophy of pseudoscience: reconsidering the demarcation problem. University of Chicago Press, Chicago, pp 165–182

Kurutz S (2018) Suspicious minds: mingling with wariness and wonder at a conference devoted to 'ancient aliens'. https://www.nytimes.com/2018/07/21/style/ancient-aliens.html. Accessed 6 Jan 2020

Laudan L (1988) The demise of the demarcation problem. In: Ruse M (ed) But is it science? the philosophical question in the creation/evolution controversy. Prometheus Books, Buffalo, pp 337–350

Mahner M (2013) Science and pseudoscience: how to demarcate after the (alleged) demise of the demarcation problem. In: Pigliucci M, Boudry M (eds) Philosophy of pseudoscience: reconsidering the demarcation problem. University of Chicago Press, Chicago, pp 29–43

Nickles T (2013) The problem of demarcation: history and future. In: Pigliucci M, Boudry M (eds) Philosophy of pseudoscience: reconsidering the demarcation problem. University of Chicago Press, Chicago, pp 101–120

Parker EA (2016) The proliferation of pseudoarchaeology through 'reality' television programming. In: Card JJ, Anderson DS (eds) Lost city, found pyramid: understanding alternative archaeologies and pseudoscientific practices. University of Alabama Press, Tuscaloosa, pp 149–166

Pigliucci M (2013) The demarcation problem: a (belated) response to Laudan. In: Pigliucci M, Boudry M (eds) Philosophy of pseudoscience: reconsidering the demarcation problem. University of Chicago Press, Chicago, pp 9–28

Pigliucci M, Boudry M (eds) (2013) Philosophy of pseudoscience: reconsidering the demarcation problem. University of Chicago Press, Chicago

Popper K (1959) The logic of scientific discovery. Routledge, London

Rejent J (2019) Friday cable ratings. https://tvbythenumbers.zap2it.com/daily-ratings/friday-cable-ratings-september-6-2019/. Accessed 6 Jan 2020

Ruse M (ed) (1988) But is it science? the philosophical question in the creation/evolution controversy. Prometheus Books, Buffalo

Sagan C (1996) The demon-haunted world: science as a candle in the dark. Ballantine Books, New York

Silverberg R (1968) Mound builders of ancient America: the archaeology of a myth. New York Graphic Society, Greenwich

Storey AA, Jones TL (2011) Diffusionism in archaeological theory: the good, the bad, and the ugly. In: Jones TL, Storey AA, Matisoo-Smith EA, Ramirez-Aliaga JM (eds) Polynesians in America: pre-Columbian contacts with the New World. AltaMira Press, Lanham, pp 7–24

Trigger BG (2006) A history of archaeological thought, 2nd edn. Cambridge University Press, New York

Turner D (2007) Making prehistory: historical science and the scientific realism debate. Cambridge University Press, Cambridge

von Däniken E (1970) Chariots of the Gods? unsolved mysteries of the past. Bantam Books, New York

Wilson JAP (2012) The cave who never was: outsider archaeology and failed collaboration in the USA. Public Archaeol 11(2):73–95

Zaitchik A (2018) Close encounters of the racist kind. https://www.splcenter.org/hatewatch/2018/01/02/close-encounters-racist-kind/. Accessed 22 Jan 2020

Chapter 3
Mortar and Pestle or Cooking Vessel? When Archaeology Makes Progress Through Failed Analogies

Rune Nyrup

Abstract Most optimistic accounts of analogies in archaeology focus on cases where analogies lead to accurate or well-supported interpretations of the past. This chapter offers a complementary argument: analogies can also provide a valuable form of understanding of cultural and social phenomena when they fail, in the sense of either being shown inaccurate or the evidence being insufficient to determine their accuracy. This type of situation is illustrated through a case study involving the mortarium, a characteristic type of Roman pottery, and its relation to the so-called Romanization debate in Romano-British archaeology. I develop an account of comparative understanding, based on the idea that humans have a natural desire to understand ourselves comparatively, i.e., in terms of how we resemble and differ from societies at other times and places. Pursuing analogies can provide this type of understanding regardless of whether they turn out to be accurate. Furthermore, analogies can provide a similar form of understanding even when the evidence turns out to be insufficient to determine their accuracy.

Keywords Analogies · Optimism · Mortaria · Romanization debate · Comparative understanding · Value of understanding

R. Nyrup (✉)
Department of History and Philosophy of Science and Leverhulme Centre for the Future of Intelligence, University of Cambridge, Cambridge, UK
e-mail: rn330@cam.ac.uk

© The Author(s) 2021
A. Killin, S. Allen-Hermanson (eds.), *Explorations in Archaeology and Philosophy*, Synthese Library 433, https://doi.org/10.1007/978-3-030-61052-4_3

3.1 Introduction

Many archaeological interpretations are based on analogies with known societies. That is, they draw on comparisons with phenomena (artifacts, cultural practices, societal structures, etc.) that are already familiar, e.g., from anthropological studies, textual historical sources, other archaeological evidence or modern everyday experience. Nonetheless, archaeologists have often been uneasy about their reliance on analogies. In methodological debates, archaeologists have worried that analogies are unreliable or speculative, questioning whether the past is likely to resemble any familiar society—and if so, whether we would be able to tell. And to be sure, there are plenty of cases from the history of archaeology where apparently compelling analogy-based interpretations turned out to be deeply misleading.[1]

Philosophers and archaeologists who defend analogies have pushed back against this pessimistic view (e.g., Ascher 1961; Smith 1977; Salmon 1982; Wylie 2002; Currie 2016). By developing positive accounts of the criteria for good analogical inferences, they seek to counter pessimistic arguments through one or more of the following optimistic counterclaims. First, there are methodologically sound criteria by which archaeologists can recognize more plausible analogies. Second, even if past societies are unlikely to resemble present ones, analogies can still assist in generating accurate interpretations of the past. Third, archaeologists can (at least sometimes) obtain sufficient evidence to recognize an analogy-based interpretation as accurate.

While I find these arguments compelling, as far as they go, they do not in my view exhaust the case for pursuing analogies in archaeology. Most existing accounts focus on establishing the possibility of analogies *succeeding*, in the sense of leading to accurate and evidentially well-supported interpretations. Here, I want to propose a complementary line of justification, focusing on cases where analogies *fail*, in the sense that either the available analogies turn out to be inaccurate or the evidence turns out to be insufficient to determine whether they are accurate.

A justification of this kind would significantly strengthen the case for pursuing analogies in archaeology. Previous defenses of analogy show that wholesale pessimism is untenable. Nonetheless, it is not uncommon for archaeological inquiry to reach a stage where the available and foreseeable evidence only suffices to rule out some interpretations as inaccurate but not to establish any particular positive account as (most likely to be) accurate. In this chapter, I discuss a case study from Romano-British archaeology that illustrates this type of situation. The case study concerns a characteristic type of pottery called the *mortarium*, which became increasingly prevalent in Britain after the Roman conquest, and its relation to the so-called Romanization debate, i.e., the debate over how to best characterize the cultural changes that took place in regions conquered by the Romans. Analogies with other parts of the Roman world as well as more recent times are ubiquitous in these

---

[1] For discussion of such examples, see Ascher (1961, pp. 317–318), Orme (1981, ch 1) and Wylie (2002, pp. 137–38).

debates. Yet there are few well-established conclusions regarding which of the inter-pretations suggested by these analogies are most likely to be accurate.

The aim of this chapter is to propose an account of how archaeology can achieve a significant form of progress through pursuing analogies even in cases like this. Specifically, I will argue (i) that we can gain an important form of understanding of cultural and social phenomena through analogies with already familiar societies, and (ii) that this is valuable to pursue in part because it supports a natural human desire to understand ourselves comparatively, that is, in terms of how our lives and societies resemble and differ from those at different times and places. Crucially, analogies can provide this form of understanding even (or especially) when they turn out to be inaccurate. Finally, I will argue (iii) that pursuing analogies produces a similar form of understanding even when what we learn is that the available evi-dence is insufficient to determine whether the analogies are accurate or not.

I begin with some preliminaries, explaining how I define 'analogy' and review-ing some extant accounts of analogy in archaeology (Sect. 3.2). Next, I introduce the mortarium case study (Sect. 3.3) and discuss the different ways analogies are used here (Sect. 3.4), including in what sense they count as 'failed'. I then move on to my positive account. I start by clarifying the type of account I seek to defend, through a comparison with Currie's (2018) defense of optimism about the historical sciences (Sect. 3.5). Finally, I develop my account in more detail (Sect. 3.6).

3.2 Analogies, Analogical Reasoning, and Inferential Strategies

I follow Hesse (1966) and Bartha (2010) in understanding an *analogy* to be a com-parison between two systems that highlights: (a) respects in which they are known or assumed to be similar (the *positive analogy*), (b) dissimilar (the *negative anal-ogy*), and (c) respects in which it is currently unknown to what extent they resemble each other (the *neutral analogy*). Analogical *reasoning* consists in drawing such comparisons in order to suggest the hypothesis that one or more features, already known or assumed to exist in one system (*the source*), also exist in the other (*the target*). This hypothesis may concern specific parts of the neutral analogy, or simply state that the target shares some (or many/most/...) of the source system's features (or some circumscribed subset of these), leaving it open exactly which.

The term 'suggest' here is intended to be neutral with regards to the epistemic force of the suggestion. Sometimes, the analogy is intended to provide some degree of evidence or epistemic justification for the hypothesis. However, analogies can also be used to merely suggest conjectures for further pursuit, without any commit-ment to their likeliness. Indeed (as I will argue), sometimes the point of pursuing analogy-based hypotheses is exactly to show them false. We will see examples of all of these uses of analogies below.

What is the inferential structure of analogical reasoning? Under what conditions are we justified in going from the existence of a given analogy to the suggestion (with a given epistemic force) of further similarities? I do not suspect there to be any single account which adequately captures all instances of analogical reasoning as characterized above. Analogical reasoning draws and depends on our prior knowledge and assumptions about the two systems (Weitzenfeld 1984; Wylie 1988; Norton 2018). Depending on what kinds of background assumptions are available in a given context, there is likely to be several different inferential strategies available that rely on comparisons between two systems. Here, I will briefly review some commonly discussed inferential strategies for analogies in archaeology.[2]

The most straightforward strategy is to rely on background assumptions which make it likely that the source and target share relevant similarities. If I know that A and B are generally likely to be similar with regards to F-type features, and then learn that A has a feature F_1, this would provide justification for the hypothesis that B also has F_1.[3] Call this *direct extrapolation*. Many early defenses of analogies in archaeology (as summarized by Ascher 1961) relied on this strategy: they sought to identify general criteria for reliable analogical inferences, such as the source and target being temporally proximate or historically continuous, or the inferred features being technologically constrained. A perennial challenge for direct extrapolation is whether the similarity assumptions a given inference relies on are independent of the hypothesis one is aiming to support; as Grahame Clark puts it, there is a "real danger of setting up a vicious circle and assuming what one is trying to discover" (Clark 1951, p. 52, cited from Wylie 2002, p. 139). We will see this illustrated in the mortarium case study below.

The second inferential strategy, which I will call *hypothesis-testing*, is more indirect: first, a model, M, is developed based on our knowledge or assumptions about the source, describing the dependency relations between the features of the source. Second, M is used, together with our knowledge or assumptions about the target, to infer what features we should and should not expect to find in the target domain, if M also applied there. By checking whether these features can be found, we can test the hypothesis that M also correctly describes the target. If this hypothesis is sufficiently confirmed, we can infer that the target also has any further features that play an essential role in M, even if these have not been directly observed. Hesse (1966) first proposed an account of analogical reasoning along these lines, since further elaborated by Bartha (2010). Similar accounts of analogies in archaeology have been defended by Smith (1977), Wylie (2002, ch 9) and Currie (2016). A significant challenge to this strategy concerns our ability to generate the right hypotheses. First, we can only test and confirm a hypothesis as accurate if we have generated the right

---

[2] For the sake of brevity, the following is kept fairly schematic. Nyrup (2020b) discusses these different uses of analogy in more detail and situates them within the history of debates in archaeology about analogy.

[3] This is an instance of what is sometimes known as 'direct inference' or 'statistical syllogism'. See Peden (2019) for a useful recent discussion of this type of inference and its relation to material theories of induction.

hypothesis in the first place. Furthermore, even if we observe some set of features, F, predicted by the hypothesis that M applies to the target, there may be another model that also predicts F but otherwise differs significantly from M. As archaeologists have often pointed out, it can be difficult to rule out that such an alternative exists for something as variable and malleable as human social and cultural practices (Smith 1955; Freeman 1968; Gould 1980).

However, archaeologists can also use analogies to overcome, or at least ameliorate, these challenges to hypothesis-testing. The idea behind this strategy—advocated for example by Ucko and Rosenfeld (1967)—is to seek out as a broad and varied a range of potential analogies as possible in order to "widen the horizons of the interpreter" (Ucko 1969, p. 262), that is, to generate different models one might use to suggest potential interpretations of a given target. While analogies are not the only method for generating alternative hypotheses, and cannot be guaranteed to generate all possible alternatives, they nonetheless provide a fruitful method for broadening one's interpretative horizons (Wylie 1988; Currie 2018, ch 8).

Two variants of this use of analogies can be distinguished. In what I will call *generation-for-confirmation*, the underlying aim of generating interpretations is still to end up with a hypothesis that can be confirmed as true (or at least evidentially well-supported). For instance, Ucko and Rosenfeld claim that: "The more varied and the more numerous the analogies that can be adduced, the more likely one is to find a convincing interpretation for an archaeological fact" (1967, p. 157). The strategy suggested here is to first cast the net wide, in order to eventually whittle down the set of potential interpretations to a single plausible hypothesis (or at least, no more than a few), for instance through hypothesis-testing. Being able to rule out a wide and variable range of alternative hypotheses can be a powerful form of evidential reasoning (Reiss 2015; Nyrup 2020b, pp. 19–20). However, it of course requires that the alternative hypotheses can in fact be ruled out.

In the final strategy that I want to outline here, *generation-as-criticism*, using analogies to generate alternative interpretations instead serves a critical purpose. Rather than seeking to test (and ideally confirm) one of the hypotheses generated, their purpose is merely to highlight that there, in fact, exists plausible alternatives, hitherto unconceived, to some apparently well-confirmed interpretation. The reason for using analogies with known phenomena (rather than, e.g., fictional stories) is that this shows that these alternatives are more than mere logical or theoretical possibilities. Highlighting an alternative model which is already known to apply to a real case shows that the relations postulated by this model already exist (or existed) somewhere in the world, and thus represents a real possibility for other cases too, consistent with how human socio-cultural practices work in the actual world (as opposed to a mere logical or epistemic possibility).[4]

Compared to the other strategies, generation-as-criticism is the odd one out. Extant optimistic accounts of analogies in archaeology usually focus on cases where

---

[4] Bartha (2010) calls this modality 'prima facie plausibility' and defends an account of how analogies can provide this type of support, inspired by Whewell's and Herschel's concept of *vera causa* (Bartha 2010, pp. 302–304). See also Nyrup (2020a, pp. 7–11) for discussion.

archaeologists succeed in using analogies to provide positive support for an interpretative hypothesis. Generation-as-criticism, by contrast, aims to throw doubt on
some currently accepted interpretation. Perhaps for this reason, this use of analogies
is rarely discussed by optimists. However, as the case study will illustrate, archaeologists do in fact often use analogies in this way. To be clear, my point here is not
to disparage the other three inferential strategies. While I have highlighted some
challenges faced by these strategies, I do not mean to suggest these uses of analogies arc therefore untenable. As optimists have highlighted (e.g., Wylie 2002; Currie
2016), archaeologists can and do manage to overcome these challenges, often by
combining several different inferential strategies. There is no general reason that
analogies in archaeology cannot, in this sense, succeed. What I am interested in
here, however, is what progress in archaeology looks like when this does not happen.

3.3   The Mortarium and the Romanization Debate

The mortarium is a characteristic type of pottery found across the Roman world (see
Fig. 3.1), which can be described as a bowl or basin with a prominent rim, typically
with a spout and embedded grits to create a rough inner surface (Tyers 1996, p. 116;
Cramp et al. 2011, pp. 1339–1340). These characteristics, together with the fact that
some mortaria sherds show signs of heavy wear, is often taken to suggest that its
function was as grinding or mixing bowl. This interpretation is also supported by
Roman sources describing their use for mixing and pounding ingredients (Cramp
et al. 2011). Furthermore, as one archaeologist dryly notes, "Some of the vessels
even helpfully have graffiti specifically stating that they are mixing bowls (*pelves*)
or mortars (*mortaria*)" (Cool 2004, p. 30).

The mortarium is particularly interesting to archaeologists of Iron Age and
Roman Britain, as it is rarely found on British sites from the (pre-Roman) Iron Age,
but becomes increasingly common after the Roman conquest. In particular, mortaria

Fig. 3.1 Roman pottery
mortarium. Reproducible
under the Creative
Commons Attribution-
Share Alike 3.0 Unported
license. (Source: https://
upload.wikimedia.org/
wikipedia/commons/a/a8/
Roman_pottery_
mortarium.jpg)

have been found not just in relation to military sites or urban centers, where they might be assumed to have been consumed by incoming soldiers and settlers, but from the later second century also on rural farmsteads (Cramp et al. 2011, p. 1340). In fact, in some areas, such as the Northwest of England, mortaria are significantly more common on rural sites than any other form of 'Roman' pottery (Peacock 2016, p. 24; cf. Evans 1995; Rush 1997; Cool 2004). Given its apparently well-understood function, the spread of mortaria in Roman Britain is sometimes interpreted as evidence of the pre-existing populations in Britain adopting more Roman ingredients or methods of cooking (Evans 1995; Tyers 1996, p. 116; cf. Cool 2006, p. 43). However, many archaeologists have argued for other possible uses, more continuous with pre-existing practices, including that it was used to grind or dehusk cereals (Phelps 1923), as a basin for curdling milk into cheese (Oswald 1943; Alcock 2001, p. 61), or simply as a convenient large bowl (Cool 2006, ch 6; Peacock 2016). We will look at some of the evidence for and against these interpretations below.

The question of how mortaria were used is significant due to its relation to the so-called Romanization debate. 'Romanization' is a name used for traditional accounts, developed in the early- to mid-twentieth century, of the changes brought about by the Roman conquest of Britain (and other provinces). These accounts construe such changes in terms of native populations 'becoming Roman' through a process of acculturation, by gradually adopting practices and conceptions more similar to the Mediterranean core of the Empire. Within this paradigm, the prevalence of 'Roman' material culture was usually seen as indicative of increased 'Romanization' (Ghisleni 2018, p. 139). Starting in the 1990s, the Romanization debate consisted of archaeologists extensively criticizing this approach. The Romanization model was seen to embody several problematic or overly simplistic assumptions, including among others (a) a reified, uniform conception of 'Roman culture', (b) a one-way transmission of ideas and practices to native populations, which (c) neglected or downplayed the role of these populations in shaping these changes.

While such assumptions were quickly rejected, the ongoing debate concerns what a more adequate framework might be, with archaeologists often drawing on post-colonial theory and analogies with more recent colonial experiences. For instance, Webster (2001) adapts the concept of 'creolization' from North American colonial archaeology, where it is used to describe processes in which slaves combined African and European cultural elements to create novel cultural forms—e.g., using European cooking utensils but African foodstuffs and cooking techniques. This is assumed to be often "a process of *resistant* adaption" (p. 218, original emphasis), where the resulting cultural blends cannot be understood independently of the unequal power relations under which they emerged. Similarly, Webster argues that Roman objects may have been used to "negotiate with, resist, or adapt Roman styles to indigenous ends" (p. 219), rather than signifying any aspiration to become Roman.

The Romanization debate is complex and ongoing. Other alternative frameworks have been proposed, centered around concepts such as 'imperialism', 'globalization' (or 'glocalization') and 'colonialism' (e.g., Ghisleni 2018; Versluys 2014; Pitts and Versluys 2015; Hingley 2011, 2015, 2017). I will not attempt to do justice to these here. What is interesting for my purposes is the role of analogies in some of these frameworks. Specifically, I will look at a few of the ways these questions play out in debates about the mortarium.

3.4 Analogies and the Interpretation of Mortaria

In this section, I highlight some of the ways analogies have been used in debates about the use and significance of mortaria in Roman Britain. My discussion is structured according to the inference strategies outlined in Sect. 3.2.

3.4.1 Direct Extrapolation

Some parts of the debate can be analyzed in terms of direct extrapolation. For instance, Cool (2006, p. 43) criticizes some discussions of the mortarium for simply assuming the function of mortaria to be food preparation, presumably in ways similar to those known from Roman sources. As mentioned above, several archaeologists have questioned this, emphasizing that this cannot be taken as given (e.g., Cool 2004, p. 31, citing Reece 1988, pp. 30–32). One way to understand this criticism is by viewing this assumption as resting on an implicit direct extrapolation from the Italian contexts described in the written sources to British contexts. What the critics are pointing out is that there is not sufficient background knowledge to license the assumption that these two contexts are likely to resemble each other in terms of how a given type of pottery was used.

To be sure, there are some contexts for which direct extrapolations of this type do seem well-supported. For instance, soldiers from the Mediterranean would plausibly have brought with them culinary practices from their home regions. Based on this assumption, we can infer that mortaria found at military forts occupied by such soldiers were likely used in ways similar to these Mediterranean regions. However, since the Romanization debate focuses on whether the existing population adapted similar practices, this type of assumption is not available: it would amount to assuming what the presence of mortaria was supposed to demonstrate.

3.4.2 Hypothesis-Testing and Generation-for-Confirmation

For this reason, most critical discussions of mortaria rely on other inferential strategies, such as hypothesis-testing and generation-for-confirmation. While not all hypotheses tested and generated are analogy-based, some of them are. Recall from Sect. 3.2 that, apart from serving as inspiration for the generation of hypotheses, analogies can play at least two further roles. First, through their knowledge of a given source, archaeologists can reason about the types of evidence we should expect if a mortarium was used analogously to that source (hypothesis-testing). Second, the case for a given interpretation can be strengthened through generating and providing evidence against prima facie plausible alternative interpretations (generation-for-confirmation).

One example is Oswald's (1943) arguments for the cheese-curdling hypothesis. Oswald discusses two analogies. First, he argues against the idea that mortaria were used analogously to a mortar and pestle, among other things by pointing out that the walls are too thin and breakable to be pounded with a pestle.[5] Second, he considers another potential analogy, a similar-looking modern type of milk bowl used in Northern France and Savoy to curdle milk. Oswald suggests that the rough surface could have acted as a reservoir for curdling bacteria, while the spouts could have been used for draining off whey. Finally, he points out that since heating helps separate the whey from the curd, "it is noteworthy that many mortaria show signs of heat [i.e., sooting] on their exterior" (p. 46). The latter argument is an example of how background knowledge about the analogical source can be used to support hypothesis-testing. By drawing on his knowledge of the cheese-making process, Oswald is able to argue that the presence of sooting on mortaria sherds would be expected evidence if the cheese-curdling hypothesis were true.

Cool (2006, pp. 42–46) also discusses different potential analogies for the use of mortaria. First, she points out that mortaria are in fact unusually common in Britain compared to Italy (as an illustration, she cites an extensive exhibition of food-related objects from Pompeii which did not include mortaria). By highlighting the rarity of mortaria in the source of the analogy (Italy), she argues that the high prevalence of mortaria in Britain should be seen as evidence *against* them being used to produce 'Romanized' food (p. 45). Second, she points out that rural sites in first and second century Roman Britain also tend to have disproportionate amount of large glass bowls and samian ware bowls, leading her to suggest that "Perhaps on these sites mortaria were regarded simply as large bowls" (p. 45; Peacock 2016 suggests a similar interpretation for farmsteads in Cumbria).

Finally, under the heading of hypothesis-testing, it is also worth highlighting the residue analyses carried out by Cramp et al. (2011). In this study, researchers analyzed the nature of organic residues that had been absorbed into mortaria sherds, using methods such as gas chromatography and stable isotope analysis. They were able to obtain samples of sufficient quality from sherds from seven sites (six in Britain, one in Germany) and compared these to results from Roman and Iron Age cooking vessels. Thus, one of the hypotheses tested in this study is whether mortaria were used in a way analogous to these cooking vessels.

The study produced several interesting findings, including the following. First, while the mortaria and cooking vessels had contained or been used to process both

---

[5]Note, however, that this does not rule out mortaria being used to grind ingredients by hand or using a stone rubber (Alcock 2001, p. 69).

plant- and animal-based products, plant-derived residues were detected with much higher frequency in mortaria compared to the cooking vessels. Second, most of the animal fats were from the carcass, although at one site, Stanwick (an agricultural village in Northamptonshire), several sherds contained residues from dairy fat. However, plant-based lipids were still more common, ruling out the hypothesis that the mortaria at Stanwick were used as single-purpose cheese-curdling bowls. Third, although mortaria show a higher *proportion* of plant-based fats, they contain the same *types* of residues as the cooking vessels. Finally, the frequency of components is comparable in Iron Age and Roman cooking vessels. For instance, both Iron Age and Roman cooking vessels from Stanwick also contain dairy fat residues, indicating some continuity in diet at this site. Overall, the authors conclude that the mortarium had some "unique role compared with other domestic vessels", different from any known metallic or ceramic Iron Age utensil, which "suggests a shift in cultural practice involving either new commodities, especially plants, or new apparatus or new recipes" (Cramp et al. 2011). However, they are unable to determine whether this role was for culinary or some other purpose, such as preparing cosmetics or pharmaceuticals.

While Cramp et al. thus provide evidence relevant to the function of mortaria, it should be noted that their study only looks at six British sites, four of which are either Roman forts, urban or high-status buildings (e.g., a villa). It is possible that mortaria played different functions for populations in other parts of Roman Britain.

3.4.3 Generation-As-Criticism

Finally, analogies have also been used, through the generation-as-criticism inferential strategy, to challenge interpretations of the significance or meaning of the cultural changes (if any) represented by the spread of the mortarium.

While it has not been applied directly to the mortarium case, the creolization framework can be seen as an example of this: the analogy to the material culture associated with African slaves in North America serves to generate an alternative account of the cultural changes in Britain during the Roman occupation. Assuming Cramp et al. are correct that the mortarium represents the introduction of new practices (at least in the contexts their samples came from), how should we interpret those changes? On a traditional Romanization account, these would be interpreted as the population adopting Roman culture, perhaps driven by an aspiration to become Roman. By contrast, the analogy with slaves in North America would suggest a different interpretation, where the changes are instead interpreted as resulting in a new cultural blend, distinct from both Roman and pre-existing culture, involving a more ambivalent or even resistant attitude to the cultural connotations of the mortarium.

A more recent example is Ghisleni's (2018) challenge to assumptions about what constitutes continuity and change in the Romanization debate. Drawing on work in the archaeology of colonialism, Ghisleni argues that earlier alternatives to the

Romanization account, such as the creolization framework, still tend to presuppose a dichotomy between 'continuity' with pre-existing Iron Age culture vs. 'change', represented by Roman material culture. Even if we focus on more complex notions of cultural transformation beyond simply adopting Roman culture, this still "presupposes that we can recognize when something has been transformed" (p. 139). In Romano-British archaeology, Ghisleni argues, continuity tends to be defined statically by reference to how things were at the time of the Roman conquest, thereby "allying change with what is Roman or what appears to be a result of Roman occupation" (p. 142). This, however, ignores the fact that the culture of a given community is dynamic and itself continually evolving. Some changes may be natural extensions of previous practices; some forms of persistence may represent a break from previous trajectories (pp. 140–141).

Ghisleni cites several examples of these ideas from the archaeology of colonialism. One example is Silliman's (2009, 2010) discussion of European-produced pottery in colonial New England. Silliman criticizes the tendency to identify these unproblematically as 'European' items and as evidence of cultural change when found in Native American households (2010, p. 40): if used as part of daily life across generations, a given type of pottery may have been seen by people in these households as a part of their own life and history, rather than an external cultural influence. Even when found in British/Euro-American households, these often employed Native American servants who would have handled the items at least as often as their owner, making it problematic to interpret them as exclusively 'European'. Based on this observation, Silliman highlights different possible interpretations of how these servants might have viewed such items: perhaps they "cared nothing for the artifacts and resented their symbolic role in marking oppression, perhaps they handled them with care as part of their economic strategies for well-being, or perhaps they or their family members used the same kind of dishes in their own homes and felt some affinity for them" (Silliman 2010, p. 42).

Ghisleni applies this framework, among other things, to the interpretation of mortaria found near Durnovaria (a Roman town founded in Dorset around 65–70 CE), which were part of the sample studied by Cramp et al. (2011). Noting the evidence of changes in preparation techniques together with continuity in diet, Ghisleni argues:

> As mortaria changed hands and uses in different contexts, as they were worked into practice through facilitating the consumption of familiar foods, and as they began to be manufactured in Britain, mortaria-as-introductions may not have always acted as a primary factor in the production of their social meanings. Glossing the category of mortarium as already and only a new object elides potential ambiguities of meaning (Ghisleni 2018, p 149)

As the above example involving European-produced pottery highlights, there are interpretative possibilities for the cultural identity and meaning of the mortarium, beyond those suggested by the Romanization and creolization accounts. Rather than seeing the mortarium and the associated changes in food preparation as a transformation of previous traditions, they may have been perceived as a natural continuation. The mortarium may not have represented cultural changes, adopted either aspiringly or resistantly, but they may instead have been associated with familiar foods or simply viewed indifferently.

3.4.4 Summary: Success and Failure in the Interpretation of Mortaria

In this section, I have highlighted some ways that analogies are used within the Romanization debate, focusing on interpretations of the function and cultural meaning of mortaria. As we have seen, analogies are used here to generate interpretative hypotheses, to reason about evidence for and against these, and to throw doubt on received interpretations. In one sense, we might see all of these as examples of analogies succeeding, insofar as they contribute to the inferential strategies within which they are employed. However, these 'successes' are of a different kind to the ones usually highlighted in optimistic accounts of analogy. Archaeologists have not so far been able to unambiguously confirm any comprehensive account of how mortaria were used in Roman Britain.

Regarding its use, some negative conclusions can plausibly be drawn. First, there seems good reason to doubt that mortaria in Britain generally played the same role as in Mediterranean regions. Second, some of the alternative hypotheses, such as them being single-purpose cheese curdling bowls, also seem doubtful, at least on the sites studied by Cramp et al. To the extent that any positive conclusions can be drawn, these are fairly vague—e.g., it "fulfilled a unique role compared with other domestic vessels ... whether this was for culinary or non-culinary (e.g., cosmetic or pharmaceutical) purposes" (Cramp et al. 2011, p. 1349)—and the evidence so far only comes from a limited range of sites.

Regarding its cultural meaning, the situation is even more ambiguous. While there is ample archaeological evidence that the Roman conquest brought about changes in Britain, for instance in terms of the types of pottery that was used, archaeologists have not reached any consensus on the significance or cultural meaning these might have held for different people living there. Instead, as illustrated by Ghisleni's arguments, the role analogies play in relation to these questions is often mainly to suggest new possible interpretations, not in order to confirm one of them as more likely, but rather to indicate a broader range of interpretative possibilities. Here, 'successful' uses of analogy neither involve confirming some interpretation as accurate nor to show it inaccurate, but rather to highlight how open these questions still are.

The current evidential situation may only be temporary. As I have emphasized, my aim is not to argue that these uncertainties are insurmountable. Further advances in techniques such as residue analysis might yield further evidence which, together with more detailed theoretical analyses, might allow archaeologists to narrow down the range of plausible interpretations of how mortaria were used or perceived in a given context. However, this cannot be guaranteed. In the remainder of this chapter

I want to argue that even if the current situation proves to be persistent, this still represents a significant form of epistemic progress, worth achieving for its own sake.

3.5   Optimism, Progress, and Pursuit Worthiness

Adrian Currie (2018) energetically defends optimism about the historical sciences, including archaeology. Since his view bears some resemblances to the one I want to defend vis-à-vis analogies, it worth spelling out the differences between the two.

Currie defines optimism and pessimism as *predictive stances* concerning the probability that a given activity—in this case historical inquiry—will succeed: "The pessimist predicts that our attempts to reconstruct the past will often fail; the optimist predicts that we will often succeed" (p. 13). Currie emphasizes that success here should be construed in a broad, pluralistic sense. It does not just include discovering true or accurate theories, but also many other types of epistemic goods: "good explanations, adequate representations, precise predictions, new technologies, successful techniques, telling interventions, effective cures, and so forth" (p. 14). Furthermore, truth is not necessary for success: first, idealized models can sacrifice truth for the sake of producing better understanding; second, speculative hypotheses, even if they fail, are an important driver of progress in historical sciences, by generating 'scaffolds' and 'inferential tools' which can increase the epistemic reach of historical inquiry (Currie 2018, ch 10–12).

In some ways, this aligns with my view: I want to provide an argument for optimism about the use of analogies in archaeology that goes beyond them suggesting true or well-established interpretations. However, Currie's optimistic arguments still give more prominence to truth or truth-related notions than what I am after here. First, while idealizations involve falsehoods, they are usually still assumed to 'get things right' in some ways—i.e., have some truth-content—in virtue of which they generate understanding. Second, Currie's account of how failed speculative hypotheses drive progress rely on them providing resources for further stages of inquiry to generate knowledge. By contrast, I want to argue that analogy-based hypotheses can also provide an important form of understanding *in virtue* of them being rejected as false or simply shown to be less certain than previously thought.

A way to further elucidate the differences between Currie's view and mine is by considering how each justifies the pursuit of analogy-based hypotheses. Optimism is closely related to pursuit worthiness: the likelier we think an activity is to succeed, the more reason we have to invest time and resources in pursuing it. Expected utility models, along the lines of Nyrup (2015, 2017, ch 2), provide a useful framework for analyzing notions of pursuit worthiness.

In general, these models are constructed by distinguishing a set of states of the world, $\{s_1, ..., s_n\}$, and a set of epistemic states, $\{e_1, ..., e_m\}$. Each pair of these represent a distinct epistemic outcome. For instance, in a simple model, focused on a single hypothesis h, we might distinguish two states of the world, $\{h, \neg h\}$, representing h being either true or false, and three epistemic states, $\{acc_h, rej_h, sus_h\}$, representing either that we accept, reject or suspend judgement on h. This results in six possible epistemic outcomes (accepting a truth, rejecting a truth, accepting a falsehood, suspending judgement, etc.). For each epistemic outcome, there will be a utility, $u(e_j, s_i)$, representing how valuable achieving that outcome would be, and a probability, $\Pr(e_j, s_i \mid a) = \Pr(s_i) \times \Pr(e_j \mid s_i, a)$, that pursuing an activity a will result in that outcome. The pursuit worthiness of the activity a can then be modelled as the expected utility given by:

$$EU(a) = \sum_i \Pr(s_i) \times \sum_j \left(\Pr(e_j \mid s_i, a) \times u(e_j, s_i) \right) \tag{3.1}$$

For instance, in the simple model, $u(acc_h, h)$ is the utility of a true positive, $\Pr(acc_h \mid h, a)$ is the probability that pursing a will correctly determine that h is true, and so on.

Within this framework, there are two obvious ways to argue for the pursuit worthiness of some activity. One can either show that the *probability* of achieving some positive epistemic outcome (or avoiding some negative epistemic outcome) is higher than previously thought; or one can argue that the *utility* of some epistemic outcome is higher than previously thought.

Typically, optimists about analogies in archaeology have focused on probabilities: they argue that pursuing analogies is more likely to result in us learning the truth of some hypothesis than pessimists assume. In terms of the formal model, if p_{analog} denotes the activity of pursuing analogies, traditional optimists argue for higher values for probabilities of the form $\Pr(acc_h \mid h, p_{analog})$ or $\Pr(rej_h \mid \neg h, p_{analog})$. As long as $u(acc_h, h) > 0$ and $u(rej_h, \neg h) > 0$, i.e., true positives and true negatives have some positive value, this increases $EU(p_{analog})$.

Currie's optimist view augments this in two ways. First, the argument from idealization points out that there are some epistemic outcomes that involve accepting a strictly speaking false hypothesis, but nonetheless produces a valuable form of understanding. Thus, some utilities of the form $u(acc_h \mid \neg h)$ are higher than pessimists assume. Second, the argument from scaffolding highlights that for some hypotheses, the epistemic outcome of rejecting a falsehood can still have an indirect value, by increasing the probability of determining the truth of *other* hypotheses.

Like the argument from idealization, my account also focuses on raising utilities, but instead targets the utilities, $u(rej_h, \neg h)$, $u(sus_h, h)$ and $u(sus_h, \neg h)$, of rejecting false analogy-based hypotheses or suspending judgement. Furthermore, unlike Currie's argument from scaffolding, I want to argue that these epistemic outcomes (at least sometimes) have higher value in a more direct sense, in that their value can be realized even in the absence of any further archaeological inquiry.[6]

---

[6] As will become clear below, realizing the value of this epistemic outcome may still require further reasoning of some sort. Thus, the distinction here is not strictly speaking between extrinsic and intrinsic value.

3.6   Making Progress Through Failed Analogies

In this section, I outline a positive account of the value of failed analogies.[7] It proceeds in three steps. First, I argue that one important epistemic good produced by archaeology is *comparative understanding*—that is, understanding of how social and cultural practices at different times and places resemble and differ from each other—and that analogies are well-suited for producing this type of understanding. Second, I propose an account of why this type of understanding seems particularly valuable in archaeology. Finally, I argue that this account also provides an account of the value of learning how ambiguous and uncertain the answers to some of these questions are.

3.6.1   Comparative Understanding

Many of the questions archaeologists seek to address are comparative. We are not just interested in understanding what life was like for people living on farmsteads in Roman Britain, but also how similar or different this was to life at Roman forts and, in turn, how this compared, say, to life in Italian cities. Understanding the Roman world involves, not just understanding the different times and places that it comprised in isolation, but also patterns of similarity and difference between these. Pushing this argument further, we are not just interested in understanding the Roman world in isolation either. We, including many archaeologists, also naturally wonder how the Roman conquest of Britain compares to other imperial projects, both contemporaneous (e.g., Imperial China) and more contemporary (e.g., the British Empire). Thus, one important epistemic good that archaeology seeks to produce is understanding of social and cultural similarities and differences across different times and places, including our own.[8]

Recall from Sect. 3.2 that I define analogies as comparisons of similarities and differences between two systems. Given this definition, it should be obvious why the pursuit of analogies in archaeology is an effective strategy for producing comparative understanding: it consists exactly in a comparison of the similarities and differences between social and cultural practices (or materials relating to these) at different times and places. Notice that this also accounts for why we are interested in analogies with societies that do or did actually exist, as opposed to analogies with merely hypothetical or fictional societies. Counterfactual history notwithstanding,

---

[7] The argument in this section develops and expands on remarks made in Nyrup (2020b, pp. 22–23, pp. 28–29).

[8] In a paper arguing that archaeology can and should be relevant to anthropology, Binford (1962) expressed this idea thus: "it must be asked, 'What are the aims of anthropology?' Most will agree that the integrated field is striving to *explicate* and *explain* the total range of physical and cultural similarities and differences characteristic of the entire spatial-temporal span of man's existence" (p. 217, original emphasis).

the type of comparative questions I have highlighted above concern similarities and differences across *actual* time and space.

Crucially, analogies are conducive to addressing questions of both similarities *and differences*. While analogical reasoning, on my definition, consists in suggesting further similarities between the source and the target, rejecting such a hypothesis amounts to accepting the hypothesis that they have certain differences. Thus, rejecting a false analogy-based hypothesis directly provides a significant epistemic good, worth pursuing for its own sake.

3.6.2   The Value of Comparative Understanding in Archaeology

In the argument just given, I relied on the intuitive idea that many significant questions in archaeology concern similarities and differences between societies. Why is it valuable to answer such questions?

Notice first that the above observations do not translate straightforwardly to the natural sciences. It is arguably valuable in physics to discover positive analogies—say, between atoms and the solar system or liquid drops and the atomic nucleus.[9] However, it is not obvious that negative analogies are significant in the same way. To be sure, the discovery that the atom is importantly different from a solar system was indirectly valuable, since it allowed physicists to develop more adequate models of the atom. But learning about such dissimilarities does not, in and of itself, seem a particularly significant discovery. In most cases, learning about differences— e.g., that the atom differs from a gyroscope, from a biomolecule or from a crystal lattice—seems of limited value.

By contrast, learning that the mortarium played a different role in Britain than in other parts of the Roman world does seem an interesting archaeological discovery, even if we do not have a positive account of what that role was. Similarly, when it comes to analogies with our own time, learning how the past differs from the present seems at least as valuable as learning of similarities: often, the value of learning about archaeology is an appreciation of how different life in the past would have been from our own. It is this phenomenon I want to provide an account of; when we are concerned with human beings, and thus ultimately ourselves, understanding differences seems to be particularly significant.[10]

My starting point will be the account of the value of historical understanding recently proposed by Grimm (2017). Grimm argues that there is no deep difference in the 'epistemic profile' of understanding produced by natural science and history: in both cases, understanding consists in grasping how different parts of a system

---

[9] See Nyrup (2020a) for further discussion of different accounts of the value of positive analogies in physics.

[10] I do not mean to suggest that understanding differences cannot also be significant in other scientific domains. The account I propose here is specific to the human and social sciences, but is compatible with understanding differences being valuable for other reasons as well.

relate to and depend on each other. For instance, he rejects the idea that things such as narrative understanding or holistic understanding of context are absent from the natural sciences. Rather, he argues that what makes historical understanding distinctive is the extrinsic benefits that come from understanding human affairs, namely that it puts us in a better position to understand how to *live well*, which we do not get in the same way from an understanding of non-human natural phenomena.

Specifically, Grimm argues that historical understanding can help us achieve two kinds of benefits. First, it can help us evaluate the legitimacy of our current institutions by better understanding how and why they were created in the first place. For instance, if we discover that the creation of a given institution was driven by "darker forces such as power, or oppression, or privilege, this gives us a powerful lens through which we can assess our practices not just as rational or irrational, but as just or unjust, as oppressive or discriminatory, and so on" (p. 20). Second, historical understanding can give us a better sense of the different possible ways of being human, thereby offering us "different, and perhaps in some ways superior, models for living well" (p. 21). Grimm gives one concrete example of this: learning more about what goods were regarded as important at other times can help us either to realize that "our own ordering of goods might be misguided or in some ways stunted" (p. 21) or to "appreciate the merit of our own arrangement of goods, insofar as we come to think that a prior society was lacking in various ways" (p. 21, fn. 34).

The second of these, that learning about life in the past gives us new models for how to live well, provides a straightforward account of the value of understanding differences between human groups, and thus of the value of rejecting analogies. Take comparisons between the past and present first: obviously, it is by learning how a past society differs from our own, as opposed to how they resemble each other, that we can obtain a new model. Second, once we have an understanding of the similarities and differences between our own life and life in a given past society, learning about differences between this society and other past societies can provide us with further models, without having to draw analogies directly with the present.

However, the first of the benefits proposed by Grimm is also relevant. While our current institutions are not themselves directly descended from the Roman state, archaeologists and historians have highlighted that many of the concepts we use to understand and justify existing social arrangements—e.g., 'empire', 'civilization', 'just war'—have emerged from Roman concepts and been shaped by the reception of these through various points in history (e.g., Morley 2010; Hingley 2011). Better understanding the genealogy of these concepts can thus help us better evaluate these social arrangements. Furthermore, social institutions or cultural practices are often articulated and justified based directly on analogies with the past. For instance, as Morley (2010, pp. 1–13) details, analogies with the Roman Empire have variously been invoked to justify imperialism ("British rule in India [was described as] 'an Empire similar to that of Rome, in which we hold the position not merely of a ruling but of an educating and civilizing race'", p. 3), to criticize current policies ("the Romans' tolerance of diversity among their subjects [was contrasted] with the actions of the French in Algeria", p. 3), and occasionally to outright denounce institutions (e.g., comparisons between the European Union and the Roman Empire, p. 8).

While Morley focuses on Empire, this points to a more general phenomenon. Whether for better or worse, humans have a natural tendency to understand our social and cultural practices comparatively. We appreciate and evaluate these, at least in part, based on how they compare with other individuals or societies, highlighting similarities with those we admire and distinguishing them from those we despise. In a culture that admires the Romans, highlighting ways in which a set of institutions resemble those of the Romans will often lend them some legitimacy. Conversely, highlighting the role violent repression played in the Roman occupation may lead us to reassess other colonialist projects (Hingley 2017, p. 90).[11] In all of these cases, comparisons with the Romans provide more than just an abstract model, describing one possible way of organizing a given set of social and cultural practices. Rather, the analogies play a more direct role in determining their value and legitimacy.

It would go beyond the scope of this chapter to give a more detailed account of this type of moral reasoning.[12] For present purposes, it is sufficient that it constitutes one significant way that analogies are used to make decisions about how to live well. Since both positive and negative analogies are relevant to this usage, this provides an additional account of the value of pursuing analogies in archaeology.

3.6.3   The Value of Uncertainty

So far, I have argued that pursuing analogies in archaeology can provide a valuable form of understanding, regardless of whether the result is to accept or reject analogy-based hypotheses. The final step of my argument is that even when the result is an increased uncertainty about which hypothesis (if any) is most accurate, this still represents a worthwhile form of progress.

My account builds on certain methodological ideas which have recently been defended within Roman history and archaeology by Morley (2010) and Hingley (e.g., 2011, 2015, 2017).

Having highlighted how analogies with the Roman Empire have been used politically, Morley (2010) goes on to point out that these analogies usually assume an unproblematic knowledge of what the Roman Empire was like. However, he notes:

---

[11] Furthermore, the inferences can run in both directions: the role of many of the analogies invoked within the Romanization debate is to highlight interpretative possibilities according to which Roman cultural influence in Britain was less benign than otherwise supposed. This may in turn inform our assessment of more recent forms of cultural colonialism (Hingley 2017, pp. 99–102, p. 106).

[12] E.g., on the one hand, some instances of this type of reasoning may seem little more than ascribing guilt by association. On other hand, it is usually taken to be an axiom of moral reasoning that cases should be treated alike unless some morally relevant difference can be identified. How to tell apart legitimate and fallacious uses of this type of moral reasoning is not a question I will try to answer here.

> Our actual knowledge of Rome is fragmentary and sometimes contradictory ... the labor of scholars since the eighteenth century has tended to multiply uncertainties rather than establish certainties as growing understanding of the way modern societies work has highlighted our ignorance about the operations of Roman society (Morley 2010, p. 9)

In my terminology, what Morley describes is modern analogies being used within the generation-as-criticism strategy to highlight our uncertainty about the Roman world. This is a valuable contribution, since the political uses of Roman analogies draw their strength from the reality of the (supposed) historical facts they cite. As Morley puts it:

> This is one reason why the Roman Empire is worth studying: not as a means of understanding better how to run an empire and dominate other countries, or of finding a justification for humanitarian or military intervention, but as a means of understanding and questioning modern conceptions of empire and imperialism, and the way they are deployed in contemporary political debates (Morley 2010, p. 10)

Hingley defends a more general argument. Building on observations such as Morley's, he argues that studies of the ancient world are inevitably intertwined with contemporary political ideas and concerns. Because of the "highly fragmentary nature of our understanding of the Roman Empire" (2011, p. 105), it is necessary to draw on contemporary ideas and concepts to "interpret the empire" (p. 105) and "give meaning to the past" (p. 109). He criticizes approaches to classical studies based on avoiding all modern concepts in favor of "immersing oneself in classical texts" (p. 104). Far from avoiding anachronism, the result is rather that "ideas are reproduced in an anachronistic manner, often without any form of conscious acknowledgement" (p. 104). Instead, Hingley advocates a methodological approach where the influence of contemporary ideas and political concerns on archaeological interpretation are explicitly acknowledged and studied. His point is not to undermine or reject all such interpretations, but rather to understand them better in order to ensure that they "will not be misinterpreted, or even misused" (p. 112).[13]

Building on these ideas, then, my argument is this: on the account given above, confirming and rejecting analogies is valuable in part because it helps us understand social arrangements comparatively, in terms of how they compare to societies at other times and places, thereby helping us understand how to live well. However, doing this successfully requires that we are able to *correctly* accept and reject such analogies. Re-evaluating our attitude, say, towards cultural transformation brought about by colonialism based on a mistaken analogy with the Romans is not valuable; we want comparative understanding that is based on how things actually were. Thus, getting a clearer understanding of the uncertainties and ambiguities in our understanding of the past in itself contributes to our ability to live well. It helps us better understand our *epistemic* relationship to the past, thereby enabling us to better evaluate analogy-based reasons for or against the legitimacy and value of current

---

[13] Cf. Hingley (2017, p. 105): "one of the main purposes of this field of study is to enable archaeologists to continue to focus critical attention on the entanglement of the past and the present and to ensure that powerful classical concepts are not used simplistically to justify modern political and military actions".

social and cultural practices. Furthermore, as we saw above, analogies deployed in service of the generation-as-criticism strategy can play a crucial role in helping us obtain this type of understanding.

To be clear, this type of archaeological meta-understanding (i.e., understanding of the uncertainties and ambiguities in our first-order understanding of the past) is only valuable if it is itself accurate. While I follow Gero (2007) in arguing that archaeologists should 'honor' the uncertainties inherent in their interpretations, only those uncertainties which actually exist are worth honoring. Spuriously creating unnecessary ambiguity or abandoning any attempt to determine which interpretations are most plausible in light of the available evidence is as at least as problematic as uncritically ignoring or glossing uncertainty.

3.7 Conclusion

In this chapter, I have used debates about the use and cultural significance of the mortarium, and its relation to the Romanization debate more generally, to provide a broader view of the reasons for using analogies in archaeology. Firstly, there are several different inferential strategies in the service of which analogies can be deployed, including what I have called direct extrapolation, hypothesis-testing, generation-for-confirmation, and generation-as-criticism. All four strategies are illustrated in the mortarium case study. Since the purpose and challenges of each of these differ, the justification and adequacy criteria for using them will differ as well (cf. Nyrup 2020b). Secondly, I have highlighted that the result of archaeological inquiry is often to give us a deeper understanding of the uncertainties and ambiguities we face when trying to understand the past, and that analogies—in particular when used for the purpose of generation-as-criticism—can play a key role in securing this. In this way, archaeology can make progress even through failed analogies.

My account of the value of failed analogies is to a certain extent explorative; it still contains a number of loose ends which will need to be tied up. In particular, I have relied on some intuitive ideas about the comparative nature of our understanding of human phenomena in the past and about the role of analogies in helping us to understand how to live well. Further exploring the nature and value of archaeological understanding will be crucial to obtaining a broader and richer account of the reasons to be optimistic about archaeological inquiry.

Acknowledgements I'm grateful to Jennifer Peacock for helping me come to grips with mortaria and the Romanization debate. This work was supported by the Wellcome Trust [213660/Z/18/Z] and the Leverhulme Trust through the Leverhulme Centre for the Future of Intelligence [RC-2015-067].

References

Alcock JP (2001) Food in Roman Britain. Tempus, Stroud

Ascher R (1961) Analogy in archaeological interpretation. Southwestern J Archaeol 17:317–325

Bartha P (2010) By parallel reasoning: the construction and evaluation of analogical arguments. Oxford University Press, New York

Binford L (1962) Archaeology as anthropology. Am Antiq 28:217–225

Clark JG (1951) Folk-culture and the study of European prehistory. In: Grimes EF (ed) Aspects of archaeology in Britain and beyond. HW Edwards, London, pp 49–65

Cool HEM (2004) Some notes on spoons and mortaria. In: Croxford B, Eckardt H, Meade J, Weekes J (eds) TRAC 2003: proceedings of the thirteenth annual theoretical Roman archaeology conference, Leicester 2003. Oxbow Books, Oxford

Cool HEM (2006) Eating and drinking in Roman Britain. Cambridge University Press, Cambridge

Cramp LJE, Evershed RP, Eckhardt H (2011) What was a mortarium used for? organic residues and cultural change in Iron Age and Roman Britain. Antiquity 85:1339–1352

Currie A (2016) Ethnographic analogy, the comparative method and archaeological special pleading. Stud Hist Phil Sci 55:84–94

Currie A (2018) Rock, bone and ruin: an optimist's guide to the historical sciences. MIT Press, Cambridge, MA

Evans J (1995) Later Iron Age and 'native' pottery in the north-east. In: Vyner B (ed) Moorland monuments: studies in the archaeology of north-east Yorkshire in honour of Raymond Hayes and Don Spratt (CBA research report 101). Council for British Archaeology, York, pp 46–68

Freeman LG (1968) A theoretical framework for interpreting archaeological materials. In: Lee RB, DeVore I (eds) Man the hunter. Aldine, Chicago, pp 262–267

Gero J (2007) Honoring ambiguity/problematizing certitude. J Archaeol Method Theory 14:311–327

Ghisleni L (2018) Contingent persistence: continuity, change, and identity in the Romanization debate. Curr Anthropol 59:138–166

Gould R (1980) Living archaeology. Cambridge University Press, Cambridge

Grimm SR (2017) Why study history? On its epistemic benefits and its relation to the sciences. Philosophy 92:399–420

Hesse M (1966) Models and analogies in science. University of Notre Dame Press, Notre Dame

Hingley R (2011) Globalization and the Roman empire: the genealogy of 'empire'. SEMATA: Ciencias Sociais e Humanidades 23:99–113

Hingley R (2015) Post-colonial and global Rome: the genealogy of Empire. In: Pitts M, Versluys MJ (eds) Globalisation and the Roman world: world history, connectivity and material culture. Cambridge University Press, Cambridge, pp 32–46

Hingley R (2017) The Romans in Britain: colonization on an imperial frontier. In: Beule CD (ed) Frontiers of colonization. University of Florida Press, Gainesville, FL, pp 89–109

Morley N (2010) The Roman Empire: roots of imperialism. Pluto Books, New York

Norton J (2018) Analogy. https://www.pitt.edu/~jdnorton/papers/material_theory/4.%20Analogy. pdf. Accessed 25 June 2020

Nyrup R (2015) How explanatory reasoning justifies pursuit: a Peircean view of IBE. Philos Sci 82:749–760

Nyrup R (2017) Hypothesis generation and pursuit in scientific reasoning. Doctoral thesis, Durham University. http://etheses.dur.ac.uk/12200/

Nyrup R (2020a) Of water drops and atomic nuclei: analogies and pursuit worthiness in science. Br J Philos Sci 71(3):881–993. https://doi.org/10.1093/bjps/axy036

Nyrup R (2020b) Three uses of analogy: a philosophical view of the archaeologist's toolbox. In: Marila M, Ahola M, Mannermaa K, Lavento M (eds) Interarchaeologia 6: archaeology and analogy. University of Helsinki, Helsinki

Orme B (1981) Anthropology for archaeologists: an introduction. Ducksworth, London

Oswald F (1943) The mortaria of Margidunum and their development from AD 50 to 400. Antiqu J 22:45–63

Peacock J (2016) When is a mortarium not a mortarium? analogies and interpretation in Roman Cumbria. In: Erskine G, Jacobsson P, Stetkiewicz S (eds) Proceedings of the 17th Iron Age research student symposium. Archaeopress, Oxford, pp 20–27

Peden W (2019) Direct inference in the material theory of induction. Philos Sci 86:672–695

Phelps JJ (1923) The culinary use of mortaria. Trans Lancashire Cheshire Antiq Soc 39:1–15

Pitts M, Versluys MJ (2015) Globalisation and the Roman world: perspectives and opportunities. In: Pitts M, Versluys MJ (eds) Globalisation and the Roman world: world history, connectivity and material culture. Cambridge University Press, Cambridge, pp 3–31

Reece R (1988) My Roman Britain. Cotswold Studies, Cirencester

Reiss J (2015) A pragmatist theory of evidence. Philos Sci 82:341–362

Rush P (1997) Symbols, pottery and trade. In: Meadows K, Lemke C, Heron J (eds) TRAC 96: proceedings of the sixth annual theoretical Roman archaeology conference. Oxbow Books, Oxford, pp 55–64

Salmon M (1982) Philosophy and archaeology. Academic, New York

Silliman SW (2009) Change and continuity, practice and memory: native American persistence in colonial New England. Am Antiq 74:211–230

Silliman SW (2010) Indigenous traces in colonial spaces: archaeologies of ambiguity, origin, and practice. J Soc Archaeol 10:28–58

Smith B (1977) Archaeological inference and inductive confirmation. Am Anthropol 79:598–617

Smith MA (1955) The limitations of inference in archaeology. Archaeol Newsl 6:3–7

Tyers P (1996) Roman pottery in Britain. Routledge, London

Ucko P (1969) Ethnography and archaeological interpretation of funerary remains. World Archaeol 1:262–280

Ucko P, Rosenfeld A (1967) Paleolithic cave art. World University Library, London

Versluys MJ (2014) Understanding objects in motion: an archaeological dialogue on Romanization. Archaeol Dialog 21:1–20

Webster J (2001) Creolizing the Roman provinces. Am J Archaeol 105:209–225

Weitzenfeld JS (1984) Valid reasoning by analogy. Philos Sci 51:137–149

Wylie A (1988) 'Simple' analogy and the role of relevance assumptions: implications of archaeological practice. Int Stud Philos Sci 2:134–150

Wylie A (2002) Thinking from things: essays in the philosophy of archaeology. University of California Press, Berkeley

Chapter 4
Scaffolding and Concept-Metaphors: Building Archaeological Knowledge in Practice

Bruce Routledge

Abstract Scaffolding and concept-metaphors have emerged as key terms within alternative approaches to the epistemological analysis of archaeological practice. Each term contains definitional ambiguities, including distinctly broad and narrow definitions in the case of scaffolding. I argue that the broad application of the scaffolding metaphor, most closely associated with the work of Alison Wylie, allows one to understand concept-metaphors as a specific category of scaffolding. At the same time, the broad application of the scaffolding metaphor provides a dynamic and flexible way of organizing the epistemological analysis of archaeological practice because it is acting as a concept-metaphor.

Keywords Scaffolding · Concept-metaphors · Epistemology · Archaeology

The recent revival of interest in epistemology within archaeology has seen a decisive turn towards the analysis of knowledge formation in practice. This shift has brought with it a number of terms and phrases that do sufficient intellectual work and contain sufficient definitional ambiguity as to warrant further focused attention. In certain cases, these terms benefit not only from individual exposition and clarification but also from juxtaposition and dialogue.

Two such terms are 'scaffolding' and 'concept-metaphors'. Both terms refer to practices, ideas and phenomena that serve to organize, support, and enable the construction of archaeological knowledge. Both terms are also intentionally broad and flexible in their definition and application. I will argue that this allows 'scaffolding' and 'concept-metaphors' to play the same organizing and enabling role within epistemological analysis that is played by the phenomena they are meant to describe in

B. Routledge (✉)
Department of Archaeology, Classics and Egyptology, University of Liverpool, Liverpool, UK
e-mail: bruce.routledge@liverpool.ac.uk

© Springer Nature Switzerland AG 2021
A. Killin, S. Allen-Hermanson (eds.), *Explorations in Archaeology and Philosophy*, Synthese Library 433, https://doi.org/10.1007/978-3-030-61052-4_4

the construction of archaeological knowledge more generally. I will also argue that 'scaffolding' and 'concept-metaphors' are closely related, with each at least partially incorporating the other.

4.1 Scaffolding

'Scaffold' and 'scaffolding' are semantically overlapping and syntactically flexible words in English, referring to an object of ambiguous number and specificity ('a scaffold', 'the scaffolds', 'the scaffolding'), the component parts of that object ('scaffolding') and an action or process ('to scaffold'). Scaffolds and scaffolding are also common metaphors in English that have come to be used widely in academic and technical writing, retaining their original flexibility by referring to both processes and structures (Caporael et al. 2014, p. 1).

In archaeological discourse, scaffolding has been deployed as a metaphor most consistently by archaeologists interested in extended cognition and 'human-thing relations' and by philosophers interested in the epistemology of archaeological knowledge formation. As we shall see below, these two metaphorical uses of scaffolding share a common genealogy, although in this chapter my attention will be focused primarily on epistemological analysis. Within this epistemological analysis one can identify both broad and narrow applications of the scaffolding metaphor. Hence, in what follows we will consider what these distinct applications share, how they differ and what might constitute an appropriate and effective understanding of the scaffolding metaphor when it comes to the epistemology of archaeological knowledge formation.

The word 'scaffolding' appears 78 times in the text and notes of Bob Chapman and Alison Wylie's important book *Evidential Reasoning in Archaeology* (Chapman and Wylie 2016). It is clearly a *leitmotif* within their work and, as its prevalence suggests, has wide application in the organization and presentation of their arguments. Chapman and Wylie define scaffolding in archaeological knowledge formation in several ways. Most broadly, they refer to scaffolding as the:

> …ladening theory, background knowledge (tacit and explicit), technical skill, social networks, institutional infrastructure, and vigilant reflexive critique—required to make archaeological observation possible and to put the resulting data to work as evidence (Chapman and Wylie 2016, p. 6)

Chapman and Wylie also define scaffolding more narrowly in places. In relation to fieldwork, scaffolding is the:

> ….technical expertise and community norms of practice which are internalized by individual practitioners as embodied skills and tacit knowledge, and externalized in the material and institutional conditions that make possible the exercise, and the transmission, of these skills and this knowledge (Chapman and Wylie 2016, p. 55)

In discussing Stephen Toulmin's (1958) schema of the structure of argumentation in practice, Chapman and Wylie equate 'inferential scaffolding' with the warrants, backing and rebuttals that Toulmin says connect evidence to knowledge claims or support and defend those connections. In this sense, inferential scaffolding is constituted by "…the gap-crossing assumptions, auxiliary hypotheses, [and] background knowledge that constitute middle-range theory in an archaeological context" (Chapman and Wylie 2016, p. 35).

This broad understanding of scaffolding as something that contributes to both the definition and description of archaeological data and to their deployment as evidence in support of knowledge claims is maintained by Alison Wylie in her more recent work. For example, in a paper focused particularly on the uses of legacy data in archaeology, she defines scaffolding as the assumptions (about the cultural-historical subjects under study), background knowledge and technical resources that facilitate the use of material traces as evidence (Wylie 2017a, p. 204). According to Wylie (2017a, pp. 207–208), rather than only warranting evidence in an argument, the taken-for-granted nature of some scaffolding (e.g., traditions of identifying, recording and describing data) can render some data illegible until the relevant scaffolding is reconfigured.

Similarly, in a paper that places archaeological modeling into a more expansive philosophical context, Wylie (2017b) defines three types of models in archaeology; *phenomenological* models, which systematically redescribe archaeological data or interpretive analogs, *reconstructive/explanatory* models, which seek to reconstruct and/or explain target archaeological contexts or cultural processes, and *scaffolding* models which are "…auxiliary hypotheses that mediate the interpretation of archaeological data as evidence relevant for positing and testing hypotheses about the archaeological target of interest" (Wylie 2017b, p. 995). While it is tempting to equate scaffolding models narrowly with inferential scaffolding within a Toulmin schema of practical argumentation, Wylie is clear that scaffolding models are phenomenological models used to scaffold arguments and that explanatory models are assemblages of smaller scale phenomenological and scaffolding models. Hence, what counts as the descriptive versus the warranting or explanatory resources of archaeology are not easily teased apart.

Independently, Marcos Llobera (2012, pp. 503–505) has also used the phrase 'scaffolding models and/or methods' in the context of GIS applications that might serve as 'middle-ground' solutions to the problem of bridging between interpretive narratives and concrete archaeological landscapes. Llobera describes these in-between links as "…the different measures, techniques, constructs, and strategies archaeologists may mobilize when constructing and exploring possible arguments" (Llobera 2012, p. 500). Again, this could be understood narrowly as inferential scaffolding, however, the examples given by Llobera (2012, pp. 504–505) include systematic redescriptions of data that could serve in concept exploration or narrative construction, as well as in the warranting of arguments.

Wylie's broad definition of scaffolding as encompassing the material, institutional, and conceptual resources of domain specific knowledge formation is seen by some scholars as a distinctly original contribution (e.g., Monteiro et al. 2018). Wylie herself, however, makes no such claim to originality, citing John Norton (2014) and William Wimsatt (2014) as sources for her understanding of scaffolding as a metaphor (Wylie 2017a, p. 221, n. 5; Chapman and Wylie 2016, p. 53, n. 32). As neither of these scholars provide a straightforward model or exemplar of 'scaffolding analysis' of which Wylie's work could be said to be an application, it is worth looking at each of these sources more closely.

John Norton (2014) defends and expands his well-known 'material theory of induction' (Norton 2003) in part by deploying the metaphor of scaffolding. Norton's core argument is that inductive inferences escape 'Hume's problem' of infinite regress because in a mature science such inferences are warranted not by asserting a universal rule but by the deployment of well-supported facts within a given domain. As he states:

> Facts are inductively grounded in other facts; and those in yet other facts; and so on. As we trace back the justifications of justifications of inductions, we are simply engaged in the repeated exercise of displaying the reasons for why we believe this or that fact within our sciences (Norton 2003, p. 668)

Importantly, "…all inductive inference is local" (Norton 2003, p. 647) and this local induction avoids an infinite regress because its domain specific facts are densely intertwined rather than hierarchically related in a chain of dependence. Indeed, Norton compares scientific knowledge to an arch built with mutually supporting stones rather than a tower where each course is dependent on its predecessor in a sequence (Norton 2014, pp. 685–686). This image of knowledge construction shares key attributes with Wylie's long-standing metaphor of cables to describe the robustness of evidential reasoning in archaeology when it involves the consilience of multiple, independent (relative to each other), lines of evidence (Wylie 1989). However, there is an important difference. Whereas Wylie classifies all of this warranting activity as scaffolding, along with the institutions, materials, practices, etc., that make this warranting possible, Norton deploys the metaphor much more narrowly.

For Norton, 'scaffolding' refers to the initial conjectures or hypotheses that shape the early stages of research programs before domain specific facts are sufficiently well-established to justify inductions. This scaffolding of preliminary hypotheses allows research to be conceived and executed in a given domain but will eventually be replaced by well-supported facts in a mature science (Norton 2014, p. 687).

In an archaeological context, Norton's use of the scaffolding metaphor is best compared with that of Adrian Currie's, a philosopher who has analyzed archaeological inferences as one component of his larger research program investigating the epistemology of the historical sciences (Currie 2018). Like Wylie, Currie emphasizes what he calls the 'methodological ominvory' of historical scientists, who opportunistically make use of a diverse range of evidence in building knowledge about the past. Unlike Norton, Currie gives detailed attention to exactly how material postulates license inferences in domain specific ways within the historical sciences through, for example, middle-range theories, analogies, and causal models. However, like Norton, Currie (2015, 2018, pp. 266–273) deploys the scaffolding metaphor narrowly around what he terms 'investigative scaffolding'. As Currie notes, research in the historical sciences often proceeds in a piecemeal or incremental fashion. Frequently, the evidential relevance of data only becomes evident when a knowledge claim is already on the table. These initial hypotheses serve as scaffolds supporting the recognition of newly relevant data sources, more refined hypotheses and more sharply discriminating empirical tests. Currie argues that investigative scaffolding is often characterized by idealizations, 'one-shot' (single cause) hypotheses (Currie 2019), gap-filling narratives (Currie and Sterelny 2017) and obvious simplifications or conflations as the granularity of initial hypotheses must match that of the available data. Once new data has been recognized as relevant, hypotheses can be modified, made more precise or replaced through a process that Currie terms de-idealization.[1]

One should not exaggerate the differences between Currie and Wylie in their use of the scaffolding metaphor. For example, Chapman and Wylie (2016, p. 45) use the term 'epistemic iteration' (citing Chang 2004) to describe what Currie has called 'investigative scaffolding', incorporating it into their account of how the overall scaffolding of archaeological knowledge is constructed. However, there are at least two significant differences in how each is deploying the metaphor of scaffolding. Whereas both Norton and Currie stress the temporary nature of conceptual scaffolding, Wylie stresses its entrenchment. Indeed, in a review of Chapman and Wylie (2016) Currie admits:

> …some disquiet with 'scaffolding' as an analogy…. An architectural scaffold isn't such simply because it supports, but because it is not a proper part of the completed structure— often, a scaffold is removed at or before completion. However, for Chapman and Wylie it's not obvious that there is a completed product for archaeology…. Moreover, it isn't clear that archaeological scaffolds are removable, or separate to the 'building': although the scaffolds might be removed or at least altered, it isn't part of their function to be so (Currie 2017, p. 785)

The second clear difference lies in Wylie's more expansive use of the scaffolding metaphor to incorporate the material and institutional resources of archaeological knowledge formation as well as its conceptual resources. This difference is recognized to some extent by Marco Tamborini (2020) who uses Chapman and Wylie (2016), amongst others, to argue that in the case of palaeontology, Currie's definition of investigative scaffolding should be expanded to incorporate the technological infrastructure of research practice (e.g., initially paper-based and now digital technologies). Tamborini (2020, p. 65) argues that new technological settings play as significant a role as preliminary hypotheses in scaffolding the recognition of previously unimaginable data, modes of analysis and hypotheses.

Both an expansive definition of scaffolding and its entrenchment can be found in the work of William Wimsatt (Wimsatt 2019, 2014; Wimsatt and Griesemer 2007), Wylie's other source for her understanding of the metaphor. However, despite Wimsatt being a philosopher of science (e.g., Wimsatt 2007), the epistemological

---

[1] Walsh (2019) provides an elegant analysis of this process in relation to Newton's experiments on the periodicity of light, while Janssen (2019) describes the history of the development of relativity and quantum theory through the metaphor of arches built on scaffolds of earlier theories that were then wholly or partially discarded.

implications of his use of the scaffolding metaphor are somewhat opaque and have been actively extracted by Wylie. Wimsatt's larger project is concerned with constructing a non-reductionist model of cultural evolution that is nonetheless compatible with evolutionary developmental biology (so-called 'evo-devo'). To this end, Wimsatt focuses on the role of scaffolding and generative entrenchment in the formation of cultural environments within which human beings develop and which themselves change over time. For individual humans, the development of competencies (e.g., language, social skills, motor skills, etc.) must be scaffolded by a variety of relational and environmental means, creating dependencies that entrench such skills so as to scaffold future performative elaborations and innovations. At the same time, new cultural phenomena are scaffolded on a base of existing phenomena linked through networks of dependencies.

Wimsatt defines scaffolding most succinctly as:

> ...structures or structure-like dynamic interactions among performing individuals that are the means through which other structures or competencies are constructed or acquired by individuals or organizations...; something scaffolds an action or class of actions for an individual or group of individuals, often in a larger system of interactions, in a characteristic environment or set of environments relative to a goal. Material or ideational entities that contribute to achieving this goal are scaffolds (Wimsatt 2019, p. 22)

What counts as scaffolding for Wimsatt is very broad indeed, although he defines three general categories (Wimsatt and Griesemer 2007, pp. 276–281; Wimsatt 2014, p. 81): 'artifact scaffolding'; 'infrastructure scaffolding'; and 'developmental agent scaffolding' or more simply 'agent scaffolding' (i.e., agents who facilitate skill development in other agents). Scaffolding that supports a wide range of other phenomena can be said to be entrenched insofar as the disruptive 'downstream' implications of changes to that scaffolding acts as an incentive for its conservation. For Wimsatt, 'generative entrenchment' occurs when entrenched scaffolding is used to generate new phenomena principally because it is already available in the environment. A simple archaeological example would be the way in which the 'born digital' coordinate data generated by total station surveying instruments has scaffolded the introduction of further digital recording methods that employ this coordinate data, such as GIS and digital photogrammetry, as well as the introduction of alternative technologies for collecting digital coordinate data such dGPS or laser scanning. All of this entrenches digital surveying instruments into the infrastructure of archaeological fieldwork in ways that are rarely the subject of explicit reflection (but see Huggett 2017).

The idea of development is essential for Wimsatt since it provides the elements of inheritance and change necessary to make his descriptive model evolutionary. Indeed, Wimsatt borrows the metaphor of scaffolding directly from developmental and educational psychology where it has deep roots, albeit transformed in this borrowing through its pairing with entrenchment.

In his comprehensive review of the use of the scaffolding metaphor in developmental and educational psychology, C. Addison Stone (1998) traces its introduction to the paper "The role of tutoring in problem solving" by Wood et al. (1976). This paper analyses the role played by a tutor in teaching a group of 3–5 year olds to build a three dimensional structure with blocks that requires skills initially beyond those that the children possess. The authors define six kinds of interactions between the tutor and the children that help the children master the task, referring to these interactions as "the scaffolding process" (Wood et al. 1976, p. 98). This study is linked closely to the translation and reception in the English-speaking world of the work of the early Soviet psychologist Lev Vygotsky, especially his concept of the Zone of Proximal Development (Vygotsky 1978, pp. 84–91). The Zone of Proximal Development (ZPD) for any given child "…is the distance between the actual developmental level as determined by independent problem solving and the level of potential development as determined through problem solving under adult guidance or in collaboration with more capable peers" (Vygotsky 1978, p. 84). While Vygotsky does not use the word 'scaffolding' (i.e., строительные леса) and Wood et al. (1976) do not cite Vygotsky, Stone (1998, p. 345) shows that the metaphor of scaffolding was quickly attached to Vygotsky's ZPD in English language studies within developmental and educational psychology. Indeed, scaffolding became (and seems to have remained) the primary means of describing the methods, resources, and settings (e.g., classrooms) used to facilitate skills development in children.

Importantly, amongst the four generally accepted characteristics of scaffolding in developmental psychology is the assumption that it is temporary, such that the initial scaffolding will be withdrawn as responsibility for learning is transferred from the adult to the child (Stone 1998, p. 349). Brendan Larvor (2018) notes that the use of scaffolding to refer to more permanent, or indeed entrenched, features arises when the metaphor 'jumps' from developmental psychology to cognitive science through its key role in Andy Clark's influential model of extended cognition. Clark is also influenced by Vygotsky and links Vygotsky's ZPD to the scaffolding metaphor in his early work on extended cognition, especially his book *Being There* (Clark 1997). Clark's work appears to be the primary source of the scaffolding metaphor as deployed by archaeologists interested in extended cognition and 'human-thing relations' (e.g., Coward 2016; Hodder 2011, pp. 35–36; Knappett 2005, pp. 58–62; Malfouris 2013).

Clark's idea of human cognition as being embodied and extended via external augmentation makes extensive use of the scaffolding metaphor:

> We may call an action scaffolded to the extent that it relies on some kind of external support. Such support could come from the use of tools, or the knowledge and skills of others; that is to say, scaffolding (as I shall use the term) denotes a broad class of physical, cognitive and social augmentations—augmentations which allow us to achieve some goal which would otherwise be beyond us (Clark 1998, p. 163)

While initially appearing rather similar to its application in developmental psychology, Clark's use is distinct in that many of the augmentations that scaffold human actions are not temporary—one can learn how to drive a car under instruction and ultimately remove the instructor, but remove the car and one cannot drive (cf. Larvor 2018).

From here it is only a short step to Wimsatt's entrenched scaffolding, although one further transformation is needed, namely the shift from Clark's focus on scaffolded individual cognition to what Kim Sterelny (2010, p. 471) terms 'environmentally scaffolded intelligence' (see also Sterelny 2012). The collective, distributed, and social nature of cognitive scaffolding implied by a scaffolded environment (as against extended individual cognition) is necessary for Wimsatt's focus on cultural evolution and better suited to Wylie's interest in archaeology as a community of practice.

Wylie's expansive use of the scaffolding metaphor to describe all of the material, institutional and conceptual resources that make possible the creation of archaeological knowledge fits within a genealogy that stresses the multifarious forms of external support that enable cognitive activity and shape it in path-dependent ways. To be clear, Wylie has sharpened and transformed Wimsatt's broad evolutionary program into a focused tool for epistemological analysis. However, her analysis of knowledge formation in archaeology retains both the open-ended definition of scaffolding as all forms of epistemic support, and the dynamic of its entrenchment through 'downstream' path-dependencies. The question remains, however, as to whether scaffolding is in fact a good metaphor for the complex and heterogenous set of resources that Wylie highlights in her analysis?

Brendan Larvor (2018), for example, thinks that scaffolding is a very poor metaphor for similar aspects of mathematics if evaluated on a literal and point-by-point basis (e.g., scaffolds are temporary, extrinsic, rigid, non-responsive, etc.). Larvor states:

> ...the scaffolding metaphor radically misdescribes the help that we get from mathematical inscriptions and other elements of mathematical material culture (such as cardboard models, computer-generated images and shapes drawn in the air). It does not readily express the to-and-fro between inward cogitation and the manipulation of symbols and diagrams, nor the process of internalizing shared, materially mediated mathematical practices (Larvor 2018, no pagination)

However, metaphors are not productively evaluated in this strictly literal sense. A metaphor works in terms of what we gain from the juxtaposition of the tenor and the vehicle, especially the insights that this juxtaposition makes possible that might not otherwise cohere. In this sense, metaphors can still work when they are partial or incomplete. As Stone notes in his discussion of the scaffolding metaphor in developmental psychology, the real danger lies in the dissipation of a metaphor's impact through its overextension.

> Part of the power of a metaphor derives from the richness of the image it evokes and from the analogy between elements of that image and the as-yet undiscovered or poorly conceptualized elements of the novel domain to be explored. To the extent that a metaphor fails to constrain our thinking about that novel domain, it loses some of its inherent power (Stone 1998, p. 351)

In this sense I would argue that Wylie's broad use of the scaffolding metaphor is productive because of the work it does in constraining our thinking regarding "as-yet undiscovered or poorly conceptualized elements" within the epistemology of archaeological knowledge formation. To demonstrate this point, I will try to explicate the epistemological work done by Wylie's use of the scaffolding metaphor. I will begin from two unlikely sources: a brief but explicit critique of Wylie's broad application of scaffolding; and V. Gordon Childe's understanding of the Three-Age (i.e., 'Stone Age, Bronze Age, Iron Age') System.

4.2   Concept-Metaphors

In an otherwise positive review of *Evidential Reasoning*, Gavin Lucas (2017a, pp. 742–743) directly questions Chapman and Wylie's broad application of the scaffolding metaphor. Lucas argues that while inferential scaffolding works quite well within some version of Toulmin's schema, its utility is dissipated when applied to settings, such as fieldwork, that are less clearly structured as an argument. Unlike Kristin Kokkov (2019) who seems to only recognize scaffolding in the narrower sense of inferential scaffolding in her critique of Chapman and Wylie, Lucas recognizes but rejects their broader application of the metaphor. He does so on grounds that are introduced in his review but most fully developed in his subsequent book, *Writing the past: knowledge and literary production in archaeology* (Lucas 2019). To summarize briefly, Lucas argues that archaeological knowledge is formed through distinct genres of writing, each characterized by distinct knowledge practices and epistemic registers. For Lucas (2019, p. 61) scaffolding works as inferential scaffolding and finds its place as a practice of argumentation, one of his four modes of archaeological knowledge formation. In contrast, scaffolding appears to play no part in Lucas's discussions of his other modes of archaeological knowledge formation; namely narrative, description and exposition.

Lucas recognizes the artificial fragmentation implied by his definition of distinct literary genres, since archaeological knowledge moves between these modes, both across and within actual texts. Hence, in the final section of *Writing the past* he explores the different means by which archaeological knowledge moves between literary modes and actual texts, offering 'paradigms'/'exemplars', 'models'/'analogies' and 'concepts'/'concept-metaphors' as common strategies by which archaeological knowledge is 'packaged' in order to make it mobile (Lucas 2019, pp. 136–159). According to Lucas, these three strategies are distinguished by their decreasing (from paradigm to concept-metaphors) ontological commitments and their concomitantly increasing mobility.

My purpose in this chapter is not to critique or develop Lucas's schema of literary modes nor his extended discussion of how archaeological knowledge so constructed moves about. Instead, I want to focus on one example that Lucas deploys in discussing the mobility of concepts and theories within archaeological knowledge formation.

Lucas (2019, pp. 156–157) uses Julian Thomas's (1993) analysis of the changing referents of 'the Neolithic' to illustrate what he terms mobile concepts (following Bal 2002) or concept-metaphors (following Moore 2004). To summarize, the term 'Neolithic' was originally coined in the late nineteenth century as a universal technological stage, it was transformed by V. Gordon Childe into a functional-economic mode ('food-producing societies') and by the mid-twentieth century had moved decisively away from a technological definition towards a regional/cultural one. Since the 1990s, 'the Neolithic' has come to refer also to a *discursive field* with intersecting themes, concepts, and issues, albeit one that still carries some taxonomic implications within relative chronological sequences. Lucas's argument is that the Neolithic has become a *concept-metaphor*, that is to say a flexible concept with relatively weak ontological commitments that can be deployed across a wide range of text types and data sets to facilitate the creation and movement of archaeological knowledge. Thomas's (1993) intellectual history of Neolithic studies is contestable, but for our purposes the details of his sequence of developments are secondary to Lucas's suggestion that the Neolithic so-conceived is a concept-metaphor. Here it is helpful to quote Henrietta Moore's definition of concept-metaphors at length:

> Concept-metaphors like global, gender, the self and the body are a kind of conceptual short-hand.... They are domain terms that orient us towards areas of shared exchange.... Their exact meanings can never be specified in advance—although they can be defined in practice and in context—and there is a part of them that remains outside or exceeds representation. One of their very important roles is to act as a stimulus for thought...and to act as domains within which apparently new facts, connections or relationships can be imagined (Moore 2004, p. 73)

What makes Moore's definition interesting for our purposes, is that V. Gordon Childe himself repeatedly deployed the metaphor of scaffolding when discussing the relationship of his own work on the Neolithic to Thomsen's Three-Age system, and he does it in a manner that is very compatible with Moore's definition and Lucas's analysis. How then might Childe's understanding of the Neolithic as scaffolded by the Three-Age System relate to Lucas's understanding of the Neolithic as a concept-metaphor?

4.3   V. Gordon Childe on Scaffolding

Throughout the 1940s and 1950s V. Gordon Childe repeatedly deployed the metaphor of scaffolding to describe the role of Thomsen's Three-Age System in constructing then current understandings of relative chronology, technological change and social evolution (Childe 1944, p. 7; 1946, p. 249; 2004 [1947], p. 89; 1953, p. 88; 1956, p. 93). Childe's statements were succinct but very focused and consistent. While the Three-Age system was no longer an accurate representation of specific archaeological sequences, either in terms of chronology or technological development, it "...did give a scaffolding within which a more coherent structure could be, and has been, reared" (Childe 1944, p. 7). For our purposes, it is secondary that Childe's representation of the Three-Age system as stages based on the isolated traits of material and technique was misleading (cf. Rowley-Conwy 2007, pp. 48–81). Childe understood himself to be working within Thomsen's system but using it as a vehicle to build something new, namely a functional-economic understanding of the Three-Ages that fused the evolution of the forces and modes of production (e.g., Childe 1935). For Childe, "Nowadays, archaeologists merely use the scaffolding provided by Thomsen's classification as a frame for describing cultures and their succession and finding out how their relics functioned in a working economy" (Childe 1946, p. 249).

Childe uses the metaphor of scaffolding to describe something that is useful and provisionally necessary for the creation of new knowledge but is, at the same time, extrinsic to that knowledge. Like scaffolding on a building site, Childe even argues for the Three-Age System's ultimate redundancy. Chronologically he wished that "…the editors of *Danske Oldsager* had had the courage to discard the scaffolding constructed by their illustrious predecessor, Thomson [sic]….[in order] to divide the whole of prehistory from Bromme on into a single series of periods, numbered consecutively and undistorted by superfluous technological adjectives" (Childe 1953, p. 88; see also Childe 1956, p. 92). In terms of social evolution, he argued that "If archaeological data are to be really serviceable in the social sciences, they must be presented classified on a new and less superficial basis" (Childe 2004 [1947], p. 90). However, in tension with this disposability, Childe also recognized the deep entrenchment of the Three-Age System, such that "Various attempts have been made to give these hallowed terms some other content…" (Childe 2004 [1947], p. 89). Hence, like actual scaffolding, Childe recognized that the Three-Age System might also be usefully reconfigured as a bridge between the old and the new rather than simply discarded.

4.4   Concept-Metaphors as Scaffolding

In presenting his functional-economic Neolithic as scaffolded by the technological Neolithic which emerged from the Three-Age System, Childe makes rather effective use of the scaffolding metaphor. The Three-Age System may have been wrong in detail and emphasis, but it organized a fuzzy set of sequences, sites, assemblages, types, and technologies that were interrelated in time and space. As such, it could scaffold a reimagining of the relationship between members of this set, while also making relevant new forms of evidence and new programs of research. The subsequent iterations of the Neolithic as detailed by Thomas (1993) could be understood in similar scaffolding terms. Each transformation in the referents of the Neolithic was scaffolded on an already entrenched understanding of the Neolithic while at the same time making possible the recognition of new, or newly relevant, sources of data (e.g., the increasing importance of monuments as a component of the Neolithic in Europe). But if the Neolithic is a form of scaffolding can it also be a concept-metaphor?

Importantly, each iteration of the Neolithic was scaffolded by its predecessors in very particular ways. Superficially, there is some resemblance between the succession of scaffolded reinterpretations of the Neolithic and Currie's 'investigative scaffolding', especially the iterative process and sequential revelation of new sources of evidence. However, any given version of the Neolithic is conceptually broader, ontologically thinner, and nominatively more durable than the preliminary hypotheses that scaffold a specific program of research. This nominative durability is important. Unlike a preliminary hypothesis, the Neolithic is not replaced, only redefined. Lucas (2019, pp. 136–159) has stressed the 'deracinated' nature of concept-metaphors, which minimizes the ontological commitments they demand and thereby facilitates their movement between disciplines and research programs. However, he also recognizes that (much as we have already noted for metaphors in general) if concept-metaphors become too elastic, too 'hollowed out', they lose their analytical potential (Lucas 2019, p. 155). For example, Lucas suggests that the wide application of the concept of 'landscape' within archaeology in the 1990s and early 2000s had just such a dissipating effect on its interpretive value (Lucas 2019, p. 155). This raises an important point that is otherwise underemphasized by Lucas, to maintain their creative analytical potential concept-metaphors must retain certain focal points. These focal points justify the nominative durability of concept-metaphors and provide the coherence necessary for them to act as scaffolding. Perhaps, as Thomas (1993) suggests, the Neolithic is best approached as a field of discourse rather than as a totality, but it is a discursive field that retains certain focal points. In Britain, for example, the Neolithic is a field of discourse that continues to include Grooved Ware and Stonehenge while excluding *Terra sigillata* and *The Mary Rose*.

Archaeology makes extensive use of concept-metaphors to scaffold new research programs in this sense. For example, Lucas (2017b, p. 187) has already suggested that 'assemblage' can be thought of as a concept-metaphor. Traditional uses of this term in archaeology typically include two seemingly distinct referents, assemblages are heterogenous sets of artifacts associated via their depositional proximity and assemblages are homogenous sets of artifacts associated via their typological similarity (see Lucas 2012, pp. 193–198). The classic debate between Lewis Binford (Binford and Binford 1966; Binford 1973) and François Bordes (Bordes 1953; Bordes and de Sonneville-Bordes 1970) over functional versus cultural/ethnic explanations for variability between Mousterian lithic assemblages ultimately hinges on distinct understandings of what an artifact assemblage represents. Most recently, the unusual choice of the English word 'assemblage' to translate Deleuze and Guattari's (1987) use of '*agencement*' in French (Philips 2006) has reinvigorated theoretical discussions of assemblages within archaeology (Hamilakis and Jones 2017). Despite these seemingly fundamental changes in research programs, 'assemblage' remains an essential archaeological term. Consider V. Gordon Childe's normative view that:

> …artifacts hang together in assemblages that recur repeatedly not only because they were used together in the same 'age', but also because they were used by the same people, made or executed in accordance with techniques, rites or styles prescribed by a social tradition, handed on by precept and example (Childe 2004 [1947], p. 83)

in comparison with Yannis Hamilakis's Deleuzean view that assemblages are:

...temporary co-presences, deliberate arrangements and articulations of things, beings, enunciations, memories and affects brought together and enacted as such by sensorality (Hamilakis 2017, p. 176)

Whilst representing radically different views of assemblages, they also intersect around the interpretive importance of association, evident in words and phrases like 'hang together', 'co-presences' and 'articulations'. This focal point is what gives 'assemblage' its nominative durability and allows it to act as a concept-metaphor, repeatedly scaffolding quite distinct programs of archaeological research, each of which accepts that associations are important in order to focus on what, how and why phenomena are associated in specific contexts. In this sense, Moore's definition of concept-metaphors highlights a specific kind of scaffolding, one characterized by domain terms that: (1) point to areas of shared exchange; (2) exceed representation; but (3) can be defined in practice and context; and (4) thereby allow new facts, connections and relationships to be imagined.

4.5   Scaffolding as a Concept-Metaphor

Trowels and Munsell Soil Color Charts, field schools and teaching traditions, investigative scaffolding, inferential scaffolding and concept-metaphors constitute a heterogenous mix of material, infrastructural and conceptual resources. In the broad application of the scaffolding metaphor these all constitute scaffolding insofar as they support and shape the production of archaeological knowledge. I suggested that this stretching of the scaffolding metaphor was justified because it did significant epistemic work. So, what is this work and how does the scaffolding metaphor get it done?

Here it is important to note that in Wylie's broad use of the scaffolding metaphor one cannot define what constitutes scaffolding *a priori*; her definitions are composed of open-ended lists of exemplars rather than parameters. Instead our attention is drawn in a very focused manner to two key points. First, knowledge is not self-warranting, it depends upon a potentially vast array of material, infrastructural and conceptual supports and, second, these supports can engender dependencies with downstream implications for the shape and direction of future knowledge formation. From these limited foci, key issues for epistemological research emerge. Evoking the scaffolding metaphor immediately requires one to specify for a given context what is the scaffolding, what is being scaffolded, by what means and with what downstream path-dependent effects. Scaffolding as a noun, if you will, can only be revealed by first paying attention to what it means 'to scaffold' as a verb in a given context. In other words, scaffolding reveals itself in the analysis of practices in context. Wylie has expressed this more generally as one of the implications of the work of post-positivist philosophers of science such as Ian Hacking and Andrew Pickering, stating:

> The upshot then, is that for those engaged in philosophical science studies, now more than ever before the questions of just what sorts of factors contingently shape the practices, goals, standards, regulative ideals and products of the sciences is genuinely open-ended— an empirical, a posteriori question. Philosophy is thus returned to active engagement with the sciences on several dimensions (Wylie 2002, p. 12)

The sharp focus, but limited specification, of Wylie's scaffolding metaphor is therefore potentially generative of new research programs and forms of evidence that were not necessarily imagined in the metaphor's original formulation. In this way, the scaffolding metaphor itself scaffolds the epistemological analysis of knowledge formation in practice and I would suggest it does this because the scaffolding metaphor is a concept-metaphor.

Much as Lucas (2019, pp. 136–159) suggests, the scaffolding metaphor has undergone some degree of 'deracination' as it has moved from developmental psychology, to cognitive science, to cultural evolutionary theory, to epistemological analysis. In so doing, its key attributes have narrowed such that scaffolding is now a domain term that provides a focal point for research, highlighting phenomena that support and shape knowledge production, while exceeding any representation of those phenomena outside of the analysis of specific contexts and practices. Indeed, it is because the constitution of scaffolding cannot be defined wholly in advance, that the scaffolding metaphor invites evidential and analytical innovation.

4.6   How Entrenched Are our Scaffolds? How 'Emptied' Are our Concept-Metaphors?

One further strength of Alison Wylie's broad application of the scaffolding metaphor is the recognition that any epistemic scaffolding can become entrenched via the path-dependencies it engenders. As such, scaffolding analysis demands a critical/reflexive orientation as well an exploratory/analytical one. Entrenched scaffolds need regular re-examination to uncover their downstream effects and evaluate their fitness-for-purpose. Concept-metaphors present a particular challenge due to their flexibility and changing referents. How 'emptied' of ontological commitments are our concept-metaphors? Do these concepts really move 'baggage-free' from iteration to iteration? How do earlier understandings adhere with continuing effects? It is well-known, for example, that regardless of what archaeologists make of terms like 'the Neolithic', the progressivist underpinnings of the Three-Age system continue to impact popular understandings of these terms (e.g., the use of 'Stone Age' in the writings of Jared Diamond). As Harry Allen notes, with regards to the impact of our progressivist taxonomic terms on representations of Australian Aboriginal history and culture:

> The archaeologist's dilemma lies in the fact that they continue to use the same terms but argue that these have new meanings. The rub, however, is the singular lack of success archaeologists have had in convincing the public to accept new and technical meanings for

long familiar terms. Archaeology cannot easily free itself from concepts which represent a nineteenth-century metaphysic and episteme (Allen 2015, p. 190)

The point is not that we should (or should not) cease to talk about the Neolithic, but rather that the critical analysis of its downstream effects arises naturally from Wylie's dual focus on scaffolding and its entrenchment. The same applies to Wylie's scaffolding analysis itself. If, for example, we were to agree with Larvor (2018) that scaffolding was a poor metaphor for our domain specific epistemic supports, or with Lucas (2017a, p. 742) that the metaphor was being stretched too far, our critique would need to be justified on these same grounds; the path-dependencies that the scaffolding metaphor engendered and the routes to understanding that these dependencies obscured. This potential for auto-critique highlights the fruitfulness of Wylie's approach. As a concept-metaphor, entrenched scaffolding invites evidential and analytical innovation in terms of the analysis of knowledge formation in practice, but it does so with a demand for vigorous, on-going, reflexive critique.

Acknowledgements I would like to thank participants in the University of Exeter's Zoom workshop "Archaeology Works" for their helpful comments on this paper, especially Adrian Currie, Alison Wylie, and Assaf Nativ.

References

Allen H (2015) The past in the present? archaeological narratives and aboriginal history. In: McGrath A, Jebb M (eds) Long history, deep time: deepening histories of place. Australian National University Press and Aboriginal History Inc , Canberra, pp 171 202

Bal M (2002) Travelling concepts in the humanities: a rough guide. University of Toronto Press, Toronto

Binford L (1973) Interassemblage variability-the Mousterian and the 'functional' argument. In: Renfrew C (ed) The explanation of culture change: models in prehistory. Duckworth, London, pp 227–254

Binford L, Binford S (1966) A preliminary analysis of functional variability in the Mousterian of the Levallois facies. Am Anthropol 68:238–295

Bordes F (1953) Essaie de classification des industries 'moustériennes'. Bulletin de la Société Préhistorique Française 50:457–466

Bordes F, de Sonneville-Bordes D (1970) The significance of variability in Paleolithic assemblages. World Archaeol 2(1):61–73

Caporael L, Griesemer J, Wimsatt W (eds) (2014) Developing scaffolds in evolution, culture and cognition. MIT Press, Cambridge, MA

Chang H (2004) Inventing temperature: measurement and scientific progress. Oxford University Press, Oxford

Chapman R, Wylie A (2016) Evidential reasoning in archaeology. Bloomsbury Academic, London

Childe VG (1935) Changing methods and aims in prehistory. Proc Prehist Soc 1:1–15

Childe VG (1944) Archaeological ages as technological stages. J R Anthropol Inst G B Irel 74(1/2):7–24

Childe VG (1946) Archaeology and anthropology. Southwest J Anthropol 2(3):243–251

Childe VG (1953) Review of Danske Oldsager, III. Ældre Bronzealder. By HC Broholm. Antiquaries J 33(1/2):86–88

Childe VG (1956) Piecing together the past: the interpretation of archaeological data. Routledge and Keegan Paul, London

Childe VG (2004 [1947]) Archaeology as a social science. In: Patterson T, Orser C (eds) Foundations of social archaeology: selected writings of V.Gordon Childe. Berg, Oxford

Clark A (1997) Being there: putting brain, body, and world together again. MIT Press, Cambridge, MA

Clark A (1998) Magic words: how language augments human computation. In: Carruthers P, Boucher J (eds) Language and thought: interdisciplinary themes. Cambridge University Press, Cambridge, pp 162–183

Coward F (2016) Scaling up: material culture as scaffold for the social brain. Quat Int 405:78–90

Currie A (2015) Marsupial lions and methodological omnivory: function, success and reconstruction in paleobiology. Biol Philos 30:187–209

Currie A (2017) Review of evidential reasoning in archaeology. Philos Sci 84:782–790

Currie A (2018) Rock, bone and ruin: an optimist's guide to the historical sciences. MIT Press, Cambridge, MA

Currie A (2019) Simplicity, one-shot hypotheses and paleobiological explanation. Hist Philos Life Sci. https://doi.org/10.1007/s40656-019-0247-0

Currie A, Sterelny K (2017) In defence of story-telling. Stud Hist Philos Sci 62:14–21

Deleuze G, Guattari F (1987) A thousand plateaus (trans: Massumi B). University of Minnesota Press, Minneapolis

Hamilakis Y (2017) Sensorial assemblages: affect, memory and temporality in assemblage thinking. Camb Archaeol J 27(1):169–182

Hamilakis Y, Jones A (2017) Archaeology and assemblage. Camb Archaeol J 27(1):77–84

Hodder I (2011) Entangled: an archaeology of the relationships between humans and things. Wiley-Blackwell, Oxford

Huggett J (2017) The apparatus of digital archaeology. Internet Archaeol. https://doi.org/10.11141/ia.44.7

Janssen M (2019) Arches and scaffolds: bridging continuity and discontinuity in theory change. In: Love A, Wimsatt W (eds) Beyond the meme: development and structure in cultural evolution. University of Minnesota Press, Minneapolis, pp 95–199

Knappett C (2005) Thinking through material culture: an interdisciplinary perspective. University of Pennsylvania Press, Philadelphia

Kokkov K (2019) Warrants, middle-range theories, and inferential scaffolding in archaeological interpretation. Perspect Sci 27(2):171–186

Larvor B (2018) Why 'scaffolding' is the wrong metaphor: the cognitive usefulness of mathematical representations. Synthese. https://doi.org/10.1007/s11229-018-02039-y

Llobera M (2012) Life on a pixel: challenges in the development of digital methods within an 'interpretive' landscape archaeology framework. J Archaeol Method Theory 19:495–509

Lucas G (2012) Understanding the archaeological record. Cambridge University Press, Cambridge

Lucas G (2017a) Review of Robert Chapman and Alison Wylie, evidential reasoning in archaeology. Eur J Archaeol 20(4):740–744

Lucas G (2017b) Variations on a theme: assemblage archaeology. Camb Archaeol J 27(1):187–190

Lucas G (2019) Writing the past: knowledge and literary production in archaeology. Routledge, London/New York

Malfouris L (2013) How things shape the mind: a theory of material engagement. MIT Press, Cambridge, MA

Monteiro E, Østerlie T, Parmiggiani E, Mikalsen M (2018) Quantifying quality: towards a posthumanist perspective on sensemaking. In: Schultze U, Aanestad M, Mähring M, Østerlund C, Riemer K (eds) Living with monsters? Social implications of algorithmic phenomena, hybrid agency, and the performativity of technology, IFIP advances in information and communication technology, vol 543. Springer, Cham, pp 48–63

Moore H (2004) Global anxieties: concept-metaphors and pre-theoretical commitments in anthropology. Anthropol Theory 4(1):71–88

Norton J (2003) A material theory of induction. Philos Sci 70:647–670

Norton J (2014) A material dissolution of the problem of induction. Synthese 191:671–690

Philips J (2006) Agencement/assemblage. Theory, Cult, Soc 23(2/3):108–109

Rowley-Conwy P (2007) From genesis to prehistory: the archaeological three age system and its contested reception in Denmark, Britain, and Ireland. Oxford University Press, Oxford

Sterelny K (2010) Minds: extended or scaffolded? Phenomenol Cognit Sci 9:465–481

Sterelny K (2012) The evolved apprentice: how evolution made humans unique. MIT Press, Cambridge, MA

Stone CA (1998) The metaphor of scaffolding: its utility for the field of learning disabilities. J Learn Disabil 31(4):344–364

Tamborini M (2020) Technoscientific approaches to deep time. Stud Hist Phil Sci 79:57–67

Thomas J (1993) Discourse, totalization and 'the Neolithic'. In: Tilley C (ed) Interpretative archaeology. Berg, Oxford, pp 357–394

Toulmin S (1958) The uses of argument. Cambridge University Press, Cambridge

Vygotsky L (1978) Mind in society: the development of higher psychological processes. Harvard University Press, Cambridge, MA

Walsh K (2019) Newton's scaffolding: the instrumental roles of his optical hypotheses. In: Vanzo A, Anstey P (eds) Experiment, speculation and religion in early modern philosophy, Routledge studies in seventeenth-century philosophy. Routledge, London/New York, pp 125–157

Wimsatt W (2007) Re-engineering philosophy for limited beings: piecewise approximations to reality. Harvard University Press, Cambridge, MA

Wimsatt W (2014) Entrenchment and scaffolding: an architecture for a theory of cultural change. In: Caporael L, Griesemer J, Wimsatt W (eds) Developing scaffolds in evolution, culture and cognition. MIT Press, Cambridge, MA, pp 77–105

Wimsatt W (2019) Articulating babel: a conceptual geography for cultural evolution. In: Love A, Wimsatt W (eds) Beyond the meme: development and structure in cultural evolution. University of Minnesota Press, Minneapolis, pp 1–41

Wimsatt W, Griesemer J (2007) Reproducing entrenchments to scaffold culture: the central role of development in cultural evolution. In: Sansom R, Brandon R (eds) Integrating evolution and development: integrating theory and practice. MIT Press, Cambridge, MA, pp 227–323

Wood D, Bruner JS, Ross G (1976) The role of tutoring in problem solving. J Child Psychiat Psychol 17:89–100

Wylie A (1989) Archaeological cables and tacking: the implications of practice for Bernstein's 'options beyond objectivism and relativism'. Philos Soc Sci 19:1–18

Wylie A (2002) Thinking from things: essays in the philosophy of archaeology. University of California Press, Berkeley

Wylie A (2017a) How archaeological evidence bites back: strategies for putting old data to work in new ways. Sci, Technol Human Values 42(2):203–225

Wylie A (2017b) Representational and experimental modeling in archaeology. In: Magnani L, Bertolotti T (eds) Springer handbook of model-based science. Springer, Cham, pp 989–1002

Part II
Interdisciplinary Connections

Chapter 5
Human Curiosity Then and Now: The Anthropology, Archaeology, and Psychology of Patent Protections

Armin W. Schulz

Abstract Recent anthropological, archaeological, and psychological findings support the view that humans have long been driven by a deep sense of curiosity that needs little if any special external material reward. Apart from being inherently interesting, these findings also turn out to have some wide-ranging consequences for central debates in other social sciences—such as economics. In particular, it is still often thought that, without extensive patent protections, economic actors lack the incentive to engage in innovative activity. However, as this chapter makes clear, the recent anthropological, archaeological, and psychological findings concerning human curiosity provide compelling reasons for thinking that this material interest-based view of innovative activity is unconvincing. In this manner, the chapter has two key upshots: first and most importantly, it shows how and why paying attention to anthropological, archaeological, and psychological findings can have wide-ranging implications throughout the social sciences. Second, it develops reasons for being skeptical about the need for the kinds of extensive patent regimes seen in many countries.

Keywords Curiosity · Innovation · Patent · Music · Play · Motivation

5.1 Introduction

Humans are highly curious creatures. On an abstract level, this claim does not need much defense. What is less obvious, however, is that humans are *deeply* curious. In particular, recent anthropological, archaeological, and psychological findings show that humans have long been driven, to a major extent, by an inherent sense of curiosity that needs little if any special external material reward (see, e.g., Glickman and

A. W. Schulz (✉)
Department of Philosophy, University of Kansas, Lawrence, KS, USA
e-mail: awschulz@ku.edu

© Springer Nature Switzerland AG 2021
A. Killin, S. Allen-Hermanson (eds.), *Explorations in Archaeology and Philosophy*, Synthese Library 433, https://doi.org/10.1007/978-3-030-61052-4_5

Sroges 1966; Kidd and Hayden 2015; Kang et al. 2009; Oudeyer and Smith 2016; Perlovsky et al. 2010; Reader and Laland 2003).[1] Apart from being inherently interesting, these findings turn out to have some wide-ranging consequences for central debates in other social sciences—such as economics. Making this clearer is the aim of this chapter.

In particular, in economics, it is still quite common to see extensive patent rights as a necessary motivator of innovation. That is, it is often thought that, without patent protections, economic actors lack the means to profit from their innovative activity, are thus not incentivized to engage in this innovative activity, and will therefore not in fact engage in this activity (Aghion et al. 2001, 2005; Aghion and Howitt 1992; Kesan 2015; Karbowski and Prokop 2013). Call this the 'material interest-based view' of innovative activity. However, as I show in this chapter, the recent anthropological, archaeological, and psychological findings concerning human curiosity provide a compelling reason for thinking that this material interest-based view of innovative activity is unconvincing.[2]

In this manner, the chapter has two key upshots: first and most importantly, it shows how and why paying attention to anthropological, archaeological, and psychological findings can have wide-ranging implications *throughout* the social sciences. This is important, as while it is true that economic scholars and other social scientists are increasingly turning towards the biological sciences—broadly understood to include anthropology, archaeology, and psychology—to make progress in understanding human economic behaviors (see, e.g., Santos and Rosati 2015; Kenrick et al. 2009; Li et al. 2012; Robson 2001; Ofek 2013; Saad 2017; Nelson et al. 2018; Nelson and Winter 1982; see also Veblen 1898; Schumpeter 1942), this strategy is still quite controversial (Sober 1992; Camerer 2007; Aldrich et al. 2008). To a significant degree, this is due to the fact that it is not clear how these very disparate subjects can be linked. This chapter looks to overcome that challenge by making clearer exactly how it is possible to bring anthropological, archaeological, psychological, and other biological considerations to bear on debates in economics and other social sciences.

The second upshot of the chapter is more concrete. It develops reasons for being skeptical about the need for the kinds of extensive patent regimes seen in many countries right now. In this way, it also makes a direct contribution to a live economic debate.

The chapter is structured as follows. Section 5.2 lays out some of the major recent anthropological, archaeological, and psychological insights about human curiosity. In Sect. 5.3, the debate about the need for extensive patent protections is laid out. Section 5.4 brings the previous two sections together to argue that the

---

[1] The literature on curiosity goes back to the beginnings of psychology as a subject (see, e.g., James 1890; Pavlov 1927; Skinner 1938). Here, though, the focus is on more recent work.

[2] The nature of economic motivation in general has of course long been a point of contention (Hausman 1992, 2012; Angner 2016, 2018; Fehr and Camerer 2007; Fehr and Fischbacher 2003; Fehr and Gaechter 2000; Fehr and Schmidt 1999; Henrich et al. 2005; Skyrms 2004). However, the question of what motivates *innovation* raises its own set of issues; these will be in focus here.

anthropological, archaeological, and psychological findings about human curiosity have major implications for the defensibility of existing patent regimes. Section 5.5 concludes.

5.2 Human Curiosity

'Curiosity' is often used in a narrow way to refer to purely internally-rewarded information-seeking behavior (Kidd and Hayden 2015; Loewenstein 1994; Oudeyer and Kaplan 2007). That is, organisms are said to be curious if they engage in potentially costly behaviors that result in their learning something about the world, despite the fact that they do not obtain any (major) external rewards for this learning.[3] However, following some other authors (see, e.g., Kidd and Hayden 2015), I here use this term in a slightly broader way.

On the one hand, I shall understand 'information' broadly to include both information about the state of the world and information about how to do certain things (i.e., 'know-how' as well as 'know-that'). So, I consider organisms curious not just if they seek to find out about where in the local environment noise is most easily amplified, but also if they try to find out how to build a tool that allows them to make more noise at any given location. This is reasonable for two reasons. First, it is controversial to what extent 'know-how' is different from 'know-that' (Clark 1997; Stanley 2011; Hawley 2003). Second, as will become clearer momentarily, the broader reading fits better to the most widely accepted views about the factors that have shaped the cultural and evolutionary history of humans.

On the other hand, in the literature on the evolution of curiosity, there is dispute over exactly how to distinguish external and internal rewards (Kidd and Hayden 2015).[4] If an organism seeks information because doing so leads to increased access to an *external* resource (food, say), but where access to this external resource is found *internally* rewarding (as would often be the case), this would seem to satisfy the above characterization of curiosity—though also be a paradigm example of what this characterization is meant to exclude.

To overcome this kind of problem, curious organisms will here be seen as ones that engage in information-seeking behavior (broadly understood) even when they are only provided with external resources that they generally find only mildly rewarding—if at all. So, if providing a grape to a human is generally not sufficiently motivating to lead them to make major changes in their behavior, then to the extent that they engage in information-seeking behavior even if doing so promises them only a grape as external reward, they are considered curious. By contrast, the fact

---

[3] A related phenomenon with similar evolutionary features is that of *play* (Kidd and Hayden 2015; Schank 2015; Allen and Bekoff 1994).

[4] This point is an instance of a general issue of how to distinguish internal and external motivation that also affects, among other, accounts of conditioning and prosocial behavior (Kidd and Hayden 2015; Piccinini and Schulz 2019; Rosenberg 2012).

that a human engages in information-seeking behavior if they think that doing so makes them famous and thus provides them with significant amounts of social capital does not clearly speak to them being curious: after all, obtaining significant amounts of social capital is a known *strong external* motivator for humans (Barkow 1992; Henrich 2015; Sugiyama and Sugiyama 2003). This characterization also helps with cases of mixed motivations, where an organism is motivated both internally and externally (e.g., where someone invents something both because they are interested in inventing for its own sake and because doing so provides them with a monetary bonus). If the externally-based motivator is not sufficient to lead the agent to engage in the relevant activity, but the purely internally-based motivator is, the agent can be considered curious in the sense of that term relevant here.

With this in mind, the following points about human curiosity need to be noted. First, not all organisms are equally curious (Glickman and Sroges 1966; Kidd and Hayden 2015; Reader and Laland 2003). Indeed, it has long been known that mammals appear particularly curious, and that among mammals, primates are further standouts—with humans being further standouts among the primates. However, exactly how curious humans are, and exactly what distinguishes human curiosity from that of other animals, has only recently become clearer.

In the first place, recent work in developmental psychology has shown that human infants are strongly and frequently driven by curiosity. From an early age onwards, humans are happy to spend significant periods of time searching for insights about the world and how to manipulate it, all without getting much in the way of reward for it. For example, even 4-month-old children have been observed to spend much time playing with puzzle toys that lack clear function, and appear happy when they figure out how to solve the puzzle (and frustrated if they do not) (Kidd and Hayden 2015; Oudeyer and Smith 2016; Kidd et al. 2012; Gopnik and Schulz 2004; Cook et al. 2011).

Importantly, moreover, work in archaeology suggests that curiosity being a strong and frequent motivator is not something that is restricted to young kids, but seems to characterize human living more generally.[5] So, for example, on a conservative estimate, intricately designed and decorated musical instruments appear in the archaeological record from about 40,0000 years ago (Conard et al. 2004; Cross 2012; Lawson and d'Errico 2002; Adler 2009; Conard et al. 2009; Morley 2013; Killin 2018). Now, it is possible that some of these were built for signaling purposes: by building complex instruments, an instrument-maker may have displayed her motoric and cognitive skills, and thus have hoped to obtain external rewards in the form of pay or social capital (Hayden 1998; Sterelny 2012; Skyrms 2010; Premo and Kuhn 2010; Renfrew and Scarre 1998; Kohn and Mithen 1999). However, it is far from clear that this was always—or indeed ever—the case: there is little (if any) further evidence that musical instruments were indeed taken as marks of cognitive

---

[5] There is of course also reason to think that early humans collected information so as to be better able to obtain external rewards at a later date (see, e.g., Bergemann and Välimäki 2008). However, since this does not concern *curiosity*—i.e., largely internally-rewarded information-seeking behavior—this is not so relevant here.

or motoric skill. It is just as plausible—if not more so—that these instruments are simply the result of curious experimentation: early humans may just have been interested to see if they could make a bone (say) make certain sounds (Cross 2012; Adler 2009; Morley 2013; Killin 2018). They might have wondered: if I drill a hole here and then blow in there, what happens? Why does this happen? Figuring this out, plausibly, was the entire reward of this often very laborious exercise.

Indeed, this idea can be extended even further back in time. Functionally useless, decorative handaxes appear in the archaeological record from about 500,0000 years ago (Mithen 2005). Again, it is *possible* that these were designed to signal group membership, motoric and cognitive skills, or social status (Kohn and Mithen 1999). However, again, there is little to support this external-reward-based view further. It is just as plausible—if not more so—that these artifacts were also built simply because their makers were interested in seeing if they could shape a rock just so, or how a rock shaped just so would feel if held or thrown (Wynn 1993; Currie 2011). In short: the act of making decorative handaxes—another very laborious exercise— plausibly was its own reward.

The idea that humans are inherently strongly curious is further supported by the fact that contemporary adult humans innovate even when by themselves and when doing so is not strongly socially rewarded (Kidd and Hayden 2015). For example, humans appear to feel pleasure—an internal reward—when engaging in unrewarded information-seeking tasks (Kidd and Hayden 2015; Kang et al. 2009; Perlovsky et al. 2010). It is plausible that this has been the case since very early times in hominin evolution.

Importantly also, the gene-cultural foundation of the exceptionally curious nature of humans is increasingly well-understood. While some forms of curiosity and a general drive towards innovation are adaptive in a wide variety of circumstances (Reader and Laland 2003; Kang et al. 2009; Sober 1994), the extensive curiosity of humans requires a different explanation. This explanation is likely to be provided by the human disposition towards cultural learning.

In particular, there is good reason to think that human curiosity coevolved with the human propensity for cultural learning. In the terms of Heyes (2018), curiosity is a 'cognitive gadget' that was adaptive in the presence of another cognitive gadget: the disposition to learn from others. In general, being very curious is *not* adaptive: seeking information even in the absence of external pressures to do so tends to be risky, as it is likely to be a waste of time and resources (Lee 2003). However, in environments in which cultural learning is adaptive—which are not universally instantiated (Boyd and Richerson 2005; Henrich and McElreath 2011)—curiosity can be adaptive. There are two reasons for this.

On the one hand, environments in which cultural learning is adaptive tend to be environments characterized by significant degrees of cooperation (Henrich 2015; Boyd and Richerson 2005). In turn, this buffers the individual from the potential costs of curiosity. Individuals that spend time and resources seeking out information can rely on others' help in compensating for the costs of the (potentially entirely fruitless) time thus spent.

On the other hand and relatedly, tolerating curious individuals in a society of cultural learners can be adaptively advantageous to the members of this society, as this kind of curiosity can lead to successive, but unpredictable improvements in the cultural information being transmitted (Boyd and Richerson 2005; Henrich 2015; Sterelny 2012). Being purely internally driven to seek information is adaptively valuable in situations in which such information can have unforeseeable, but biologically useful consequences at other times and places. Being motivated to drill holes in bones, just to see how this changes the sounds that are coming out of these bones when one blows into them, can be adaptive in cases where such knowledge sufficiently often turns out to have useful applications for *some* people—even if these people are not the inventor/discoverer (or anyone related to them), and even if there is no way to predict when and why this knowledge will turn out to be useful (Anderson 1993; Boyd et al. 2011; Cross 2012; Killin 2018).

All of this matters here, as it further supports and explains the idea that humans should be seen to be exceptionally curious. Humans are known to be standouts in their capacity for cultural learning (Henrich 2015; Boyd et al. 2011; Henrich and McElreath 2011; Heyes 2018), which suggests that they are also standouts for their curious psychology. Humans have evolved in conditions under which innovation for its own sake was encouraged to run wild (Loewenstein 1994; Kang et al. 2009).

This is undoubtedly interesting in and of itself. However, it is also important from the point of view of economics—and, especially, the defensibility of institutions of extensive patent protections.

5.3   Patent Protections: Disputes and Justifications

Patent protections are monopoly rights to sell a given type of good or service—typically for a fixed time period (Horowitz and Lai 1996; Nordhaus 1969; Klemperer 1990). They are initially awarded to the inventors or discoverers of the good or service in question, but they are increasingly treated as commodities to be traded (an important point to which I return momentarily) (Kesan 2015; Sichelman 2010; Teece 1986). Since their inception and development, a number of issues surrounding the issuance of patent protections have been raised.

The first of these issues is that there is no question that patent protections have downsides. On the one hand and most directly, in virtue of the fact that they create monopolies, patent protections create inefficiencies (Mas-Colell et al. 1996). By definition, such protections reduce the supply of a given good or service and thus raise its price. In turn, this implies that (a) some economic actors that would otherwise have access to the good or service are now prevented from doing so, and (b) the economic actors that are still getting access to the good or service now have to pay a higher price for doing so. All in all, this creates a 'deadweight loss': while some of the losses in (a) and (b) are compensated for by gains in the profits of the patent holders, not all of them are. Hence, possible gains from trade are lost through patent protections (Nordhaus 1969; Klemperer 1990; Boldrin and Levine 2013;

Gifford 2004). On the other hand, patent protections—especially if they are very broad—can make it too difficult for new inventors to enter into a given product market, and thus prevent beneficial, new inventions from getting made (Aghion et al. 2001, 2005; Aghion and Howitt 1992; Horowitz and Lai 1996; Nordhaus 1969; Nelson et al. 2018; Muthukrishna and Henrich 2016).[6]

Given these downsides of patent protections, why then are they thought to be useful at all? Several different—not necessarily mutually inconsistent—answers have been given to this question (Karbowski and Prokop 2013). Some have suggested that patent protections are a consequence of the fact that discoverers or inventors have natural (intellectual) property rights over their discoveries or inventions (Shrader-Frechette 2006; Shiffrin 2001; for discussion, see, e.g., Otsuka 2003; Damstedt 2002). Others have noted that patents can lead to an economy functioning more efficiently, as inventions become more easily tradable or commercializable (Kesan 2015; Sichelman 2010; Teece 1986; Penin 2005).

However, the central justification of the need for extensive patent regimes—and the one that is key in this context—is that patents are often seen as a major motivational tool. Specifically, patent protections are typically thought to be necessary to incentivize innovation (see, e.g., Kesan 2015; Torrance and Tomlinson 2009; Nordhaus 1969; Karbowski and Prokop 2013). The term 'innovation' here includes all kinds of R&D activity: anything from the discovery of new medicines to the invention of a new process of building some of the components that go into the assembly of a particular kind of machinery. The core idea behind this argument for patent rights is that R&D activities can be very expensive to conduct and tend to be uncertain in terms of their payoff: a company might need to expend significant resources in the development of a new drug, even though it is quite likely that this drug ends up not getting regulatory approval or failing to come to market for other reasons. If it then were the case that companies would face the possibility that even successful innovations would almost immediately face direct competition, thus driving their price down to just cover their marginal costs (which are also often near-zero), they would have no way to recoup the costs of failed innovations. In turn, this would mean that they have no profit incentive to engage in the innovation activities to begin with. This argument is stated very clearly and explicitly throughout the literature. Consider:

> But clearly the most important reason that firms and private investors put in the resources and time to advance a technology is that they hope and expect to profit from their efforts. Their ability to gain returns from their work then depends to a considerable degree on their ability to control the use of their inventions. (Nelson et al. 2018, p. 51)

Or this:

---

[6] This was part of what led Schumpeter (1942) to worry about patents leading to the collapse of the entrepreneurial system. Of course, as will also be made clearer below, it is also the case that patents are thought to spur innovation (Aghion et al. 2005; Schumpeter 1959); the point in the text is just that, if patents prevent an innovator from producing anything at all like a given product, it may make the creation of incremental improvements of this product overly difficult.

The theory of patent law is straightforward. Society benefits from new technology. Yet in the absence of patent protection, invention would often go unrewarded. Unless a portion of this newly created economic value can be captured by its inventor, there is no incentive to innovate. Indeed, in certain cases there would be a negative incentive: Invention often requires the expenditure of substantial resources in research and experimentation. This failure of the market to supply the incentive to invent is a result of a crucial absence of property rights. (Gifford 2004, p. 81)

Similarly:

Formal IP is designed to provide *ex ante* incentives to innovate by providing a reward system that makes it easier for innovators to make *ex post* profits if their innovation is successful, by allowing them to exclude imitators for a finite period. (Hall et al. 2014, p. 376)

Now, it needs to be noted immediately that a number of challenges to this motivation-based argument exist. Some of these are historical: they note that changes in a country's patent regime often do not lead to concomitant increases in the country's growth or innovation rate (Branstetter and Sakakibara 2001; Qian 2007). However, these historical challenges (so far) failed to dethrone the motivational argument for patent regimes from its central position in the literature. On the one hand, this is due to the fact that the historical situation is in fact quite ambiguous: for example, while some studies fail to find a positive effect of patent regimes on innovation rates, other studies do find such positive effects (Hu and Png 2013; Moser 2005). On the other hand, these studies can be hard to interpret, as changes in the patent regimes of a given country often come with other changes as well (such as changes in the tax regimes or broader legal environment). In turn, this makes it hard to disentangle whether and to what extent changes in the country's growth rate (or a lack thereof) are due to the changes in the patent regimes or to some of the other changes that occurred.

Somewhat similar points can also be made concerning the other major challenge towards the above classic argument for extensive patent regimes. This other challenge looks to gauge innovators' motivations directly—especially through survey studies. Here, it has been found that researchers often seem to be motivated for reasons *other than* patent protections of their inventions. For example, being the first in a given market is often thought to be an important motivator of innovative activities (Arundel et al. 1995; Cohen et al. 2000; Mazzoleni and Nelson 1998; Bessen and Meurer 2008).

However, this challenge has also failed to fully resolve the debate here.[7] In the main, this is because survey studies are also famously hard to interpret. Since subjects may not be able to introspect their motivations accurately, and thus misrepresent the importance of patent regimes in incentivizing the R&D activities in which they engage, these kinds of surveys cannot easily settle the issues here (Hall et al. 2014; Torrance and Tomlinson 2009, p. 138).

---

[7] Note also that this challenge anyway still supports the idea that innovation is motivated by profit; it is just that being first-to-market is seen as ensuring such a profit. As noted below, there are reasons to question the basic premise underlying this argument.

This point is strengthened by the fact that a more charitable reading of the traditional argument for patent protections requires acknowledging that not *all* the motivational sources of innovative activities need to lie a profit motive. The claim made by defenders of the classic argument for patent protections should really just be seen to be that a *significant* proportion of the motivation for innovation has its source in a search for material profits. After all, it can be taken for granted that humans (like most other organisms) are motivated *both* by external, physical rewards—such as monetary gains—and by other considerations—such as social status or curiosity.[8] The question really is how important each of these is. Defenders of the traditional argument for patent protections think the answer is: the profit motive is crucial. Without it, only insignificant amounts of R&D would be completed—humans are *mostly* driven by the promise of direct or indirect external rewards.

However, assessing whether the defenders of the traditional argument for patent protections are right in this is not straightforward. Even assuming that people could accurately determine whether they are motivated to innovate by the promise of monetary rewards or by something else, it is far from obvious that they would be able to accurately assess *how strong* each of these motivations really is (Wilson 2002; Nisbett and Wilson 1977; Machery 2005).

Putting all of this together, therefore, it needs to be concluded that a compelling assessment of the traditional, motivational argument for patent protections still requires an evaluation of the *relative importance* of non-monetary—or even non-external—rewards. In the rest of this chapter, I therefore seek to do two things: first, I aim to take steps towards such an evaluation by leveraging the anthropological, archaeological, and psychological findings sketched in the previous section. Second, I aim to make clearer the methodology of this way of bringing anthropological, archaeological, and psychological considerations to bear on economic debates.

5.4 Human Curiosity and Patent Protection: Some Implications

What can archaeological interpretations of early material culture (Hayden 1998; Adler 2009; Morley 2013; Killin 2018; Klein and Edgar 2002; Mellars 1989), model-based investigations of cultural evolution (Boyd and Richerson 2005), experiments about infant play (Gopnik et al. 2004), and studies of the neural reward

---

[8] Social status, for example, is widely accepted to be a major human motivator, too (Barkow 1992; Henrich 2015; Sugiyama and Sugiyama 2003; Hayden 1998; Sterelny 2012; Skyrms 2010). This is especially noteworthy here, since significant aspects of the institution of contemporary science are based on the fact that people are keenly driven by social approval (Zollman 2018; Strevens 2003). (That said, the issues here are complex as scientists can also obtain patents on some of their inventions and might choose to become university researchers rather than industry ones for external, monetary reasons—such as a job opening near family, a better benefits package, a better work-life balance, etc.; see also Nelson et al. 2018; Torrance and Tomlinson 2009).

circuitry involved in exploratory behavior (Kidd and Hayden 2015) contribute to the evaluation of whether patent protections are the key motivator of R&D activities? I submit that the answer is that, collectively, these findings suggest that extensive patent protections are *not* needed to motivate even significant amounts of R&D.

In particular, as noted earlier, what we know about the importance of curiosity in human evolution (Boyd et al. 2011; Henrich 2015; Henrich and McElreath 2007, 2011; Heyes 2018), together with what we know about exploratory behavior in contemporary infants and adults (Kidd and Hayden 2015; Gopnik et al. 2004), makes it very plausible that curiosity is a *frequent and significant human motivator*. Given this, there is good reason to think that it would continue to drive significant amounts of R&D activity even in the absence of guaranteed major external rewards for successful such activity. Of course, this is not to say that external considerations are not also very motivating for humans. However, since cultural learning is universally regarded as very important to human evolution, and since curiosity does seem to drive much behavior of contemporary human infants and adults, there is reason to think that a *strong and frequent* drive for innovation is likely to be a significant part of human psychology.

This matters, as it suggests that the traditional argument for the plausibility of extensive patent regimes is built on unconvincing motivational assumptions. Given that humans plausibly are very curious by nature—that is, given that humans have evolved so as to be internally driven to frequently engage in information-seeking behavior—innovative activity does *not* need to be incentivized by patents, and would still remain about as abundant and creative as before. Paying attention to the above anthropological, archaeological, and psychological findings about human motivation thus does away with much of the attraction of a key argument for extensive patent regimes.

As noted earlier, this conclusion fits well to some other recent work on the plausibility of patent regimes (see, e.g., Hall et al. 2014; Cohen et al. 2002; Arundel et al. 1995; Nelson et al. 2018, p. 51; Root-Bernstein 1989). However, the present conclusion is more wide-ranging than these other arguments, as the latter still tend to assume some kind of material self-interest. The point here, though, is that, from an anthropological, archaeological, and psychological viewpoint, there is reason to think that humans would continue to engage in innovative activities even if that yielded no guaranteed material gains whatsoever. That is, given what else we know about the motivations underlying human innovation, it is reasonable to think that significant amounts of economically important innovative activities would remain even if they were not rewarded with extensive patent regimes.

More than that: the anthropological, archaeological, and psychological evidence above is also relevant here for providing an *explanation* for why humans do not need to be promised major material rewards to be motivated to engage in R&D activities. The fact that humans evolved in a rich cultural setting that made it culturally adaptive for them to have a deeply curious psychology explains *why* there is no need for the promise of major material rewards—whether in the form of patents or in some other way—to motivate their inventive and innovative behavior. So, besides adding evidence to other studies about the need (or lack thereof) for patents to motivate research, the anthropological, archaeological, and psychological perspective laid out in this chapter aids in coming to a deeper understanding of why these other studies turned out the way that they did. What underlies the fact that humans can be expected to innovate even in the absence of guaranteed external monetary rewards for successful inventions is the fact that humans evolved in conditions that favored the spread of a particularly curious psychology.

However, to fully understand this argument, it is important to get clearer on its methodological foundations. There is no question that there is much uncertainty surrounding the interpretation of early material culture (Currie and Killin 2019; Pain 2019; Sterelny and Hiscock 2017; Sterelny 2017; Hodder 1994), the ways in which ancient motivational systems operate in contemporary technological environments (Buller 2005; Richardson 2007; Garson 2014; Downes 2013), and the details of the distribution of psychological traits within human populations (Lee 2003; Buss and Hawley 2010; Brandon 1990). For example, the interpretation of the material remains of ancient cultures involves making assumptions about what is useful for what and how difficult it was to create which artifacts. Such assumptions can be tricky to validate and may be influenced by the biases of the researchers conducting them, thus weakening the trust we can place on them (Wylie 1994; Sterelny 2017). Similarly, even if there is reason to think that humans have had certain psychological dispositions for a long time, this does not mean that these dispositions necessarily need to display themselves in contemporary, real-life, non-laboratory situations. The features of contemporary society may well counterbalance or sidestep such psychological dispositions: for example, the continual reinforcement or suggestion of rival profit-based motivations may crowd out curiosity-based ones (Promberger and Marteau 2013; Reeson and Tisdell 2008). Lastly, it is not at all obvious whether psychological dispositions—even assuming they can be shown to have evolved and to be active in the relevant situations of contemporary society—are evenly distributed in the population, or whether they are more strongly represented in some groups. So, perhaps dispositions towards curiosity only display themselves in times of severe stress, or only among younger humans (Lee 2003).

These methodological constraints need to be emphasized and acknowledged. However, they should not be overstated either. Even if uncertainties remain about the interpretation of the anthropological and archaeological record, the impact of contemporary culture on older psychological dispositions, and the distribution of these dispositions in human populations then and now (among other such uncertainties), this does not mean that our currently best interpretation of key anthropological, archaeological, and psychological findings cannot provide *evidence* for or against some social scientific hypotheses. That is, even if we cannot be sure whether ancient musical instruments or decorative handaxes were not useful for something, and even if we cannot be sure to what extent an intrinsically rewarded drive towards innovation is not permanently crowded out by externally-derived motivations for many or all humans in contemporary society, this does not mean that the findings sketched in Sect. 5.2 do not provide *a reason* for thinking that extensive patent protections are not necessary to motivate significant innovation.[9]

To be sure, for an account that rationally compels *full acceptance* of the view that humans are strongly driven to be curious, and so need little in the way of external, material incentive to engage in exploratory activities, the questions surrounding the interpretation of material culture, etc. *do* need to be addressed. However, this does not mean that there is no epistemic value in a more modest, evidential account that acknowledges these limitations. In particular, such an account can still provide *a reason* for thinking that a significant drive towards curiosity is present in humans, and that therefore extensive patent protections are not needed to incentivize innovation. While this is less than a full confirmation of the latter conclusion, it is social scientifically valuable nonetheless (see also Schulz 2018; Wylie 1994).

5.5 Conclusion

This chapter showed how the increasingly widely accepted anthropological, archaeological, and psychological finding that humans are deeply curious—i.e., strongly internally rewarded to seek novel information—is relevant beyond its direct relevance specifically to anthropological, archaeological, and psychological investigations. In particular, the plausibility of deep and strong human curiosity *supports and explains* the economic claim that extensive patent protections are *not* necessary to motivate significant amounts of R&D activity, contra to the widely held and influential material interest-based view of innovative activity. While this does not speak to the existence of other reasons for why patents may be useful—for example, patents may be useful for making the commercialization or trade of inventions easier (Kesan 2015; Sichelman 2010; Teece 1986; Penin 2005), and they may flow from the very nature of property rights (Shrader-Frechette 2006; Shiffrin 2001)—this conclusion is noteworthy. On the one hand, the kinds of patent regimes that can be justified with non-motivational arguments may be quite different from the ones that can be justified with motivational arguments (see also Otsuka 2003). On the other hand, it is widely accepted that the motivational argument holds a central position in the debate surrounding extensive patent regimes (Kesan 2015; Torrance and Tomlinson 2009; Nordhaus 1969; Karbowski and Prokop 2013). By calling this

---

[9] Of course, it is possible that humans are only significantly inherently motivated to innovate when it comes to aesthetic matters (such as building musical instruments and decorative handaxes). However, as yet, there is no reason to think that this is true—especially in the light of the further facts cited in Sect. 5.2 above, such as the fact that human kids spend significant amounts of time puzzling out things without getting external rewards for doing so. In the absence of such a reason, the above archaeological findings retain evidential value—nothing more, but also nothing less.

argument into question, therefore, the need for extensive patent regimes is dealt a relatively heavy blow.[10]

Apart from its inherent interest, this conclusion is also important for bringing out the value of linking biological considerations (broadly understood) and social science. As a number of other authors (Nelson et al. 2018; Wilson et al. 2013; Hodgson and Knudsen 2010) have recently suggested, bringing these two seemingly disparate subjects together can be very useful. The present chapter illustrates this usefulness further by showing how extra evidence for or against given economic policies can be obtained through the appeal to considerations from anthropology, archaeology, and psychology—as long as this appeal is properly understood.

References

Adler D (2009) The earliest musical tradition. Nature 460:695–696
Aghion P, Howitt P (1992) A model of growth through creative destruction. Econometrica 60(2):323–351
Aghion P, Harris C, Howitt P, Vickers J (2001) Competition, imitation and growth with step-by-step innovation. Rev Econ Stud 68(3):467–492
Aghion P, Bloom N, Blundell R, Griffith R, Howitt P (2005) Competition and innovation: an inverted-u relationship. Q J Econ 120(2):701–728
Aldrich HE, Hodgson G, Hull D, Knudsen T, Mokyr J, Vanberg VJ (2008) In defence of generalized Darwinism. J Evol Econ 18(5):577–596
Allen C, Bekoff M (1994) Intentionality, social play, and definition. Biol Philos 9(1):63–74
Anderson JR (1993) Problem solving and learning. Am Psychol 48(1):35–44
Angner E (2016) A course in behavioral economics, 2nd edn. Palgrave Macmillan, London
Angner E (2018) What preferences really are. Philos Sci 85(4):660–681
Arundel A, Paal G, Soete L (1995) Innovation strategies of Europe's largest industrial firms: results of the PACE survey for information sources, public research, protection of innovations and government programmes; final report. Maastricht Economic Research Institute of Technology, Maastricht
Barkow J (1992) Beneath new culture is old psychology: gossip and social stratification. In: Barkow J, Cosmides L, Tooby J (eds) The adapted mind: evolutionary psychology and the generation of culture. Oxford University Press, Oxford, pp 627–637
Bergemann D, Välimäki J (2008) Bandit problems. In: Durlauf SN, Blume LE (eds) The new Palgrave dictionary of economics: volume 1–8. Palgrave Macmillan, London, pp 336–340
Bessen J, Meurer MJ (2008) Patent failure. Princeton University Press, Princeton
Boldrin M, Levine D (2013) The case against patents. J Econ Perspect 27:3–22
Boyd R, Richerson P (2005) The origin and evolution of cultures. Oxford University Press, Oxford
Boyd R, Richerson P, Henrich J (2011) The cultural niche: why social learning is essential for human adaptation. Proc Natl Acad Sci 108(suppl-2):10918–10925
Brandon R (1990) Adaptation and environment. Princeton University Press, Princeton
Branstetter L, Sakakibara M (2001) Do stronger patents induce more innovation? Evidence from the 1988 Japanese patent law reforms. Rand J Econ 32(1):77–100
Buller D (2005) Adapting minds. MIT Press, Cambridge, MA

---

[10] Of course, since patent rights are deeply embedded in the workings of most contemporary economies, any shift away from the currently existing extensive patent regimes will need to be carefully managed.

Buss DM, Hawley PH (eds) (2010) The evolution of personality and individual differences. Oxford University Press, Oxford

Camerer C (2007) Neuroeconomics: using neuroscience to make economic predictions. Econ J 117:C26–CC4

Clark A (1997) Being there. MIT Press, Cambridge, MA

Cohen W, Nelson R, Walsh J (2000) Protecting their intellectual assets: appropriability conditions and why US manufacturing firms patent or not. NBER, discussion paper 755

Cohen W, Nelson R, Walsh J (2002) Links and impacts: the influence of public research on industrial R&D. Manag Sci 48(1):1–23

Conard N, Malina M, Münzel S, Seeberger F (2004) Eine Mammutelfenbeinflöte aus dem Aurignacien des Geißenklösterle: Neue Belege für eine Musikalische Tradition im Frühen Jungpaläolithikum auf der Schwäbischen Alb. Archäologisches Korrespondenzblatt 34:447–462

Conard N, Malina M, Münzel S (2009) New flutes document the earliest musical tradition in southwestern Germany. Nature 460:737–740

Cook C, Goodman ND, Schulz LE (2011) Where science starts: spontaneous experiments in preschoolers' exploratory play. Cognition 120(3):341–349

Cross I (2012) Music as an emergent exaptation. In: Bannan N (ed) Music, language and human evolution. Oxford University Press, Oxford, pp 263–276

Currie G (2011) The master of the Masek beds: handaxes, art, and the minds of early humans. In: Schellekens E, Goldie P (eds) The aesthetic mind: philosophy and psychology. Oxford University Press, Oxford, pp 9–31

Currie A, Killin A (2019) From things to thinking: cognitive archaeology. Mind Lang 34(2):263–279

Damstedt BG (2002) Limiting Locke: a natural law justification for the fair use doctrine. Yale Law J 112(5):1179–1221

Downes S (2013) Evolutionary psychology is not the only productive evolutionary approach to understanding consumer behavior. J Consum Psychol 23(3):400–403

Fehr E, Camerer C (2007) Social neuroeconomics: the neural circuitry of social preferences. Trends Cogn Sci 11(10):419–427

Fehr E, Fischbacher U (2003) The nature of human altruism. Nature 425:785–791

Fehr E, Gaechter S (2000) Fairness and retaliation: the economics of reciprocity. J Econ Perspect 14:159–181

Fehr E, Schmidt KM (1999) A theory of fairness, competition, and cooperation. Q J Econ 114(3):818–868

Garson J (2014) The biological mind. Routledge, London

Gifford DJ (2004) How do the social benefits and costs of the patent system stack up in pharmaceuticals? J Intellect Prop Law 12:75–124

Glickman SE, Sroges RW (1966) Curiosity in zoo animals. Behaviour 26(1/2):151–187

Gopnik A, Schulz LE (2004) Mechanisms of theory-formation in young children. Trends Cogn Sci 8(8):371–377

Gopnik A, Glymour C, Sobel DM, Schulz LE, Kushnir T, Danks D (2004) A theory of causal learning in children: causal maps and Bayes nets. Psychol Rev 111(1):3–32

Hall B, Helmers C, Rogers M, Sena V (2014) The choice between formal and informal intellectual property: a review. J Econ Lit 52(2):375–423

Hausman DM (1992) The inexact and separate science of economics. Cambridge University Press, Cambridge

Hausman DM (2012) Preference, value, choice, and welfare. Cambridge University Press, Cambridge

Hawley K (2003) Success and knowledge-how. Am Philos Q 40(1):19–31

Hayden B (1998) Practical and prestige technologies: the evolution of material systems. J Archaeol Method Theory 5(1):1–55

Henrich J (2015) The secret of our success: how culture is driving human evolution, domesticating our species, and making us smarter. Princeton University Press, Princeton

Henrich J, McElreath R (2007) Dual-inheritance theory: the evolution of human cultural capacities and cultural evolution. In: Dunbar R, Barrett L (eds) The Oxford handbook of evolutionary psychology. Oxford University Press, Oxford, pp 555–570

Henrich J, McElreath R (2011) The evolution of cultural evolution. Evol Anthropol 12:123–135

Henrich J, Boyd R, Bowles S, Camerer C, Fehr E, Gintis H et al (2005) 'Economic man' in cross-cultural perspective: behavioral experiments in 15 small-scale societies. Behav Brain Sci 28(6):795–815

Heyes C (2018) Cognitive gadgets: the cultural evolution of thinking. Harvard University Press, Cambridge, MA

Hodder I (1994) The interpretation of documents and material culture. In: Denzin NK, Lincoln YS (eds) Handbook of qualitative research. Sage Publications, London, pp 393–402

Hodgson G, Knudsen T (2010) Darwin's conjecture. University of Chicago Press, Chicago

Horowitz AW, Lai ELC (1996) Patent length and the rate of innovation. Int Econ Rev 37(4):785–801

Hu AGZ, Png IPL (2013) Patent rights and economic growth: evidence from cross-country panels of manufacturing industries. Oxf Econ Pap 65(3):675–698

James W (1890) The principles of psychology. Holt, New York

Kang MJ, Hsu M, Krajbich IM, Loewenstein G, McClure SM, Wang JT-Y, Camerer C (2009) The wick in the candle of learning: epistemic curiosity activates reward circuitry and enhances memory. Psychol Sci 20:963–973

Karbowski A, Prokop J (2013) Controversy over the economic justifications for patent protection. Procedia Econ Finance 5:393–402

Kenrick DT, Griskevicius V, Sundie JM, Li NP, Li YJ, Neuberg SL (2009) Deep rationality: the evolutionary economics of decision making. Soc Cogn 27(5):764–785

Kesan JP (2015) Economic rationales for the patent system in current context. George Mason Law Rev 22(4):897–924

Kidd C, Hayden BY (2015) The psychology and neuroscience of curiosity. Neuron 88(3):449–460

Kidd C, Piantadosi ST, Aslin RN (2012) The Goldilocks effect: human infants allocate attention to visual sequences that are neither too simple nor too complex. PLoS One 7(5):e36399

Killin A (2018) The origins of music: evidence, theory, and prospects. Music Sci. https://doi. org/10.1177/2059204317751971

Klein R, Edgar B (2002) The dawn of human culture. Wiley, New York

Klemperer P (1990) How broad should the scope of patent protection be? RAND J Econ 21:113–130

Kohn M, Mithen S (1999) Handaxes: products of sexual selection. Antiquity 73:518–526

Lawson G, d'Errico F (2002) Microscopic, experimental and theoretical re-assessment of upper Paleolithic bird-bone pipes from Isturitz, France: ergonomics of design, systems of notation and the origins of musical traditions. In: Hickmann E, Kilmer AD, Eichmann R (eds) The archaeology of sound: origin and organization. Marie Leidorf, Rahden, pp 119–142

Lee PC (2003) Innovation as a behavioral response to environmental challenges: a cost and benefit approach. In: Reader SM, Laland KN (eds) Animal innovation. Oxford University Press, Oxford, pp 261–278

Li YJ, Kenrick DT, Griskevicius V, Neuberg SL (2012) Economic decision biases and fundamental motivations: how mating and self-protection alter loss aversion. J Pers Soc Psychol 102(3):550–561

Loewenstein G (1994) The psychology of curiosity: a review and reintrepretation. Psychol Bull 116(1):75–98

Machery E (2005) You don't know how you think: introspection and language of thought. Br J Philos Sci 56(3):469–485

Mas-Colell A, Whinston MD, Green JR (1996) Microeconomic theory. Oxford University Press, Oxford

Mazzoleni R, Nelson R (1998) The benefits and costs of strong patent protection: a contribution to the current debate. Res Policy 27(3):273–284

Mellars P (1989) Major issues in the emergence of modern humans. Curr Anthropol 30(3):349–385

Mithen S (2005) The singing Neanderthals: the origins of music, language, mind and body. Weidenfeld & Nicholson, London

Morley I (2013) The prehistory of music: human evolution, archaeology and the origins of musicality. Oxford University Press, Oxford

Moser P (2005) How do patent laws influence innovation? Evidence from nineteenth-century world's fairs. Am Econ Rev 95(4):1214–1236

Muthukrishna M, Henrich J (2016) Innovation in the collective brain. Philos Trans R Soc B Biol Sci 371:20150192

Nelson R, Winter S (1982) An evolutionary theory of economic change. Belknap Press, Cambridge, MA

Nelson R, Dosi G, Helfat CE, Pyka A, Saviotti PP, Lee K et al (eds) (2018) Modern evolutionary economics: an overview. Cambridge University Press, Cambridge

Nisbett RE, Wilson TD (1977) Telling more than we can know: verbal reports on mental processes. Psychol Rev 84:231–259

Nordhaus W (1969) Invention, growth and welfare. MIT Press, Cambridge, MA

Ofek H (2013) MHC-mediated benefits of trade: a biomolecular approach to cooperation in the marketplace. In: Sterelny K, Joyce R, Calcott B, Fraser B (eds) Cooperation and its evolution. MIT Press, Cambridge, pp 175–194

Otsuka M (2003) Libertarianism without inequality. Oxford University Press, Oxford

Oudeyer P-Y, Kaplan F (2007) What is intrinsic motivation? A typology of computational approaches. Front Neurorobot 1:6

Oudeyer P-Y, Smith LB (2016) How evolution may work through curiosity-driven developmental process. Top Cogn Sci 8(2):492–502

Pain R (2019) What can the lithic record tell us about the evolution of hominin cognition? Topoi. https://doi.org/10.1007/s11245-019-09683-0

Pavlov IP (1927) Conditioned reflexes: an investigation of the physiological activity of the cerebral cortex. Oxford University Press, Oxford

Penin J (2005) Patents versus ex post rewards: a new look. Res Policy 34(5):641–656

Perlovsky LI, Bonniot-Cabanac M, Cabanac M (2010) Curiosity and pleasure. Paper presented at the 2010 International Joint Conference on Neural Networks (IJCNN), July 2010

Piccinini G, Schulz A (2019) The ways of altruism. Evol Psychol Sci 5:58–70

Premo LS, Kuhn SL (2010) Modeling effects of local extinctions on culture change and diversity in the Paleolithic. PLoS One 5(12):e15582

Promberger M, Marteau TM (2013) When do financial incentives reduce intrinsic motivation? Comparing behaviors studied in psychological and economic literatures. Health Psychol 32(9):950–957

Qian Y (2007) Do national patent laws stimulate domestic innovation in a global patenting environment? A cross-country analysis of pharmaceutical patent protection, 1978–2002. Rev Econ Stat 89(3):436–453

Reader SM, Laland KN (2003) Animal innovation: an introduction. In: Reader SM, Laland KN (eds) Animal innovation. Oxford University Press, Oxford, pp 3–37

Reeson AF, Tisdell JG (2008) Institutions, motivations and public goods: an experimental test of motivational crowding. J Econ Behav Organ 68(1):273–281

Renfrew C, Scarre C (eds) (1998) Cognition and material culture: the archaeology of symbolic storage. McDonald Institute for Archaeological Research, Cambridge

Richardson R (2007) Evolutionary psychology as maladapted psychology. MIT Press, Cambridge, MA

Robson AJ (2001) The biological basis of economic behavior. J Econ Perspect 39(1):11–33

Root-Bernstein RS (1989) Strategies of research. Res Technol Manage 32(3):36–41

Rosenberg A (2012) Philosophy of social science, 4th edn. Westview Press, Boulder

Saad G (2017) On the method of evolutionary psychology and its applicability to consumer research. J Mark Res. https://doi.org/10.1509/jmr.14.0645

Santos LR, Rosati AG (2015) The evolutionary roots of human decision making. Annu Rev Psychol 66(1):321–347

Schank JC (2015) The evolution and function of play. Adapt Behav 23(6):329–330

Schulz A (2018) Efficient cognition: the evolution of representational decision making. MIT Press, Cambridge, MA

Schumpeter J (1942) Capitalism, socialism, and democracy. Harper & Brothers, New York

Schumpeter J (1959) The theory of economic development. Harvard University Press, Cambridge, MA

Shiffrin SV (2001) Lockean arguments for private intellectual property. In: Munzer SR (ed) New essays in the legal and political theory of property. Cambridge University Press, Cambridge, pp 138–167

Shrader-Frechette K (2006) Gene patents and Lockean constraints. Public Aff Q 20(2):135–161

Sichelman T (2010) Commercializing patents. Stanford Law Rev 62(2):341–402

Skinner BF (1938) The behavior of organisms: an experimental analysis. Appleton-Century, Oxford

Skyrms B (2004) The stag hunt and the evolution of social structure. Cambridge University Press, Cambridge

Skyrms B (2010) Signals: evolution, learning, and information. Oxford University Press, Oxford

Sober E (1992) Models of cultural evolution. In: Griffiths P (ed) Trees of life. Springer, Dordrecht, pp 17–39

Sober E (1994) The adaptive advantage of learning and a priori prejudice. Ethol Sociobiol 15(1):55–56

Stanley J (2011) Know how. Oxford University Press, Oxford

Sterelny K (2012) The evolved apprentice: how evolution made humans unique. MIT Press, Cambridge, MA

Sterelny K (2017) Artifacts, symbols, thoughts. Biol Theory 12(4):236–247

Sterelny K, Hiscock P (2017) The perils and promises of cognitive archaeology: an introduction to the thematic issue. Biol Theory 12(4):189–194

Strevens M (2003) The role of the priority rule in science. J Philos 100:55–79

Sugiyama LS, Sugiyama MS (2003) Social roles, prestige, and health risk. Hum Nat 14(2):165–190

Teece DJ (1986) Profiting from technological innovation: implications for integration, collaboration, licensing and public policy. Res Policy 15:285–290

Torrance AW, Tomlinson WM (2009) Patents and the regress of useful arts. Columbia Sci Technol Law Rev 10:130–168

Veblen T (1898) Why is economics not an evolutionary science? Q J Econ 12(4):373–397

Wilson TD (2002) Strangers to ourselves. Belknap Press, Cambridge, MA

Wilson DS, Gowdy JM, Rosser JB Jr (2013) Rethinking economics from an evolutionary perspective. J Econ Behav Organ 90S:S1–S2

Wylie A (1994) Evidential constraints: pragmatic objectivism in archaeology. In: Martin M, McIntyre L (eds) Readings in the philosophy of social science. MIT Press, Cambridge, MA, pp 747–765

Wynn T (1993) Two developments in the mind of early homo. J Anthropol Archaeol 12:299–322

Zollman K (2018) The credit economy and the economic rationality of science. J Philos 115(1):5–33

Saad G (2017) On the method of evolutionary psychology and its applicability to consumer research. J Mark Res, Innovation, AI, 1300ffman Hotel

Santos LR, Rosati AG (2015) The evolutionary roots of human decision making. Annu Rev Psychol 66(1):321–347

Schmitt JC (2015) The evolution and function of play. Annu Relev 23(2):329–330

Schulz A (2018) Efficient cognition: the evolution of representational decision making. MIT Press, Cambridge, MA

Schumpeter J (1942) Capitalism, socialism, and democracy. Harper & Brothers, New York

Schumpeter J (1959) The theory of economic development. Harvard University Press, Cambridge, MA

Shariffin SV (2011) Enchanted imagination for private intellectual property. In: Munzer SR (ed) New essays in the legal and political theory of property. Cambridge University Press, Cambridge, pp 138–157

Shadmehr R, Krakauer JW (2010) Gene patents and Delicense constraints. Biolaw AR Q 20(2):115–141

Sichelman T (2010) Commercializing patents. Stanford Law Rev 62(2):341–402

Skinner BF (1938) The behavior of organisms: an experimental analysis. Appleton-Century, Oxford

Skyrms B (2004) The stag hunt and the evolution of social structure. Cambridge University Press, Cambridge

Sloutsky (2010) Similarity, category learning, and information. Oxford University Press, Oxford

Sober E (1992) Models of cultural evolution. In: Griffiths P (ed) Trees of life. Springer, Dordrecht, pp 17–39

Sober E (1994) The adaptive advantage of learning and a priori prejudice. Ethol Sociobiol 15(1):55–60

Stanley J (2011) Know how. Oxford University Press, Oxford

Sterelny K (2012) The evolved apprentice: how evolution made humans unique. MIT Press, Cambridge, MA

Sterelny K (2012) Artifacts, symbols, thoughts. Biol Theory 12(4):236–247

Sterelny K, Hiscock P (2017) The point and promise of Cognitive archaeology: an introduction to the thematic issue. Biol Theory 12(4):189–194

Stevenson M (2001) The role of the priority rule in science. J Philos 100:55–79

Sugiyama LS, Sugiyama MS (2003) Social roles, prestige, and health risk. Hum Nat 14(2):165–190

Teece DJ (1986) Profiting from technological innovation: implications for integration, collaboration, licensing and public policy. Res Policy 15:285–290

Torrance AW, Tomlinson WJ (2009) Patents and the regress of useful arts. Columbia Sci Technol Law Rev 10:130–168

Arthur T (1888) Why is economics not an evolutionary science? Q J Econ 12(4):373–397

Wilson TO (2002) Sneakers to ourselves. Belknap Press, Cambridge, MA

Wilson DS, Gowdy JM, Rosser JB Jr (2013) Rethinking economics from an evolutionary perspective. J Econ Behav Organ 90:S1–S2

Wylie A (1985) Evidential constraints: pragmatic objectivism in archaeology. In: Martin M, McIntyre L (eds) Readings in the philosophy of social science. MIT Press, Cambridge, MA, pp 747–765

Wynn T (1993) Two developments in the mind of early homo. J Anthropol Archaeol 12:299–322

Zollman K (2013) Network epistemology: communities of science. Philos Compass 1(5):574–587

Chapter 6
Music Archaeology, Signaling Theory, Social Differentiation

Anton Killin

Abstract Musical flutes (pipes) constructed from bird bone and mammoth ivory begin to appear in the archaeological record from around 40,000 years ago. Due to the different physical demands of acquiring and working with these source materials in order to produce a flute, researchers have speculated about the significance—aesthetic or otherwise—of the use of mammoth ivory as a raw material for flutes. I argue that biological signaling theory provides a theoretical basis for the proposition that mammoth ivory flute production is a signal of increasing social differentiation in Upper Palaeolithic human life.

Keywords Archaeological record · Behavioral modernity · Music · Signaling theory · Social complexity

Not only is music a highly valued feature of all known living human cultures, permeating many aspects of daily life, playing many roles. Music is ancient. The oldest known musical instruments—bird-bone and mammoth-ivory flutes (pipes)—appear in the archaeological record from 40,000 years ago and from these we can infer even earlier musical artifacts/activities, as yet unrepresented in the material record (e.g., Morley 2013; Killin 2014). Several researchers have speculated that the use of mammoth ivory for flute production (rather than bird bone) reflects some special status or significance, whether aesthetic, or otherwise. I agree that an aesthetic dimension is likely to be part of the story. But that is not what is at issue in this chapter. Rather, I argue that biological signaling theory (an increasingly common tool now of both biological theorists and philosophers of biology) provides a

A. Killin (✉)
Department of Philosophy, Mount Allison University, Sackville, NB, Canada

School of Philosophy and Centre of Excellence for the Dynamics of Language, Australian National University, Acton, ACT, Australia
e-mail: anton.killin@anu.edu.au

© Springer Nature Switzerland AG 2021
A. Killin, S. Allen-Hermanson (eds.), *Explorations in Archaeology and Philosophy*, Synthese Library 433, https://doi.org/10.1007/978-3-030-61052-4_6

theoretical basis for taking seriously a complementary suggestion: that mammoth ivory flute production can justifiably, albeit tentatively, be thought to reflect something of significance: increasing social differentiation in Upper Palaeolithic human life.[1] By adopting this theoretical framework, the methodological focus of attention is thus shifted to the natural phenomenon of signal production and use, and away from some standard approaches to signs and symbols such as structuralism, semiology, and C.S. Peirce's semiotics (Saussure 1959; Lévi-Strauss 1969; Leach 1976; Atkins 2013). The case for such a refocus is made in Godfrey-Smith (2014a, 2017) and Kuhn (2014). Here I hope to contribute to demonstrating its utility.

6.1   Musical Technologies in Prehistory

The archaeological evidence for ancient musicking is intriguing. The oldest known musical instruments are the Upper Palaeolithic flutes (pipes) from the Swabian Jura in southwestern Germany (specifically, Hohle Fels, Vogelherd, and Geißenklösterle). Most of these are made from bird bone (predominantly vulture radius or ulna; also swan bone); some from mammoth ivory. They appear from around 40 kya (40 thousand years ago) onwards. Since it was around this time—40 kya—that *sapiens* are currently thought to have arrived in western Europe, it seems that they brought the cognitive wherewithal to make and use musical artifacts with them from Africa; that is, modern *sapiens'* founder population was musical. I'll return to this thought shortly.

One of the oldest flutes so far discovered (in 2008 in Hohle Fels and reconstructed from fragments: see Adler 2009; Conard et al. 2009; Morley 2013, ch 3) is made from a griffon vulture (*Gyps fulvus*) wing bone. The preserved portion of the flute sports five fingerholes and is nearly 22 cm long, which researchers presume is virtually the complete item. The body of the flute has been scraped smooth. The fingerholes were created by piercing, with a tool, areas of the bone that were first etched into thinned-out, concave depressions. One end of the bone has been fashioned into a mouth-hole by carving into it two V-shapes. Cut-marks near the fingerholes suggest that the placements of the holes were measured. The underlying reason for that measuring is unknown (perhaps it is indicative of a pitch standard or scale, or was for reasons of physical practicality or pedagogy) but nevertheless it appears that this aspect of the flute was designed with something in mind. Reconstruction and performance experiments of prehistoric flutes from the Swabian Jura exhibit a wide range of tones possible which, according to some researchers,

---

[1] This chapter expands and develops an idea I suggested in Killin (2018). The basic idea follows Kim Sterelny's argument that symbolic culture signals not an upgrade in individual cognitive sophistication or the first appearance of self-identifying collectives, but is both partial cause and effect of increasing social hierarchy and wealth differential (Sterelny 2012, pp. 52–54).

establishes these instruments as "fully developed musical instruments" (Conard and Malina 2008, p. 14).[2]

To my mind, these flutes suggest a long history of musical technological production. They are not crude whistles. Their sophistication of design, and the pre-planning evident, is indeed indicative that they must be "several conceptual stages removed from the earliest origins, even of instrumental musical expression, to say nothing of those universal vocal, manual-percussive and dance forms which must have existed independently of—and before—any need for such tools" (d'Errico et al. 2003, p. 46). I thus side with the view that musical technology has a much older past than the Upper Palaeolithic flutes, currently (and perhaps indefinitely) hidden from the material record.

Prehistoric musical instruments enable rare and fascinating glimpses into an otherwise largely hidden culture, revealing more and more about our lineage's ancient past. The archaeologist Iain Morley (2013) offers an excellent survey and inventory of prehistoric musical instruments so far unearthed by archaeologists, including 104 bird-bone and ivory pipes/flutes, as well as pierced reindeer-foreleg phalanges (alleged whistles) and other sound-producers such as bullroarers and various forms of struck percussion such as rasps. Steve Kuhn and Mary Stiner (1998), for example, identify a modified ungulate bone from around 32–35 kya that is reminiscent of rasps found in several contemporary musical cultures. Its function as a musical artifact may be an educated guess, as is that of the alleged bullroarers and other such artifacts discussed by Morley, however, use-wear analysis might shed light. The musical status of the Upper Palaeolithic flutes, though, I take to be non-controversial. They will be my focus.

It is a striking fact, in my view, that even the oldest known flutes demonstrate such an investment of time, energy and resources. Consider the oldest known *mammoth-ivory* flute (Conard et al. 2004, 2009), dated to around 40 kya. Compared to bird-bone, the production of flutes from ivory requires greater skill, precision work, and effort. Vulture and swan radius and ulna are naturally hollow and already an appropriate size, as well as light, sturdy and thus easier to craft in comparison to ivory, which is oversized, layered, and tough to work. Ivory flute production requires that:

> ...a section of ivory must be sawn to the correct length, it must then be sawn in half along its length, the core lamellae (layers) must be removed, and then the two halves of the flute must be refitted and bound together with a bonding substance which must create an airtight seal in order for the pipe to produce a sound. (Morley 2013, p. 50)

An "ambitious challenge" in the words of Conard and Malina (2008, p. 14)! Ivory flute production indicates the maturity and sophistication of Upper Palaeolithic musical technologies. And the commitment of time, energy and valuable resources

---

[2] For example, see (hear) experimental archaeologist and flautist Fredrich Seeberger's CD *Klangwelten der Altsteinzeit* ("Soundworld of the Old Stone Age"), for Urgeschichtliches Museum Blaubeuren. Seeberger recorded experimental/explorative performances of replicas of Upper Palaeolithic flutes in Hohle Fels cave, demonstrating a range of possibilities of these flutes in one of their presumed acoustic spaces.

to musical technologies implies that music really mattered to ancient humans. Morley suggests that given that bird-bone flutes could be produced much more easily than ivory flutes, and that the ivory flutes looked more or less identical to (though a little bigger than) their bird-bone counterparts, "the choice of raw material (mammoth ivory) was itself considered highly significant" (Morley 2013, p. 51). Although Morley does not get into it, this is consistent with the idea that the use of ivory for flute production was aesthetically significant, perhaps for its timbre or pitch/range differences compared to that of the bird-bone flutes, or significant due to associations with the raw material itself, for instance whether it was supposed by ancient folk to have magical powers or the like. I strongly suspect that de facto there is an aesthetically oriented dimension to ivory flute production. But in this chapter, I propose that *signaling theory* from biology provides a theoretical basis for drawing some complementary and less speculative, though nonetheless still provisional, inferences about what the mammoth-ivory flutes might reflect. The basic strategy will be to employ *trace-based reasoning* with biological signaling theory supplying the *mid-range theory*: a theory about that which exists in the 'mid-range' between law-like generalizations and particulars. This inferential framework is described and defended by Currie (2018) and Currie and Killin (2019); see Pain (2019) for constructive discussion. The next section introduces the core methodology.

6.2   Trace-Based Reasoning

Much of archaeology involves inferring from some physical trace (or set of traces) of the past to behavioral, social, or cognitive features of past human life. For an example of this methodology, take Thomas Wynn and Frederick L. Coolidge's (2004) argument in support of the proposition that Neanderthals possessed long term working memory. Wynn and Coolidge argue for this cognitive claim by inferring *from* physical traces (here, Levallois tools) by way of a *midrange theory* (or, in this case, theories)—the long-term working memory model of expert cognition (Ericsson and Kintsch 1995) and the situated learning model of technical cognition (Keller and Keller 1996). These theories, which have independent empirical support, are taken to provide an inferential window into the past: a window through which the traces can be interpreted, and from this theoretically and empirically constrained interpretation it can be inferred that the best explanation for the production of the traces in question is that Neanderthals possessed the capacity for long term working memory.[3]

The preceding example of trace-based reasoning has been characterized as a *minimal-capacity inference* (Currie and Killin 2019): in this case, it purports to identify a prerequisite cognitive capacity. Rather than focusing on some minimal capacity, in this chapter I suggest that the musical technologies of the Upper

---

[3] For discussion within this framework of this and other relevant examples, see Pain (2019).

Palaeolithic comprise part of a larger package enabling a *causal-association infer-ence*; an inference which builds an association between lines of evidence causally connecting material traces with alleged features of past societies. It aims to give provisional license to the claim that Upper Palaeolithic musical technologies are evidence of increasing social differentiation. Before human societies made the tran-sition from egalitarian Pleistocene forager lifeways to those of the hierarchically structured social groups which later came to characterize much of the post-Neolithic Transition human world, more complex social dynamics and internal divisions than that of the standard picture of Pleistocene forager groups presumably characterized late Middle Palaeolithic and Upper Palaeolithic social life. Kuhn runs an argument to that effect by appeal to Palaeolithic adornment technologies. By adding musical artifacts to adornment technologies—broadening the evidential range—Kuhn's argument's plausibility is strengthened. This is my goal in the following three sections.

6.3 Palaeolithic Adornment Technologies

Recall that the midrange theory I have in mind is biological signaling theory. Ancient humans have been sending/receiving signs through the use of material sub-stances and objects as communicative media for a long time—perhaps at least since ochre and other pigments appear in the archaeological record, initially rarely (per-haps since 230–280 kya; see McBrearty and Brooks 2000) and more frequently and in greater supply over time (McBrearty and Stringer 2007). Perforated beads enter the record around 100 kya (Vanhaeren et al. 2006), though in a patchy fashion. Vanhaeren and colleagues describe two beads from Skhul in Israel and one from Oued Djebbana in Algeria. Analysis on one of the Skhul beads indicates that it comes from a layer containing human fossils and stone tools dated to 100–135 kya. The distance of these sites from the shore (for Djebbana, 190 km) indicates deliber-ate human selection and transport of the shells: typical natural causes of shell dis-placement such as major storms or transport due to natural predators cannot explain the presence of the shells at these sites. 41 shell beads discovered at Blombos Cave in South Africa have been dated to around 75 kya (Henshilwood et al. 2002, 2004; d'Errico et al. 2005); other beads from various sites in North Africa 40–60 km from the shoreline date to 82 kya (e.g., Bouzouggar et al. 2007; d'Errico et al. 2009). These early beads are 'simple' in the sense that they were made from easily-worked and easily-acquired local raw materials (e.g., shells of littoral gastropods, mainly *Nassarius*, and ostrich shell) and those that were humanly modified were not much modified. Use-wear analysis indicates that at least some of the beads were strung/suspended; some beads contained traces of red ochre (iron oxide) most likely trapped into the grooves of the shell by its rubbing against colored materials such as hide or thread (Bouzouggar et al. 2007, pp. 9966–9967). It is thus safe to infer that they were worn. It also appears some beads were intentionally heated in order to change their coloring (d'Errico et al. 2005). It is unlikely that the use of Middle

Palaeolithic beads was frivolous and without symbolic import. Bouzouggar and colleagues argue that "the choice, transport, coloring, and long-term wearing of these items were part of a deliberate, shared, and transmitted nonutilitarian behavior... [To] be conveyed from one generation to another over a very wide geographic area, such behavior could not have survived if they were not intended to record some form of meaning" (Bouzouggar et al. 2007, pp. 9969). Add to this the Middle Palaeolithic evidence of ostrich eggshell fragments and ochre pieces both with engraved geometric designs (Texier et al. 2010; Henshilwood et al. 2009); "The engraved eggshell fragments were probably originally parts of canteens or other storage vessels, and so could have been used for displaying a variety of messages" (Kuhn 2014, p. 44).

Middle Palaeolithic beads typically appear in the archaeological record among rubbish, presumably lost or discarded, not in large concentrations or storage caches. After a period of stasis, more complex ornaments enter the record as time goes on. The Upper Palaeolithic record contains ostrich-eggshell ornaments, pierced animal teeth, carved ornaments made from ivory, bone, and stone, in addition to more littoral-shell beads appearing more ubiquitously and in greater quantity, throughout European and western Asian sites as well as throughout Africa (Álvarez Fernández and Jöris 2007). Ornaments as grave goods appear from 30 kya. Although the actual information encoded in the signs sent by way of the use of these ornaments *qua* communicative technologies remains obscure, it is possible to make inferences about the types of messages—their more abstract or general role.

With that goal in mind, archaeologist Steve Kuhn (2014) argues that these accumulative innovations in Palaeolithic adornment technologies—from mineral pigments (ochre and other metallic oxides), to simple beads, to more complex ornaments and ornaments appearing as grave goods—reflect changes in social dynamics and complexity; that they point to the solving of coordination problems in ancient social life, and resolving of conflicts in human groups with increasingly larger populations where the interests of individuals may be less 'aligned'. The idea receives its theoretical license from biological signaling theory, an influential framework for the study of animal communication systems that emphasizes the evolutionary consequences of various signaling strategies and their adaptive potential.[4]

6.4   Signaling Theory

Signaling theory distinguishes senders from receivers (though being a sender does not exclude one from also being a receiver, and vice versa). A sender produces a signal reflecting some perceived state of the world which a receiver reads or interprets and acts upon (Lewis 1969). The meaning of a signal is typically taken to be a

---

[4] For philosophical treatments of signaling theory, see work by Brian Skyrms (e.g., Skyrms 2004, 2010) and Peter Godfrey-Smith (e.g., Godfrey-Smith 2014a, b).

function from the state of the world to the act via two 'rules': the sender's rule and the receiver's rule (Godfrey-Smith 2014a). Speaking in this way is not supposed to imply that the rules are conscious or volitional, although they could be. The sender's rule maps world-states to signs; the receiver's rule maps signs to acts. Famously, the vocal alarm calls of vervet monkeys guide the predator-evasion strategies of other members of the group (Seyfarth et al. 1980), which provides an excellent example of a sender-receiver system in action. Schematically, the vervet who sees a leopard approaching produces a species-typical vocalization, sign L; the vervet who detects an eagle produces sign E; the vervet who spots a snake produces sign S. Upon receiving sign L, the other vervets climb up high into the treetops where the leopard cannot follow. Upon receiving sign E, they look up and hide under thick bushes exploiting the protection of branches and foliage. Upon receiving sign S, they stand up on their hind legs and scan the area surrounding them for snakes, which they can usually drive off with the threat of a mob attack.

Signaling theory distinguishes costly from non-costly signals. As the saying goes, "talk is cheap". There are many contexts in which it is in the sender's interests to send low cost but false messages and it's in the receiver's interests to ignore false messages; consequently, signs that are low cost typically become established only in low stake contexts where those party to the sign have aligned interests.[5] Such signs usually function to coordinate action—food-associated vocal calls of chimpanzees and bonobos, for example, guide the foraging of other members of the group. The bow of a male adult lion signals an invitation to a cub to engage in play fighting behavior; the worker honeybee's waggle dance signals the location of a nectar source, recruiting a team of carriers. And consider again the alarm calls of vervet monkeys. Idealizing somewhat and assuming common interest, the vervet signaling system is a Nash equilibrium: neither senders nor receivers have incentive to unilaterally change their behavior. In more demanding contexts, or contexts in which the individuals' interests diverge, signaling theory predicts that senders of signs should pay some nontrivial cost and/or produce a sign that is hard-to-fake in order for them to be taken as honest, credible signs.[6] Gazelle stotting is a typical example. Stotting expends more energy than, say, vocalizing would, and is a reliable signal to the predator that not only has the gazelle noticed the predator, the gazelle is fit and capable of escaping, for otherwise it should not risk expending valuable energy before a potential chase, or risk revealing a tired leap. Stotting is a costly, hard-to-fake display of energetic resources. If the signal is heeded, both animals avoid a pointless, energy-expensive chase. (They have that interest in common! And it is that very interest which sustains informative communication via the sign.)

Other typical examples include male-male conflict signals arising in contexts of sexual selection: the loud bellows of male red deer (*Cervus elaphus*) during roaring

---

[5] Actual cases of *complete* common interest are of course rare. However, cases of partial common interest can be more or less aligned, and this graded conception is all we need.

[6] Not all high-cost signals are hard-to-fake and vice versa; for reasons of exposition I do not explicitly distinguish these here as that is a task not without difficulties and is orthogonal to my aims. See Brusse (2019) for philosophical analysis.

matches over females signals body size, fighting prowess and energetic resources (Clutton-Brock and Albon 1979). Both roaring and fighting are high-energy and utilize the same set of thoracic muscles, so powerful bellows are only possible from stags in tip-top shape; weak fighters cannot fake a costly roar and would do better to withdraw from a potential fight. The important point for our purposes here is that relatively low-cost signals typically only establish in cases where sender and receiver have aligned interests; hard-to-fake, differentially costly signals typically reflect diverging interests. These examples also demonstrate a principle of natural sign emergence: senders and receivers, whether in cases of (more or less) aligned interests or diverging interests, are coadapted entities. Signals, in part, mediate their interactions (Godfrey-Smith 2017).

Kuhn infers from ethnographic studies of recent hunter-gatherer societies that "[Palaeolithic] beads and pigments decorated bodies and clothing, and that they most often carried social information, messages about an individual's identity, affiliations, social roles, and social standing"; "They tell well-informed viewers about the wearer's place in kinship networks, their marital status, group affiliations, and so forth" (Kuhn 2014, p. 43). Similarly, Stiner observes, "In recent cultures, beads are combined, often in large numbers, to express group affiliation, personal identity or state (wealth, marital, or social status), or serve as simple attention-fixing devices in social gatherings. They are also common components of symbol-laden gifts" (Stiner 2014, p. 53). Kuhn argues that the use of ochre and other pigments as material signs does not express cost very effectively, that simple ornaments express some cost effectively, and that more complex ornaments, as well as the action of adding material objects to burials, expresses more cost (see also Kuhn and Stiner 2007).

Via signaling theory, Kuhn infers that low-cost signs (ochre and other pigments; simple ornaments) are correlated with low-stakes contexts, probably "reflecting efforts at coordinating human actions, small-scale rituals, or social gestures promoting shared identity and cooperation" (Kuhn 2014, p. 46). Their use may have helped to identify an individual's specific role in some context, for example. They may have a dual function, as Godfrey-Smith points out: "one function is to convey information about the bearer to observers and another is to foster unity" (Godfrey-Smith 2014a, p. 84). The exact details are of course conjectural, but that's not what's at issue here; the idea is that they were used for something, and whatever it was, it was in a context in which individual interests were (more or less) aligned, given the low cost of the sign:

> There is no point in making the marks if their behavioral consequences in observers will not be helpful to the sender, the person marked. But a sender cannot dictate how a receiver will interpret the marks; the receiver will react to the marks with habits or rules that have been selected to serve the receiver's interests. If the marking of the body is stable in some context, this is probably because the behavioral consequences of observing the marks are beneficial, on balance, for both sides. (Godfrey-Smith 2014a, p. 84)

This fits the generally received egalitarian picture of Late Pleistocene anatomically modern human social life—a social life without formal leadership, internal structural demarcations, or accumulation of personal wealth—and suggests some nudge of an increase in social complexity and inequalities at or around the time that beads are entering the picture. Their information-carrying role is bound to be key: "they conveyed messages about an individual's identity, affiliation, and social standing… human groups used these markers in the formation and navigation of expanding social networks, which provided some degree of 'insurance' and mutualistic support" (Godfrey-Smith 2014a, p. 86; see also Stiner 2014). Beads are more durable than the application to skin of ochre and other pigments—such application is transient by its very nature.[7] Beads can be clearly transferred between individuals (and thus any message they convey has the potential to transcend single encounters/individuals) and they express quantity in a clearer way than pigment application (Kuhn and Stiner 2007). They are also recombinable: beads are small, individual units, with easily perceived repeating (or contrasting) geometries, that may be combined into larger compositions (Stiner 2014). Given these features (coupled with the potentially not-very-high costs involved in collecting, selecting/modifying, and—given their light weight—transporting beads), Kuhn argues that they reflect an increase in social complexity and diverse, but still potentially congruent, interests. Social dynamics are, gradually, on the move, but we have still not at this stage produced evidence of a substantial shift from the standard picture.

In light of signaling theory, this picture of ancient human social life changes somewhat with the introduction of the more complex ornaments. Ivory is difficult to modify, and presumably more difficult to acquire than shells or mineral pigments.[8] Ditto for carnivore teeth and the canines of male red deer—possibly signs of hunting prowess, or wealth/status. And even 'simple' beads take on a dimension of added cost as they begin to be traded over hundreds of kilometers, potentially as symbols of place, rarity, group associations. The costliness of these ornaments (and the hard-to-fake skill required for the modification of some) may still reflect coordinated action but may also reflect social competition as a release valve for conflict due to diverging interests. As social complexity and group size increased, the assumption is that ancient societies became less egalitarian, internally divided, and thus greater social differentiation emerges. Complex ornaments might reflect one way that ancient humans handled conflict management (for example, status conflict between two families or subgroups within the larger group). The action of adding ornaments and other valuable goods to burials, Kuhn suggests, is "clear evidence for a new kind of social competition, probably between the families or lineages of the deceased" (Kuhn 2014, p. 47). Note that the ornaments unearthed at Upper Palaeolithic human burials at Sunghir in Russia—one extreme example—required thousands of hours of labor (White 2003): a costly, hard-to-fake, conspicuous indicator of status or social standing, possible religious/ritual functions

---

[7]Tattooing is an ancient practice at least several thousand years old, but there is no evidence of it in Palaeolithic humans, and in any case, "tattooing does not consume much pigment, so most of what we find was probably used for other purposes" (Kuhn 2014, p. 44).

[8]The material record provides ample evidence of the big game hunting skills of Late Pleistocene humans. Even if the ivory was acquired through scavenging practices rather than active hunting it would still have to be extracted from the carcass, which is not effortless, and may well have involved deterring other predators or scavengers.

notwithstanding. Indeed it is curious that 30 kya, ancient humans would dedicate such time, energy, and resources to an impressive array of goods, only for these goods to be buried and covered over, and presumably, not intended to be seen again (at least, not by human eyes).

6.5   Adornments, Music, Sociality

According to the standard picture, Pleistocene foragers comprised mobile, loosely egalitarian bands: "They lived in what might be called societies of equals, with minimal political centralization and no social classes" (Boehm 2001, p. 4; Knauft 1991; Kaplan et al. 2009). Since the Neolithic Transition (~10 kya), many human societies, independently, in geographically isolated regions, began living in sedentary, larger scale, inequal, inegalitarian social groups. The change did not occur overnight, and the advent of agriculture in those regions presumably played a significant role in ramping up the inequalities in the accumulation and distribution of wealth (Barker 2006). However, farming and food storage practices don't provide a full explanation (Flannery and Marcus 2012; Sheehan et al. 2018) and there is considerable debate surrounding provisionally dating changes in social complexity. The preceding discussion, however, locates the seed of such a transition. That seed can be dated back to the Middle Palaeolithic in Africa prior to the final out-of-Africa exodus. African, European, and Asian sites continue, from 45 kya onwards, to provide a picture of gradual increases in social complexity and differentiation compared to the Middle Palaeolithic baseline. (Of course, that picture is nothing like that of, say, Göbekli Tepe, let alone the explicitly hierarchical social environments of other such post-Neolithic Transition human worlds as Ancient Egypt.) Signaling theory provides theoretical license for the inference that cumulative innovations in Palaeolithic adornment technologies were a material expression[9] of a shift towards greater differentiation and internal conflict in human social groups.

Kuhn's discussion of ornaments fits nicely with the preceding discussion of music archaeology. Early musical activities and musical instruments can be thought of as part of a suite of communicative media. Complex musical instruments demonstrable of high cost enter the archaeological record around the time of more complex ornaments and grave goods; also cave art and figurines. But, as suggested earlier, these almost certainly reveal a much older prehistory of music, perhaps lost to the record.[10] Turning to the ethnographic record (and considering the cross-cultural ubiquity of music, the typically ephemeral types of materials used for musical instruments by ethnographically known foragers, the emphasis of the body and

---

[9] They may also have been a direct causal mechanism of that shift, though I have not made the case for that herein.

[10] Moreover, it was only due to painstakingly intricate efforts at waterscreening, organizing, and refitting that many of the bone and ivory Upper Palaeolithic flutes have been reconstructed and identified. Other excavations may have recovered fragments that remain unidentified.

voice in musicking, and so on) motivates the idea that humans were musically active for a long time. That is, musically active by vocalizing and using their bodies (clapping, stomping, dancing), perhaps making and using 'low cost' instruments (e.g., from easily acquired and easily workable materials, quite likely materials that would not survive deposition until today were they in fact used) in cooperative contexts (given the low costs involved), and, presumably, for the expression of affect and group affiliation and bonding.

Research related to music's origins details many ways in which both existing and ethnographically known human forager groups made music and these typically reflect the scenario just posited (see Morley 2013, ch 2, for review). Note that music is present and deeply entrenched in every known human society. It is thus likely its evolutionary origins are in our pre-exodus African forebears, or it would have had to evolve independently all over the world in many different, and geographically isolated, cultural contexts. Morley surveys the traditional music and ways of musicking of several hunter-gather societies from geographically disparate regions: the Blackfoot and Sioux Native Americans, the Aka and Mbuti African Pygmies, the Pintupi-speaking Western Desert Australian Aborigines, and the Alaskan Yupik and Canadian Inuit. All these musical traditions are drawn from lifeways prior to the influence of Western settlement and agricultural modes of subsistence; lifeways within wildly diverse ecological niches—from plains to forest to desert to tundra. And, in different ways, in each culture, music deeply pervades their day to day lives. The details are fascinating, but I must pass over these here (Killin 2016, 2018 revisits these and others). Although these modern-day and historically recent foragers are of course not Palaeolithic people, it is somewhat telling that their musics are all predominantly vocal, with little if any instrumentation.[11] Non-vocal sound production, when used, comprises predominantly unpitched percussion (drums, rattles, rasps, clapsticks) or bodily sound production (stomping, hand-clapping). Some pitched instrumentation features—e.g., flutes, whistles, and bullroarers, etc., but not all of these feature in every one of these musical traditions. These instruments are made using such naturally occurring organic materials as cane and wood. Native Australians have the *didjeridu*—traditionally a termite-hollowed eucalyptus tree limb, blown to produce a drone upon which a rich variety of embellishments are produced—but these are distinctive. Although the Australian, Yupik and Inuit musics contain lyric traditions encoding, for example, geographic, ecological, or community knowledge, much Native American and African Pygmy song comprises *vocables*, not words, which contribute to the wider context of performance emotionally/affectively. In all these cultures, the ritual and ceremonial uses of music are crucial, although music is also a pleasurable and entertaining communal pastime. These last points extend well beyond the specific forager groups Morley discusses. As Wiessner points out, in her investigation of day- versus night-time activities of forager groups:

---

[11] I write in the present tense even though some traditions may no longer be practiced as they used to be prior to Western influence and other modern-day factors.

For 60 hunter-gather societies in the eHRAF [Human Relations Area Files for world cultures], there was mention of ceremonies held by firelight involving song and dance for celebration, healing, initiation, mourning, fertility, and other purposes. Song and dance bonds groups through rhythm, coaction, and in many cases, altered states of consciousness. Night was a prime time for entrance into imaginary worlds of the supernatural. (Wiessner 2014, p. 14032)

Wiessner's own case study focuses on the Ju/'hoansi (!Kung Bushmen) of southern Africa (northwest Botswana and northeast Namibia). As the flickering flames of firelight extend the day for social rather than subsistence/economic pursuits long past sundown, people would gather to talk, dance, and make music. Singing was sometimes accompanied by thumb piano, lute (pluriarcs), and musical bow, coopting a utilitarian technology for use in musical production.[12] It was also the time for celebrations and ceremonies:

On 8 of 36 evenings observed, women broke into festive song and dance that flooded night. On 6 of 36 nights, trance-healing dances were performed: three were smaller ones for healing and three nightlong dances drew in people from three or more camps. Efforts by everybody present contributed to healing the sick, closing social rifts, and restoring spiritual cohesion. (Wiessner 2014, p. 14029)

The implications of this brief ethnographic interlude for interpreting the archaeological record should be clear. The absence of clear archaeological evidence of musical technologies prior to 40,000 years ago in Europe should not dissuade us from viewing earlier humans, indeed the *sapiens* founder population, as musical:

First, the music of the hunter-gatherers described here is largely non-instrumental, with melody frequently being carried by vocalization alone and much of the percussion being produced by clapping, slapping, and foot-stamping. Instrumentation, when it is used, is almost exclusively percussive. Second, the instrumentation produced is constructed from readily occurring natural resources with relatively little modification; most use skins, wood, cane, or other vegetable matter such as gourds. Wood is used more readily than bone for scrapers, flutes, drums, and drumsticks; the use of bone is rare.... Apart from bone, none of these instruments would leave an archaeological trace, as vegetable or animal matter degrades rapidly under most natural conditions. Had they been used by early humans, there would be no evidence of it in the archaeological record. (Morley 2013, pp. 30–31)

The physiological and cognitive evidence suggests that music today utilizes evolutionarily deep capacities and that its foundations may be ancient indeed (for review, see Morley 2013, ch 5–10). And the Palaeolithic stone-tool record attests the extraordinary technological capacities of ancient humans. It was certainly within their capacity to produce technologies to accompany song and dance. Moreover, much of the instrumentation evinced in the ethnographic record requires little modification: a dried gourd and its seeds makes an excellent rattle, or can be overblown like a bottle to produce notes, for example. So indeed it appears absence of evidence, in the case of music's prehistory, is not evidence of absence. I thus side with the view that basic musicality is an ancient feature of hominin social life (Killin

---

[12] Hunting bow technologies appear in the archaeological record from 64 kya (Lombard and Haidle 2011; Lombard and Phillipson 2010).

2017, 2018). Applying Kuhn's use of signaling theory, admittedly to a 'ghost trace', we can infer with a high degree of plausibility that these prior forms of musicking, being low-cost and 'simple', would have been correlated with cooperative, low-stakes contexts: for example, ritual, group bonding, entertainment, coordinating activity, perhaps lullaby and play song, and so on. The first archaeological evidence of flutes is not evidence of the beginnings of music,[13] but rather evidence that in Upper Palaeolithic Europe ancient humans made use of durable sources in musicking. Those flutes were found in the European caves (which Upper Palaeolithic *sapiens* occupied) and may not have survived the elements were it not for their somewhat protected deposition—they almost certainly would not have been discovered and pieced together were it not for their being in the caves.

Now, extending Kuhn's application of signaling theory beyond the domain of adornment technologies—to encompass the known, already-sophisticated ancient flutes—I suggest that the mammoth-ivory flutes too reflect an incrementally growing and internally differentiated social group, with increased division of labor and specialization well beyond that of subsistence activities, and increasingly complex social affiliations and networks.[14]

Mammoth-ivory flutes are an impressive display of *costly, hard-to-fake* production skill and raw material acquisition, use and control. Yet reconstructions of these instruments and performance experiments on them suggest that although ivory and bone flutes both would have been capable of "a range of notes and a potential for musical diversity comparable to modern recorders and flutes" (Conard and Malina 2008, p. 15), ivory flutes are only equivalent or even inferior to bird-bone flutes in some respects, in particular due to the effects of moisture build-up (that occurs by playing the flute) on the adhesive connecting the two ivory halves (Conard and Malina 2008). After around 30 min of playing, the sound-quality would decline and the halves of the flute would need to be reglued (or at least the adhesive connecting the halves would need to be 'touched up') to restore an air-tight fit. Ivory flutes were thus not only costly to produce, but to maintain, if they were to be played repeatedly or for an extended period. Bone flutes, on the other hand, can be played for an indefinite duration. Of course, aesthetically relevant differences between the sounds produced by ivory and bone may well have mattered to Upper Palaeolithic humans.[15] The ivory flutes, owing to their larger size, produce a deeper sound than the bird-bone flutes. Timbral differences may have been perceived as significant too. And

---

[13] Contrary to Adler's evocative article title, "The earliest musical tradition", about the bird-bone flute discussed earlier.

[14] Some theorists have interpreted particular overwrought, finely crafted Middle Palaeolithic stone handaxes as sexual signals—the 'sexy handaxe' hypothesis (see, e.g., Kohn and Mithen 1999)—though this is controversial (e.g., Nowell and Chang 2009). A possible interpretation of the mammoth-ivory flutes, similarly, sees these as 'sexy flutes'. That possibility is compatible with the view argued here (that they reflect increasing social complexity and differentiation) and thus may be a part of the story. Ancient humans were presumably constantly scrutinizing one another for partnership making/unmaking; being a skilled craftsperson presumably would not go unnoticed by potential partners. That said, the view here does not explain them as products of sexual selection.

[15] I'm grateful to Myriam Albor for helpful discussion about this point.

there may have been religious, superstitious or 'magical' associations made with items of one raw material but not the other. Well, perhaps. Other arguments are needed to make the case for such significances or associations. But in any event, I contend it is difficult to escape the conclusion that there was some signaling role (or roles) realized by way of these items, even if non-consciously, as biological signaling theory predicts. And that at least one such role is one which pushes the boundaries of 'low-stakes', interest-aligned contexts, given the material sign's high cost and difficulty to be faked. One possible way this could have gone—signaling of unique status—is captured by the discoverers of many ancient artifacts, Conard and Malina, to whom I give the final word:

> Access to such an outstanding example of craftmanship could be seen as providing the maker and the player of the instrument with increased status within the group. While music may have been a part of daily life during the Aurignacian, certainly access to a precision made ivory flute of this kind would have been highly limited, and would thereby have lent unique importance to its owner. This unique status would have been all the more important if the owner, as we assume was the case, were a skilled musician. (Conard and Malina 2008, pp. 15–16)

6.6   Conclusion

With mammoth-ivory flutes, we have conspicuously costly and hard-to-fake signals. Making/owning/using/maintaining these 'big ticket' artifacts appears to reflect increasing social differentiation in the Upper Palaeolithic. Admittedly, like Kuhn, I paint with a very broad brush. That said, I have argued that biological signaling theory gives license to such a causal-association inference: from musical technological traces to Upper Palaeolithic social dynamics, by way of signaling theory— an independently plausible theoretical framework, and its independently plausible application to Palaeolithic adornment technologies and the lessons learned by that application.

References

Adler D (2009) The earliest musical tradition. Nature 460:695–696

Álvarez Fernández E, Jöris O (2007) Personal ornaments in the early Upper Paleolithic of western Eurasia: an evaluation of the record. Eurasian Prehist 5(2):31–44

Atkins A (2013) Peirce's theory of signs. In: Zalta EN (ed) The Stanford encyclopedia of philosophy (summer 2013 edition). http://plato.stanford.edu/archives/sum2013/entries/peirce-semiotics

Barker G (2006) The agricultural revolution in prehistory: why did foragers become farmers? Oxford University Press, Oxford

Boehm C (2001) Hierarchy in the forest: the evolution of egalitarian behavior. Harvard University Press, Cambridge, MA

Bouzouggar A, Barton N, Vanhaeren M, d'Errico F, Collcutt S, Higham T, Hodge E, Parfitt S, Rhodes E, Schwenninger J-L, Stringer C, Turner E, Ward S, Moutmir A, Stambouli A (2007) 82,000-year-old shell beads from North Africa and implications for the origins of modern human behavior. PNAS 104(24):9964–9969

Brusse C (2019) Signaling theories of religion: models and explanation. Relig Brain Behav. https://doi.org/10.1080/2153599X.2019.1678514

Clutton-Brock TH, Albon SD (1979) The roaring of red deer and the evolution of honest advertisemdent. Behaviour 69:145–170

Conard N, Malina M (2008) New evidence for the origins of music from the caves of the Swabian Jura. In: Both AA, Eichmann R, Hickmann E, Koch L-C (eds) Challenges and objectives in music archaeology. Studien zur Musikarchäologie VI, Orient-Archäologie 22. Verlag Marie Leidorf, Rahden, pp 13–22

Conard N, Malina M, Münzel S, Seeberger F (2004) Eine Mammutelfenbeinflöte aus dem Aurignacien des Geißenklösterle: neue Belege für eine Musikalische Tradition im Frühen Jungpaläolithikum auf der Schwäbischen Alb. Archäologisches Korrespondenzblatt 34:447–462

Conard N, Malina M, Münzel S (2009) New flutes document the earliest musical tradition in southwestern Germany. Nature 460:737–740

Currie A (2018) Rock, bone and ruin: an optimist's guide to the historical sciences. MIT Press, Cambridge, MA

Currie A, Killin A (2019) From things to thinking: cognitive archaeology. Mind Lang 34(2):263–279

D'Errico F, Henshilwood C, Lawson G, Vanhaeren M, Tillier A-M, Soressi M, Bresson F, Maureille B, Nowell A, Lakarra J, Backwell L, Julien M (2003) Archaeological evidence for the emergence of language, symbolism, and music—an interdisciplinary perspective. J World Prehist 17(1):1–70

D'Errico F, Henshilwood C, Vanhaeren M, van Niekerk K (2005) Nassarius kraussianus shell beads from Blombos Cave: evidence for symbolic behaviour in the Middle Stone Age. J Hum Evol 48(1):3–24

D'Errico F, Vanhaeren M, Barton N, Bouzouggar A, Mienis H, Richter D, Hublin J-J, McPherron SP, Lozouet P (2009) Additional evidence on the use of personal ornaments in the Middle Palaeolithic of North Africa. PNAS 106(38):16051–16056

Ericsson KA, Kintsch W (1995) Long-term working memory. Psychol Rev 102(2):211–245

Flannery K, Marcus J (2012) The creation of inequality. Harvard University Press, Cambridge, MA

Godfrey-Smith P (2014a) Signs and symbolic behavior. Biol Theory 9:78–88

Godfrey-Smith P (2014b) Sender-receiver systems within and between organisms. Philos Sci 81(5):866–878

Godfrey-Smith P (2017) Senders, receivers, and symbolic artifacts. Biol Theory 12:275–286

Henshilwood C, d'Errico F, Yates R, Jacobs Z, Tribolo C, Duller GAT, Mercier N, Sealy J, Valladas H, Watts I, Wintle A (2002) Emergence of modern human behaviour: Middle Stone Age engravings from South Africa. Science 295:1278–1280

Henshilwood C, d'Errico F, Vanhaeren M, van Niekerk K, Jacobs Z (2004) Middle Stone Age beads from South Africa. Science 304:404

Henshilwood C, d'Errico F, Watts I (2009) Engraved ochres from the Middle Stone Age levels at Blombos Cave, South Africa. J Hum Evol 57:27–47

Kaplan HS, Hooper PL, Gurven M (2009) The evolutionary and ecological roots of human social organization. Philos Trans R Soc B 364:3289–3299

Keller CM, Keller JD (1996) Cognition and tool use: the blacksmith at work. Cambridge University Press, Cambridge

Killin A (2014) Musicality in human evolution, archaeology and ethnography. Biol Philos 29:597–609

Killin A (2016) Rethinking music's status as adaptation versus technology: a niche construction perspective. Ethnomusicol Forum 25(2):210–233

Killin A (2017) Plio-Pleistocene foundations of hominin musicality: coevolution of cognition, sociality, and music. Biol Theory 12(4):222–235

Killin A (2018) The origins of music: evidence, theory, and prospects. Music Sci. https://doi. org/10.1177/2059204317751971

Knauft B (1991) Violence and sociality in human evolution. Curr Anthropol 32:391–428

Kohn M, Mithen S (1999) Handaxes: products of sexual selection. Antiquity 73:518–526

Kuhn S (2014) Signaling theory and technologies of communication in the Paleolithic. Biol Theory 9(1):42–50

Kuhn S, Stiner MC (1998) The earliest Aurignacian of Riparo Mochi (Liguria, Italy). Curr Anthropol 39(suppl-3):175–189

Kuhn S, Stiner MC (2007) Body ornamentation as information technology: towards an understanding of the significance of early beads. In: Mellars P, Boyle K, Bar-Yosef O et al (eds) Rethinking the human revolution: new behavioural and biological perspectives on the origins and dispersal of modern humans. MacDonald Institute of Archaeology, Cambridge, pp 45–54

Leach E (1976) Culture and communication: the logic by which symbols are connected. Cambridge University Press, Cambridge

Lévi-Strauss C (1969) The raw and the cooked. Harper & Row, New York

Lewis D (1969) Convention. Harvard University Press, Cambridge, MA

Lombard M, Haidle MN (2011) Thinking a bow-and-arrow set: cognitive implications of Middle Stone Age bow and stone-tipped arrow technology. Camb Archaeol J 22(2):237–264

Lombard M, Philllipson L (2010) Indications of bow and stone-tipped arrow use 64,000 years ago in KwaZulu-Natal, South Africa. Antiquity 84:635–648

McBrearty S, Brooks A (2000) The revolution that wasn't: a new interpretation of the origin of modern human behaviour. J Hum Evol 39:453–563

McBrearty S, Stringer C (2007) Palaeoanthropology: the coast in colour. Nature 449:793–794

Morley I (2013) The prehistory of music: human evolution, archaeology and the origins of musicality. Oxford University Press, Oxford

Nowell A, Chang ML (2009) The case against sexual selection as an explanation of handaxe morphology. PaleoAnthropology 2009:77–88

Pain R (2019) What can the lithic record tell us about the evolution of hominin cognition? Topoi. https://doi.org/10.1007/s11245-019-09683-0

Saussure F (1959) Course in general linguistics. Philosophical Library, New York

Seyfarth R, Cheney D, Marler P (1980) Monkey responses to three different alarm calls: evidence of predator classification and semantic communication. Science 210:801–803

Sheehan O, Watts J, Gray R, Atkinson Q (2018) Coevolution of landesque capital intensive agriculture and sociopolitical hierarchy. Proc Natl Acad Sci 115:3628–3633

Skyrms B (2004) Stag hunt and the evolution of social structure. Cambridge University Press, Cambridge

Skyrms B (2010) Signals. Oxford University Press, Oxford

Sterelny K (2012) The evolved apprentice: how evolution made humans unique. MIT Press, Cambridge, MA

Stiner MC (2014) Finding a common bandwidth: causes of convergence and diversity in Paleolithic beads. Biol Theory 9:51–64

Texier PJ, Porraz G, Parkington J et al (2010) A Howiesons Poort tradition of engraving ostrich egg shell containers dated to 60,000 years ago at Diepkloof Rock Shelter, South Africa. PNAS 107:6180–6185

Vanhaeren M, d'Errico F, Stringer C, James SL, Todd JA, Mienis HK (2006) Middle Palaeolithic shell beads in Israel and Algeria. Science 312:1785–1788

White R (2003) Prehistoric art: the symbolic journey of humankind. Harry N. Abrams, New York

Wiessner PW (2014) Embers of society: firelight talk among the Ju/'hoansi Bushmen. Proc Natl Acad Sci 111(39):14027–14035

Wynn T, Coolidge FL (2004) The expert Neanderthal mind. J Hum Evol 46(4):467–487

Chapter 7
The Archaeology and Philosophy of Health: Navigating the New Normal Problem

Carl Brusse

Abstract It is often taken for granted that notions of health and disease are generally applicable across the biological world, in that they are not restricted to contemporary human beings, and can be unproblematically applied to a variety of organisms both past and present (taking relevant differences between species into account). In the historical sciences it is also common to normatively contrast health states of individuals and populations from different times and places: e.g., to say that due to nutrition or pathogen load, some lived healthier lives than others. However, health concepts in contemporary philosophy of medicine have not been developed with such cross-lineage, non-human, or diachronic uses in mind, and this generates what I call the 'new normal' problem. I argue that the new normal problem shows that current naturalistic approaches to health (when based on biological reference classes) are worryingly incomplete. Using examples drawn from evolutionary archaeology and the human fossil record, I outline an alternative, function-based strategy for naturalizing health that might help address the new normal problem. Interestingly, this might also reconstruct a certain uniqueness for humans in the philosophy and science of health, due to the deep history of obligate enculturation and cultural adaptation that archaeology demonstrates.

Keywords Health concepts · Pathology · Paleopathology · Cultural evolution · Biologial function · Biological anthropology

C. Brusse (✉)
Department of Philosophy and Charles Perkins Centre, University of Sydney,
Sydney, NSW, Australia

School of Philosophy, Australian National University, Acton, ACT, Australia
e-mail: carl.brusse@sydney.edu.au

© Springer Nature Switzerland AG 2021
A. Killin, S. Allen-Hermanson (eds.), *Explorations in Archaeology and Philosophy*, Synthese Library 433, https://doi.org/10.1007/978-3-030-61052-4_7

7.1   Introduction: The New Normal Problem

The bones of our ancestors can tell us many things about the quality of their lives at different times and places. In many cases, bones and other material traces generate clear signals of transitions in health and wellbeing, or at least in the normal indicators of health and wellbeing. One example of this is life expectancy. Our hominin lineage has become unusually long-lived compared to our closest extant relatives, the great apes, who become 'elderly' by their mid-30s. Dental evidence suggests that the developmental maturation of our *Homo erectus* ancestors was inconsistent with our own (Robson and Wood 2008), and some researchers argue on the basis of skeletal evidence that modern forager life expectancies were not achieved by *Homo sapiens* populations until approximately 40,000 years ago (Caspari and Lee 2004). A more recent and archaeologically unambiguous story can be told about how the hunter-gatherers who lived just prior to the advent of agriculture were healthier than the first generations of sedentary farmers. The remains of early farmers show them to have been significantly shorter, less robust, and comparatively malnourished: smaller, weaker bones exhibiting more signs of injury and disease, with missing teeth and in generally worse condition. It seems that their forager predecessors lived healthier lives, and (in this respect at least) were straightforwardly better off. Much ink has been spilled in addressing the obvious question accompanying this evidence, i.e., if the agricultural way of life made the lives of individual people worse, why was it adopted? Regardless though, the intuitive thing to say is that early Holocene populations were less *healthy* than their ancestors in the late Pleistocene, and such talk is treated as unproblematic in the literature on archaeology and biological anthropology.

However, there is a long-standing philosophical problem for such diachronic assessments of health. As Canguilhem noted in his 1943 *Essai sur quelques problèmes concernant le normal et le pathologique* (and its subsequent editions and translations), many conditions which in modern populations are seen as pathological were extremely common facts of life for prehistoric populations (Canguilhem 1989, p. 172). One small but well-evidenced example here is the hypoplasia, or underdevelopment of tooth enamel, which afflicted about 75% of Neanderthals (Ogilvie et al. 1989). Conversely, certain common facts of life for our contemporary populations, such as crooked teeth, were far rarer and stranger for those same ancient humans (and indeed foragers in general), because diets and dental development were more closely aligned (Ungar 2004). What is normal and typical changes over time as populations and/or their environments change, which means there is a *prima face* problem in projecting the normal-pathological distinctions of one population (such as ourselves) onto discontinuous populations for whom those distinctions may be less relevant. This is the broad issue that this chapter aims to grapple with: understanding biological norms in ways that support attributions of health and pathology over evolutionary timescales.

This basic problem persists in contemporary philosophy of medicine. Health and disease are inescapably normative and evaluative concepts, in that the health/

disease states of individuals are ways of describing how well things are going for them, at some basic level. There is a voluminous literature on the nature of these judgements, roughly divided between the so-called 'naturalists', who see health or disease states as grounded in objective biological facts, and 'normativists', who argue that health or disease states should be seen as relativized (in some way) to factors external to the natural sciences, such as values, interests, medical practices, or experiences; and therefore as culturally constructed, subjective, or in some other way *irreducibly* normative. This brief sketch does not do justice to the diversity of debate within the literature (especially on the normativist side), with many nuanced theories and interpretations on the table (Boorse 2011; Cooper 2017; Kingma 2014) including hybrid positions (see, e.g., Wakefield 1992). In broad brush strokes though, we can characterize naturalism and normativism as trading off against two competing desiderata, i.e., (i) that being in a healthy or diseased state is to some degree scientific, in the sense of being a matter of mind-independent fact, and (ii) that the notion of health and disease we arrive at should be *humane*, in the sense of being relevant and immediate to the human experience of health/disease and/or medical practice. Much of the debate is centered around the acceptability of the trade-offs, especially with respect to how they deal with problem cases (such as in mental health).

Despite Canguilhem's early contributions though, diachronic comparisons have been largely absent from the debates between normativists and naturalists, which tend to focus on concepts of disease and health suitable for contemporary clinical practice.[1] Conversely, the cognate science of paleopathology is concerned with differences over time with respect to the incidence and expression of disease conditions, largely without considering the philosophical question of why conditions count as diseases (Grauer 2012, 2018; Klepinger 1983). But from such a perspective it is intuitive to side with naturalism to some degree, i.e., to assume (as in our motivating transition cases) that there are some mind-independent facts of the matter about whether this or that individual was more or less healthy in some regard. This becomes especially salient when considering other hominins and non-human animals. For example, cancer has been observed within almost every branch of the tree of life where complex multi-cellularity evolved (Aktipis et al. 2015). All sides of the debate would recognize cancer as a biological phenomenon or *condition* that these organisms suffer from, but at least some non-naturalistic approaches would struggle to recover the intuition that it is still a *pathological* one for creatures lacking the cognitive or cultural sophistication to appreciate it as such, or where the spotlight of human attention is absent. The intuition to make sense of (or explain away) is that cancers (and other 'fundamental' diseases/injuries) represent ways in which things 'go wrong' for an organism entirely at the biological level. While we can recognize that disease, injury, and health are value-laden within the human social context, it is not unreasonable to hope for something to underpin this which is more biologically

---

[1] See for example Murphy (2020), and the absence of such considerations in significant handbooks and collected volumes (e.g., Carel and Cooper 2014; Humber and Almeder 1997; Huneman et al. 2015; Marcum 2016; Schramme and Edwards 2017; Solomon et al. 2017).

general and objectively grounded. The minimal naturalistic intuition is that at least *some* of the norms of health are firmly grounded in physiology, fitness, or other biological facts.

But if biological norms are to be objective in some sense, they should still be relativized to relevant biological populations. This seems unavoidable. For example, while the maximum life expectancy for humans is decades longer than it is for even the most pampered of chimpanzees in captivity (and by more than a factor of two when considering the oldest known individuals from each species), it is absurd to treat this a *health* comparison. What constitutes a healthy lifespan for a chimp (or for our most recent common ancestor) is not a matter for human biological norms to weigh in on, any more than humans should be subject to the biological norms of 200-year-old Bowhead whales. Unless strictly figurative or otherwise un-scientific, the terms 'long' and 'healthy' with respect to the lives of individual organisms only make sense in the context of the biological norms of their relevant biological populations (at a first pass, their species). Some sort of relativization to biological populations seems like a plausibility constraint. But coupled with the fact of common descent, this means that objective biological norms must also change and diverge over time.

This generates a problem: normative change makes archaeological evidence of diachronic health differences potentially ambiguous. It might be evidence that earlier or later populations were healthier or more diseased. But without assuming invariant biological norms, the difference might instead constitute evidence that those norms have changed, such that comparative health judgements between the two populations become problematic. After all, we know that biological norms must have changed in various ways during the evolution of our lineage. The challenge then is to filter the archaeological record of hominin health changes in a principled, suitably objective manner: which of the changes we see are changes in health status, and which are changes in what it takes to be healthy? Updating Canguilhem's terminology, the difference is between normal vs pathological on one hand, and the normal vs the 'new normal' on the other. This is the new normal problem.

The phenomena to save, in the face of the new normal problem, are common sense attributions of health differences across human/hominin populations, and the challenge is to do so in a way that fits what the archaeological evidence can tell us, without circularity. Objective biological norms cannot be invariant across lineages and over time. But it is equally implausible that they change whenever the population does—this cannot be the case if they are to be useful in describing the health effects of novel innovations, mutations, environmental changes, etc. The challenge is finding a useful middle ground. And the key question raised by the new normal problem is: what *kind* of changes license drawing a line between discontinuous regimes of biological normativity? Failure to address that question leaves naturalism about health incomplete, and unable to support attributions of health and pathology over evolutionary timescales.

This chapter is an attempt explore that problem, and to consider how evidence from evolutionary archaeology might help guide formulations of biological normativity which avoid it. The ideal goal for biomedical naturalism would be a positive account that is archaeologically informed, anthropologically sound, and which accounts for human health evaluations within an evolutionary context. The more limited goal of this chapter is to make progress in this regard by (in Sect. 7.2) providing an analysis of the failure points of one well-known naturalistic account, and (in Sect. 7.3) outlining several bodies of archaeological evidence for major transitions in health/wellbeing, which will assist in carving up that problem space in a principled manner. I will argue that a naturalistic account of human biological norms must be bolstered by a robust notion of biological function, but that human archaeology cautions against over-generalizing from non-human biology. This is because *cultural* evolution plays a key role in each of the problematic transitions. The unique reliance on culture to direct the development of human phenotypes raises difficult questions about how to demarcate characters as relevant or not to health and pathology, but (as I will cautiously suggest) the answering of those questions might also allow naturalism to be more humane.

7.2   Bolstering Biomedical Naturalism Against the New Normal Problem

The new normal problem for biomedical naturalism can be better explored by considering how it manifests for the most well-known naturalistic theory of health: the biostatistical theory of health, or BST, proposed by Christopher Boorse (1977) and further defended and updated in a series of responses to critics (Boorse 1997, 2014). Though there are more recent starting points for biomedical naturalism (Chin-Yee and Upshur 2017; Griffiths and Matthewson 2018; Matthewson and Griffiths 2017; Neander 1995), the BST is a useful reference point because the new normal problem is closely related to a known criticism of it. And the way that it falls short permits lessons to be drawn for naturalism in general.

7.2.1   Reference Class Problems

By way of brief summary, the BST defines health as the absence of disease, and diseases as impairments or limitations that prevent relevantly important functionality of an organism (or 'normal functional ability') from operating at levels that are statistically typical, given the reference class of the organism in question. Which functionalities are important and relevant is determined by their typical contribution to survival and reproduction, with the reference class being "a natural class of organisms of uniform functional design; specifically, an age group of a sex of a species" (Boorse 2014). Health then is constituted by having all your biologically significant functions operating as expected for your sex and age cohort.

One line of critique here (though by no means the only one) is to question the objectivity of these reference classes, the synchronic version of which is succinctly laid out by Kingma (2007). Boorse's rationale with reference classes (again, broadly speaking) was to follow medical practice in recognizing that some variable biological characters, such as hormone levels, play different adaptive roles for different people: men and women, young and old. Having a specific level of some specific hormone might be normal and healthy, e.g., for a young man, but aberrant and pathological in women and older men. In this way, Boorse hopes to keep statistical normality in alignment with intuitive judgments about health with as little 'splitting' as possible, being ambivalent about treating race (for example) as another reference class category (Boorse 2014, p. 702). Boorse is right to be resistant, as arbitrarily fine-grained reference classes would allow an absurd gerrymandering of the BST (until we each end up in a reference class of one, where being healthy is tautological). But what Kingma points out is that Boorse's line-drawing is dangerously ad hoc. It is one thing to identify a convenient set of reference classes which generate right-sounding results, but the biological objectivity of BST's health judgments depends on these reference classes themselves having a biological rationale. Boorse, in reply, is instead happy with having a *medical* rationale, i.e., to allow current medical practice to instead precisify the reference classes for BST, fit for current day clinical application (see Boorse 2014, p. 693). Whether this gives away too much of BST's aspirations for biological objectivity will not be considered here, though it will depend on how value-laden, conventional/pragmatic the references classes of current medical practice are taken to be.

This debate becomes relevant for present purposes when it is generalized diachronically. The reference classes used in health statistics (e.g., of life expectancy or infection rates, etc.) are typically synchronic snapshots of populations, often broken down according to Boorse's envisioned reference classes, but also usually by territory.[2] Using such snapshots as reference classes for BST purposes, the average British man in his 40s (for example) will have biological norms against which his health can be assessed. Assuming his relevant functional abilities are indeed operating near the statistical mean for 40-year old British men today, the BST should rule that he is healthy. But it should also have made the same assessment (i.e., 'healthy') of the average British man of 150 years ago based on the statistical norms of that era, even if this person would be horribly unhealthy as measured by today's standards. Alternatively, if we rigidly adhere to snapshot reference classes then we cannot even make that comparison, as these two individuals simply have different biological norms, relativized to different reference classes. This recreates Canguilhem's discontinuity on an uncomfortably short timescale: there would be no objective basis for saying that British 40-somethings now are healthier than British 40-somethings 200 years ago (when an alarming proportion of them were dead or dying).

---

[2] It should be noted that this casts doubt on the *biological* objectivity of the reference classes used in medical practice, as 'you are healthy for a man/woman your age' typically carries with it an implicit '… in your country/territory/ethnic group'.

7.2.2   Functions and Norms

The full suite of problems arising from the diachronic reference class problem (for both the BST and biomedical naturalism in general) is something I explore more fully elsewhere. To be fair to Boorse, it is not clear he has any ambition to make biomedical naturalism applicable to diachronic contexts. The point is not that it is a bad idea to take the BST out of the clinic in this way (though it is a bad idea), but rather to illustrate and generalize the new normal problem in more detail.

The BST (at least the 2014 version) has three features which combine to make it unsuitable for application to evolutionary timescales:

(i) it relativizes health evaluations for individuals against defined populations,
(ii) it defines those populations synchronically and at least partially non-biologically, and
(iii) its assessments of disease/health are based on statistical typicality within the populations.

The easiest of these to point to is the statistical assessment procedure (iii). Condition prevalence can ebb and flow wildly due to contingencies like epidemics and environmental conditions, and it seems implausible that biological norms too are tossed around this way. An assessment procedure which was less hostage to such contingencies seems preferable. As suggested, we should also address (ii) here, and find some objective way of fixing the qualitative and spatial/temporal limits of reference classes that does not rely on inference from standard medical practice (though perhaps at the price of diverging from that practice). However, we should also question relativization to populations *per se*. The BST uses different reference classes to grade people on their own curve, as it were. But this means that, even synchronically, it is hard to see how it can say of two average specimens from different reference classes whether one is more healthy than the other, as we intuitively can between typical 30 year olds and typical 80 year olds. Ideally, naturalists should want biological norms that can partially overlap, leaving some (but not all) inter-class comparisons intelligible.

I would argue that the BST's main rivals in the naturalist literature can already do this.[3] Whereas the BST bases health/disease assessments on class-typical levels of class-typical biological function, these views typically attend to biological functions directly, where functions are either defined systemically or as selected effects. Rather than explore these alternatives in depth though, we can imagine functional approaches in the abstract, as a third potential relativization strategy in contrast to 'normativist' relativization and BST-style relativization to populations. The

---

[3] Approaches appealing to common accounts of biological function (such as the selected effects account) include Chin-Yee and Upshur (2017); Griffiths and Matthewson (2018); Matthewson and Griffiths (2017); Neander (2016); Wakefield (1992, 2007). The organizational account of function has also been appealed to for health-naturalist purposes (Saborido and Moreno 2015). The arguments in this chapter are intended to be broadly compatible with any of these, though I err on the side of language friendly to the selected effects account.

longevity comparisons between humans, chimps and whales show that health judgements must be relativized in *some* way, as health *simpliciter* makes less and less sense the further we range over the tree of life. Normativists of various stripes relativize to human values, conventions, interests, etc., and Boorse (despite protestations of sufficient objectivity) ends up relativizing the reference class component of his view to conventions concerning medical practice. The obvious move was to relativize the health of organisms to the natural populations they are members of. But, as Kingma points out, we all belong to many, varied natural populations defined by various combinations of (arguably) natural features—sex, species, lineage, sexuality, developmental stage, race, blood type, handedness, etc. At the very least, this requires a further naturalistic story to fix upon one of the many 'natural' groupings we fall under as normatively special for us, without over-compartmentalization.

The alternative relativization strategy I would like to suggest is relativization to functional design itself. Adaptive functional homologies might be a way of making sense of 'ideal' functional design across branching lineages, where this is independent of the statistics of contingent functional attainment,[4] and of value-laden projection. In simple terms, this would involve considering functional complexes (like longevity or cardiac function) one at a time, and assessing health *in those respects* based on achieved functionality relative to the functions as evolved. On this approach, the population subject to a biological norm would be derivative of and relative to the relevant functions and their homologies.[5] For example, chimpanzee and human biological norms of longevity presumably diverged because of selective pressures for longer lifespans in the post-divergence hominin lineage. We lack relevant functional homology with chimps in this respect, but (depending on the sequencing of the divergent selection) we share a normative clade in this respect with a smaller set of hominids with a more recent common ancestor. This will be similar for many cognitive functions (for example). But with respect to many other adaptive functions where things can go wrong, we would be part of clades with a deeper homology, for example with respect to aspects of cardiac health, or endocrine function, or vulnerability to certain cancers or mitochondrial diseases. Details will of course matter a great deal here, but function-relativization of biological norms might allow a more flexible regime of health judgments, some being very narrow and specific to human populations (e.g. as regarding dyslexia), and others with which we can make meaningful comparisons of health and pathology further across time and between lineages and morphs with shared functional heritage.

---

[4] Boorse, when pressed, makes a similar suggestion: "to the problem of typical disease, I see no solution but to retreat to a concept of ideal design which, so far, I am unable to define" (Boorse 2014, p. 707).

[5] I am using 'homology' here in a very loose sense. Functionally homologous groups (in my sense) would be the result of both common selective pressures that explain the origin of that function, and subsequent maintaining selection. They would therefore often be paraphyletic, i.e., subsequent divergent selective pressures could split some lineages away from the functionally homologous group. This may be a problematic use of the terminology (Griffiths 2006).

This is of course more of a theory schema than a theory itself, and I will not be defending it in any depth here. In any case, it can serve as a framework for what follows. We can assume that the best-pass version of biomedical naturalism will involve some appeal to biological function. To the extent that biological function is construed and constructed in evolutionary terms, naturalism will be highly dependent on the kinds of functional changes evidenced in the archaeological record: how robust they are/were, and whether they are likely to have been adaptive (as opposed to being the result of some other process of change). While the overall goal of this chapter is not to articulate a complete positive view, I will return to consider some directional recommendations in this regard in the concluding section, in light of the case studies to which I now turn.

7.3 Health Transitions in the Archaeological Record

So far, the argument has been both negative and abstract. In what follows, I consider what we can infer from the archaeological record regarding how our lineage changed during some distinctive physiological transitions, and what a philosophical account of health should ideally say about those transitions. Using modeling terminology, the goal is both to establish data points that an eventual theory of health should to try to fit, and to look at how plausible human biological functions (and their evolution) might figure into such a theory.

7.3.1 Gross Human Transitions and Discerning Adaptive Functions

Human archaeology, broadly construed (i.e., to include human teeth and bones as well as material culture, the narrow conception of archaeology), is invaluable here. The majority of our biological functions of course have their origins much deeper in time; the functional homologies we have with other primates, mammals and so forth. But fossil evidence at those timescales is likely too low resolution to provide many clear transition examples. But archaeology can supply higher temporal resolution for recent transitions, including information about context, order, and causation.

There are numerous examples available, and we can begin by considering gross characteristic changes. Human beings are replete with evolved features that are unusual in the context of our nearest extant relatives, many of which are part of the suite of adaptations associated with foraging and the diet it allowed. The genus *Homo* is broadly marked out from preceding *Australopithecus* and divergent *Paranthropus* taxa as being more gracile and behaviorally sophisticated, with an overall trend toward less robust bones, smaller teeth, shorter guts, larger brains, and complex lifeways and social interdependencies, up until anatomically modern humans. There is a (descriptive) sense in which this gross transition was an enfeeblement, especially for anatomically modern humans: we are much less robust and physically resilient than many of our ancestors and other archaic species. Is there also a normative sense in which this is true? I.e., are we less *healthy* in this respect? That seems like an implausible comparative judgement to make, at least in contrast to considering parasite load, or other incidences of common diseases, injuries, and afflictions. For example, Neanderthal skeletal evidence appears to show (i) that they were stronger with bones somewhat thicker and more robust than anatomically modern humans,[6] but (ii) that they suffered from a higher incidence of serious injury (Berger and Trinkaus 1995; Spikins et al. 2019). It seems reasonable to say that they were no more or less healthy than us for having the bones they had (i.e., that these differences are just descriptive differences), but that they were less healthy for having them damaged more often. At least, the aptness of these propositions seems like an empirical matter that a philosophical framework for health judgements should respect, rather than pre-empt. Functional relativization would seem to allow this: morphological differences in the bones of the two lineages could be signs of innocent functional specialization, but they would also share a deeper common function (i.e., not to break).

Various evolutionary narratives seek to explain human enfeeblement as the result of tool use. Tools became force multipliers, enabling hunting and other high-value food sources, and a way of outsourcing much of the food-processing work previously done by chewing. One of the best sources of data here, fossil teeth, shows a nuanced and piece-wise trend (Ungar 2017). *Homo habilis*, one of the presumed basal species for our genus, appeared to have molars which were roughly similar size to *Australopithecus* (relative to body size), and larger incisors. It is not until *Homo erectus*, about 500,000 years after the first known stone tools, that teeth are appreciably small enough to indicate that a significant degree of food-processing is being outsourced to technology.[7] Though fossil evidence at this the time-depth is by no means voluminous, it is still higher resolution than in deeper time periods, and indicates a likely sequencing of changes in environment and organism.

The usual cautions against adaptationist thinking should of course be kept in mind. But we can still draw reasonable conclusions. For example, the inference made by Ungar is that the reduction in human dental (and skeletal) robustness is adaptive, due to an alleviation of the selection pressures that previously demanded developmental investment in those structures. This is reasonable and again fits function-relative naturalism, given the evolutionary trajectories discernible across plausible reconstructed evolutionary trees for the hundreds of attested early hominin fossils. Even if we did not have modern humans as a baseline object of study, it is beyond implausible that, e.g., the hundreds of *H. erectus* specimens we have

---

[6] Though this should not be over-stated, see Pearson et al. (2006).

[7] However, tooth shape in early *Homo* species had already changed to become more capable of eating tougher and more elastic foods than the hard, brittle foodstuffs that *Australopithecus afarensis* specialized in (Ungar 2004).

found (scattered over Africa and Eurasia, and across almost two million years) were all pathological individuals whose dental development had been stunted in more or less the same way.

However, interesting debates sometimes arise between adaptive transformation and maladaptive pathology, especially when it comes to describing new species from initially small sample sizes. For example, the famous and surprising discovery in 2003 of the LB1 partial skeleton in Liang Bua cave on Flores, one of the Lesser Sunda islands in Indonesia, led to conflicting claims about whether this was a new species, or merely a pathological specimen of a known species. Because it was initially dated as living during the known settlement time of *H. sapiens* in the region, and because of its unique 'mosaic of primitive and derived traits', small stature and low endocranial volume, the describing authors considered and rejected the possibility that LB1 was a modern human afflicted with IGF-related postnatal growth retardation, pituitary dwarfism, or microcephalic dwarfism (Brown et al. 2004; Morwood et al. 2004). Their description of LB1 as a new species, *Homo floresiensis*, was challenged however by further claims of alternative, pathology-based explanations of its condition. These included secondary microcephaly (Henneberg and Thorne 2004), myxoedematous endemic cretinism caused by congenital hypothyroidism (Obendorf et al. 2008), and Down Syndrome (Eckhardt et al. 2014; Henneberg et al. 2014). These explanations have been rejected by subsequent comparative studies (Baab et al. 2016; Brown 2012). Subsequent re-dating and phylogenetic analyses of morphological characters now suggest that the *H. floresiensis* lineage originated prior to *H. erectus* and was likely extinct before *H. sapiens* arrived in the region (Argue et al. 2017). Teeth and statistical inference also come to the rescue here, as dental fragments from several other individuals were also found in the Liang Bua cave, and there is enough commonality of characteristics to make it unlikely that LB1 is radically unrepresentative of that small population (Kaifu et al. 2015). It is likely then that *H. floresiensis* is a genuine hominin lineage with distinctive small-bodied physiology and associated biological norms evolved (presumably) according to insular dwarfism (Montgomery 2013; Tucci et al. 2018).

One point to emphasize here is that scientific methods can tell the difference between functionally defined health and pathology in a population by pinning down the variation within it. There are more ways of being unhealthy than being healthy, in the same way that there are more ways of being far from the fitness maxima than being near it. The function-relative theory schema therefore sounds plausible at this level. To a first approximation, selection and adaptive optimization will tend to drive convergence to a functionally adaptive phenotype, with pathologies then impinging to divert organisms from this (healthy) phenotype in a variety of ways and degrees. It would be very difficult even for an endemic pathology resulting from persistent, detrimental environmental conditions, to mimic the distribution of a stable, more-or-less optimally adapted population. Pathology should therefore leave at least a statistical signal in the fossil record, given enough available specimens.[8] Evidence of convergence on a trait within a biological population (again, generally speaking) might then be a good sign that the trait is associated with health, if health is understood in some way as approximating adaptive well-functioning.

There will no doubt be devils in the details, but at this level of description at least, relativizing health judgments to adaptive functions allows us to draw some reasonable conclusions about health in the face of evolutionary change. Regarding the gross human transition: rather than an enfeeblement, we can see the shortening of our gut, reduction in bone mass and teeth size (etc.) as healthy adaptations which coevolved with our larger brains and the technologies (physical and social) that they made available—stone tools, fire control, cooperative complexity, paternal investment in children, etc. We are considerably less robust than Neanderthals, and weaker than chimpanzees. But given our ancestors' other advantages, including their complex suite of cultural adaptations (Henrich 2015; Laland 2017), they did not need to invest as heavily in developmentally expensive bone and muscle, and these resources were better spent furthering survival and reproduction in other ways. So, while *H. sapiens* might have been descriptively enfeebled, we are no less healthy for it, assuming that the weakening was part of overall phenotypic adaptation. We share many functional adaptations with our cousins and ancestors, and so we can recognize many of the same diseases across the clade which run counter to our common biological norms by impairing those functions. But insofar as we have our own distinctive adaptive biological functions, our biological norms (and potential pathologies) will differ in those respects.

Function-relative naturalism seems to get the basic intuitions right here, and successfully addresses the change-vs-new-normal problem. In principle, it offers a biologically grounded, objective and mind-independent way to determine the scope of distinctive human biological norms, in a way which still lets some of them extend to archaic populations. If so, valid comparisons of health and disease can still be made qua specific common biological functionality and associated pathologies. And this approach may allow us to push back against Canguilhem's concerns about the normal/pathological distinction in archaic populations. Unlike the BST, it can recognize that, as it is for us, dental hypoplasia *was* pathological for Neanderthals despite being endemic among them. For both lineages, dental hypoplasia is not a functional adaptation, but instead a developmental impairment[9] working against a functional adaptation with a deeper homology. Of course, the success of this sort of move depends on there being plausible evolutionary narratives to support it, and this would have to be defended on a case-by-case basis. But the reliance on a plausible narrative is also the point: this is an approach which leaves discernible empirical facts in the driver's seat.

---

[8] Though lack of convergence is by no means definitive proof—drift and other evolutionary processes can be sources of variation that are fitness-neutral and not pathological.

[9] As hypothesized by Ogilvie et al. (1989), the widespread underdevelopment of the tooth enamel in their sample (of 699 crowns) is indicative of nutritional stress from weaning to adolescence.

7.3.2 Recent Transitions and the Question of Culture

Before acknowledging the open questions for a function-relative approach (e.g., what 'function' is supposed to mean, and the difference between functional adaptation and adaptive function, etc.), there is another important point to acknowledge: the causal role of culture. As mentioned, we should be cautious of biasing attention toward 'human uniqueness' and human-distinctive adaptations, since the majority of biological functions and norms will have deeper origins. But it is an unavoidable feature of recent functional adaptations that they involve cultural traits as well as classically biological characters.

Technology is strongly implicated in the gross human transition, and is treated as characteristic of the later hominin phenotype, including (accurately or not) the *Homo* genus as a whole. Contemporary and ethnographically described foragers such as the Hadza and Inuit simply could not survive in their environments without specific tool kits, and detailed skills and knowledge with respect to their manufacture and use (Henrich 2015). This dependence of course evolved earlier in the *Homo* lineage, with respect to more archaic technologies such as stone axes and cleavers, and the control of fire. Fire provided material advantages in the form of heat for cooking, and for generation of further cultural adaptations, such as resins and heat-tempering. The shortening of the gut as the result of liberated calories is just one physiological change this supposedly fed into (Wrangham 2010). Another apparent consequence is our truncated sleep cycle: fire effectively extended the hours of daylight and the time available for social activity. Adult humans sleep less than our great ape relatives (who sleep from dusk to dawn), though we appear to compensate for this by sleeping more deeply (Nunn and Samson 2018; Samson and Nunn 2015).

But characters like the control of fire or mastery of stone knapping (as well as undoubtedly many other early *Homo* technological skills which did not preserve as well) are not transmitted vertically as part of our classical biological heritage—as anyone who has tried their hand at knapping will appreciate. They are culturally transmitted, sometimes through social learning and/or teaching, and sometimes through trial and error learning in an environment which scaffolds that learning (Sterelny 2012). But it is reasonable to think of stable, reliably reproduced, culturally transmitted skills and knowledge as integrated into the human phenotype—at least, if 'phenotype' is understood in the ecological sense of how well an organism can do in a given environment. A forager who does not know how to forage is not really a forager. Such skills and knowledge are straight-forwardly referred to as 'cultural adaptations' by some (Boyd et al. 2011; Henrich 2015; Perreault 2012), in that they provided unequivocal fitness advantages and were 'selected for' because of those advantages,[10] but via social learning mechanisms rather than (or in addition

---

[10] Cultural evolution theory is divided regarding whether 'selection' is the best way (or even necessary at all) to describe the mechanisms by which successful cultural traits are propagated and/or upregulated. See for example Tim Lewens' taxonomy of cultural evolutionary thinking and related discussion (Brusse 2017; Heyes 2016; Lewens 2015). Indeed, clarifying relevant notions of suc-

to) natural selection. To cover the semantic bases here of transmission and upregulation (selectively or otherwise) we can call them 'culturally propagated'.[11]

Cultural traits and social learning might therefore be quite deeply implicated in human evolution, certainly in its most recent chapters, and this poses interesting questions for the function-relative approach to health we are considering. Suppose, as seems likely, that some of our traits are i) culturally propagated, ii) selected as ecologically adaptive, and iii) developmentally integrated, in the sense that they are built up over time, especially in early life—examples here might include specialized forms of cognition (Heyes 2018). There is no obvious, non-question begging reason to rule out such traits and characters from inclusion in the suite of adaptive functions we might otherwise take as underpinning biological normativity. Therefore, some failures of adaptive enculturation or cultural development might be just as pathogenic as deleterious mutations, malnutrition, or environmental mismatch. Though this could be a theoretical minefield (for reasons alluded to in passing), it could also be an advantage for biomedical naturalism. For example, depending on the details of cultural function, it might allow naturalism to take a more nuanced approach toward (for example) some mental illnesses than under a more crudely biological 'brain malfunction' paradigm. At the very least, it makes conditions such as dyslexia more naturalistically intelligible as pathologies, given the likely reality that dyslexia is the failure of an underlying cognitive system that has only recently been exapted by cultural evolution. These are of course open questions that will depend on how biological/cultural functions (and their relationships to health) are specified.

7.3.3   Functional Adaptation and Culture in Behavioral Modernity

With these frameworks and caveats in mind, we can now return to the two candidate cases for human health transitions we began with: the (allegedly) dramatic increase in longevity in humans of the Early Upper Palaeolithic, and the Neolithic revolution. In both these cases, cultural processes are clearly implicated which may challenge the function-relative theory schema, and in both cases difficult interpretive questions are opened up, illustrating the ways in which the viability of this approach will have to be decided.

---

cess and cultural analogues of fitness and adaptiveness are non-trivial (Ramsey and De Block 2017), meaning that the idea of a cultural adaptation should be approached with reasonable caution. I am assuming that some such useful account is possible for the purposes of this discussion, though perhaps only in a very loose sense.

[11] This is not to say that the basic cognitive capacities to acquire traits did not evolve through biological natural selection, but even this is disputed (e.g., by Heyes 2018, who argues that there is a degree of cultural propagation even for the enabling capacity-traits of cultural propagation).

7.3.3.1 Longevity

In a pair of papers (Caspari and Lee 2004, 2006), Rachel Caspari and Sang-Hee Lee used dental age-at-death estimates of fossil hominids to estimate the ratio of mortality between old and young individuals for hominid groups in different times and locations. In the first study, 768 fossil hominids in four groups were analyzed: later australopithecines, Early and Middle Pleistocene *Homo*, West-Asian/European Neanderthal, and European Early Upper Palaeolithic *H. sapiens* (50–30 kya). The authors found both a chronological trend toward increased longevity (consistent with a gradual increase in longevity over human evolution), and a five-times higher old-to-young ratio in the Early Upper Palaeolithic group compared to the Early and Middle Pleistocene group. Caspari and Lee hypothesized that this striking difference was due to an increased reliance on cumulative cultural innovation in the Middle to Upper Palaeolithic transition, whereby cultural complexity meant that senescent individuals (i.e., beyond reproductive age) became more valuable and prestigious as reservoirs of skills and knowledge. In their 2006 paper, the old-to-young ratios of four populations were compared: European Early Upper Palaeolithic *H. sapiens*, and three populations were from the prior Middle Palaeolithic—West Asian *H. sapiens*, West Asian Neanderthals, and European Neanderthals. At this time (prior to known *H. sapiens* migration into Europe) the *H. sapiens* and Neanderthal populations in West Asia are culturally indistinguishable in the archaeological record. Overall, these populations had three cross-cutting domains of similarity which allowed meaningful comparisons to be made: the three Middle Palaeolithic populations were culturally similar, the two West Asian Middle Palaeolithic populations (*H. sapiens* and Neanderthal) had similar environments, and the two *H. sapiens* and two Neanderthal populations of course each had closer phylogenetic affinities. According to their analysis of comparative age-at-death ratios, the two *H. sapiens* populations (separated by time and geography) were found to be significantly different, which the authors see as attributable to either environmental or cultural differences. The contemporaneous West Asian and European Neanderthals were different, implying environmental differences (as might be expected). But the contemporaneous *H. sapiens* and Neanderthal populations in West Asia were similar, suggesting that merely biological differences between the lineages was not a strong causal factor.

These papers are intriguing both in their purported findings, and for their place in the context of two contentious debates at the time: i) concerning the so-called grandmother hypothesis and its sequencing, and ii) concerning the now mostly debunked 'human revolution' theory of a rapid transition to human behavioral modernity in the European Middle-Upper Palaeolithic (McBrearty 2007). Briefly, the grandmother hypothesis agrees that longevity increase was driven by social changes whereby the contributions of post-reproductive individuals became more valuable. Advocates of this hypothesis such as Kristen Hawkes typically attribute this effect to grandmothers contributing to the fitness of reproductive relatives (i.e., their daughters), but at a much earlier stage than the Middle-Upper Palaeolithic, and proceeding more incrementally, driven by kin selection rather than cultural evolution. Hawkes and co-authors have expressed skepticism about Caspari and Lee's results, arguing (a) that skeletal assemblages are unlikely to represent unbiased samples of the age-structure of populations, and (b) that longevity and old-to-young statistical ratios are by no means proxies for one another, as evidenced by the similarity of old-to-young ratios in living non-human primates despite very different life expectancies (Hawkes and O'Connell 2005; Hawkes and Coxworth 2013).

Regardless of the wider debates, this summary can still serve as an instructive hypothetical case study. Assume for sake of argument that Caspari and Lee are correct, and there was a dramatic increase in longevity because of a cultural change. How should a function-relative grounding of health judgments treat this case? That might depend on exactly how the culture changed and how we interpret that change. Would this be a case of better health achieved though cultural change alleviating excess mortality, or would it be an alteration of the altered *H. sapiens* phenotype via cultural adaptation, and therefore a non-valent change with respect to the relevant norms of human health? This is effectively a resurgence of the new normal problem with respect to the culturally-driven functional traits being considered as candidates for normative relevance alongside paradigmatically biological functions.

But the problem here has a somewhat different aetiology, and stems from a lack of interpretive unity within contemporary cultural evolutionary theory. On one hand we might take robust, intergenerational cultural innovations to be part of an extended human phenotype, and therefore take Boyd, Henrich et al. literally: cultural adaptations are adaptations, and cultural functions are functions. Alternatively, cultural innovations, their substratum being stored externally to the traditional organism in the form of practices, rituals, institutions or artifacts, might instead be seen as more like an engineered niche, à la cultural niche construction, in being just a conveniently, reproducibly modified part of the environment. On this reading they would scaffold behavior and development, but they are not themselves part of the phenotype, and so are not functions in the relevant sense. These two approaches would give very different answers to the new normal problem, yet at least some argue that the choice between them is effectively a modeling decision (Sterelny 2010). If so, this seems like a liability all of a sudden for function-relativism: it seems like a bad thing if a modeling decision were to decide whether a condition counted as healthy or not. Without a way to resist this sort of interpretational pluralism, the function-relative approach would fail to be clearly objective, and in a similar way as the BST, i.e., with respect to grounding out in convention or utility (in this case explanatory convention/utility rather than clinical).

This is not to say that this problem is unanswerable, but answering it is beyond the scope of the current discussion. It will hinge on the nature of cultural adaptations and how and when (if ever) to take them seriously as adaptive functions in the literal, phenotypically-constitutive sense, and it is therefore part of a much broader debate about the overall status of cultural evolution and the extended evolutionary synthesis.[12]

7.3.3.2 The Neolithic Transition

The final transition I want to revisit in more depth is the Neolithic transition from foraging life to sedentary farming (in the Middle East, from about 11 kya), which presents another interesting problem case. Prior to the work of a number of archaeologists in the 1960s,[13] the dominant explanations were self-flatteringly progressive. The assumption was that settled agricultural life was a straightforwardly rational improvement on forager life with more security, ease, and leisure (Barker 2006), and so simply inventing the technology was enough to kick off settled civilization. If that were really the case, then health and 'civilizational' cultural adaptations would seem to go hand in hand. However, ethnographic studies of hunter-gatherers like the !Kung-San put lie to this, as even in inhospitable environments foragers enjoyed relatively easy and secure lives compared to hard-toiling traditional farmers, with more healthy, diverse, and fungible food sources, and a lower parasite load. This comparison would have been much worse for early farmers with a much narrower range and quality of domesticated crops, a fact now conclusively attested to by skeletal and environmental evidence showing a marked decline in bone density, stature, and nutrition, and a corresponding increase in cavities, damage, and conditions which usually serve as markers for pathology and poor quality of life (Larsen 2006).

Explanations of this run the gamut. One is sedentism itself: once farming productivity reached a level capable of providing full subsistence (as opposed to occasional supplementation), a switch to sedentary farming could increase effective fertility by supporting raising more children than a mobile life would allow. The gradual encroachment of farmers onto forager lands might likewise be explicable economically/militarily, via the order of magnitude greater increase in carrying capacity for land under the plough (Diamond 1998). And the increased strategic value of land might also make the difference, so that hundreds of sickly farmers defending fixed investments would out-fight (in both strength and morale) dozens of relatively healthy foragers who have the option (for now) to move on and hunt elsewhere. But the flipside of this is greater vulnerability and deep planning horizons: take away everything a forager owns and they are less well-equipped foragers, but take everything from sedentary farmers and they cannot farm at all. Social solidarity and bonding mechanisms such as collective religions and ideologies, and associated

---

[12] For example, we might perhaps understand cultural adaptations as adaptations proper to the degree that they are intergenerationally stable and partake in other paradigmatic features of adaptations, in a multi-dimensional conceptual space (à la Mitchell 2000), though this obviously presents further challenges.

[13] Famously culminating in the conference and subsequent book *Man the Hunter* (Lee and DeVore 1968).

assurance mechanisms and cognitive biases may have taken on a much greater role as a result, and there are numerous theories and modeling approaches along these lines, both adaptive (Brusse 2020; Bulbulia 2004; Handfield 2020) and maladaptive (Norenzayan et al. 2016). In short, the transition to sedentary farming lifeways might have been i) an unpleasant trade-off for greater fitness, ii) a forced but still rational choice (Sterelny 2015), iii) a fitness trap, or iv) driven by maladaptive or fitness-orthogonal cultural evolution.[14]

What to make of the options here? Simplifying greatly, there are two binary judgments to be made. First, are the health differences caused by the transition validly comparative? I.e., is it fair to say that early farmers had worse health than the preceding foragers? Second, were the causes of the transition functional adaptations? Intuitively, the first answer seems like an unequivocal 'yes'—this is exactly the sort of bio-anthropological judgement we want a naturalistic theory of health to accommodate. But if the answer to the second question is also yes, this will spell trouble for the proposed theory schema, in that it would give the intuitively wrong answer to the new normal problem in this case. If the shift to farming lifeways was the saltation of a suite of cultural adaptations, then this might just be the new adaptive phenotype, which trades off a more miserable form of enfeeblement for increased overall fitness. The problem is that the form of enfeeblement here looks very much like malnourishment, or at least the prevalent environmental susceptibility to it. This prospect seems much harder to swallow than the gradual weakening of *H. sapiens* discussed earlier.

Again, (a) it might be seen as a virtue of the function-relative approach that it is so much a hostage to empirical fortune, but (b) figuring out exactly how much danger it is in here would be extremely involved. Everything hinges on the empirical details and how to interpret them in answering the second question. Over and above clarifying the nature of adaptive (bio)cultural functions, the ideal strategy for a function-relative approach, to retain its full scope and potential, would be to show that only the most plausible transition-drivers met that standard. This would be easier with a restrictive standard for cultural adaptations, though this of course would trade off the potential advantages of cultural function-relativization with respect to 'humanizing' biomedical naturalism.

7.4   Conclusions and Prospects

Time to take stock. I began this chapter with the new normal problem, which looked like a curious dilemma for both common sense uses of comparative health judgements in archaeology, and for naturalistic theories of health in general. The main point of the discussion was to demonstrate what it would take to construct a naturalistic conception of health that is *complete*, in the sense of being able to address the

---

[14] This is of course also bracketing off exogenous drivers like climatic change.

new normal problem and also being *applicable* naturalistically, outside narrow clinical applications. It is easy to make the case that naturalism should do better than the BST, and do better than relativizing health and disease to fixed biological populations. Relativization to biological functions solves many problems. And, for at least some cases of health changes in the human archaeological record, a function-based approach also seems to make plausible judgments. Speculatively, there might be advantages if cultural adaptations were taken into account, with respect to a wider scope for the content of such judgements, opening up less narrowly biological assessments of health and pathology, perhaps more in line with the judgements of normativist theories. What is currently missing from this picture are objectively justifiable *inclusion criteria*, to adjudicate such functions as either internal or external to health-relevant biological (or biocultural) normativity. But there is also no apparent justification for ignoring them, or for pretending that culture and biology can be neatly separated in this context.

In any case, this approach is extremely dependent on empirical evidence from archaeology and evolutionary anthropology, and how it is interpreted. In exploring this, a number of sticking points were bracketed off, with the conclusion here largely programmatic and several philosophical questions left wide open. Some of these are very broad, and arguably more imposing than the original problem which led to them: what are biological functions, in this context? To what degree are these functions grounded in culture? How and to what extent are these integrated into human phenotypes? If they are integrated to some degree, are there health states or pathologies which can be objectively associated with them? These questions become more acute when culture becomes a significant driver of phenotypic changes. Pinning down the answers to many of these questions better would be crucial to making progress toward a more complete conception of biological norms, and how to make sense of health and disease from any given naturalist and evolutionary perspective.

In attempting to extend the scope of health judgment beyond the strictly medical context and out to archeologically-relevant populations, it is ironic that human archaeology also makes it plain that grounding these judgements objectively will not be simple. We are likely to miss the mark if we simply use traditional notions of biological adaptation and function which downplay the distinctive contribution of culture to human phenotypes. Humans of every species and era were always cultural organisms to some degree, and our understanding of their health and well-functioning will need to respect the stories that both their bones and their material culture speak to us.

Acknowledgements My thanks to Paul Griffiths and to the editors of this collection for written comments on earlier drafts of this chapter. Thanks also for insights and feedback from Benjamin Jeffares, John Matthewson, Emily Parke, an audience at the New Zealand Association of Philosophy, and the team at Theory and Methods in Bioscience at the Department of Philosophy and Charles Perkins Centre, University of Sydney. This publication was written while a visiting fellow at the School of Philosophy and Centre for Philosophy of the Sciences at the Australian National University, and was made possible through the support of a grant from the John Templeton Foundation (Grant ID 60811). The views expressed in this publication are those of the author and do not necessarily reflect the views of the John Templeton Foundation.

References

Aktipis CA, Boddy AM, Jansen G, Hibner U, Hochberg ME, Maley CC, Wilkinson GS (2015) Cancer across the tree of life: cooperation and cheating in multicellularity. Philos Trans R Soc Lond B Biol Sci 370(1673):20140219. https://doi.org/10.1098/rstb.2014.0219

Argue D, Groves CP, Lee MSY, Jungers WL (2017) The affinities of Homo floresiensis based on phylogenetic analyses of cranial, dental, and postcranial characters. J Hum Evol 107:107–133

Baab KL, Brown P, Falk D, Richtsmeier JT, Hildebolt CF, Smith K, Jungers W (2016) A critical evaluation of the Down Syndrome diagnosis for LB1, type specimen of Homo floresiensis. PLoS One 11(6):e0155731

Barker G (2006) The agricultural revolution in prehistory: why did foragers become farmers? Oxford University Press, Oxford

Berger TD, Trinkaus E (1995) Patterns of trauma among the Neandertals. J Archaeol Sci 22(6):841–852

Boorse C (1977) Health as a theoretical concept. Philos Sci 44(4):542–573

Boorse C (1997) A rebuttal on health. In: Humber JM, Almeder RF (eds) What is disease. Humana Press, Totowa, pp 1–134

Boorse C (2011) Concepts of health and disease. In: Gifford F (ed) Philosophy of medicine. North-Holland, Amsterdam, pp 13–64. https://doi.org/10.1016/B978-0-444-51787-6.50002-7

Boorse C (2014) A second rebuttal on health. J Med Philos 39(6):683–724

Boyd R, Richerson PJ, Henrich J (2011) The cultural niche: why social learning is essential for human adaptation. Proc Natl Acad Sci 108(suppl-2):10918–10925. https://doi.org/10.1073/pnas.1100290108

Brown P (2012) LB1 and LB6 Homo floresiensis are not modern human (Homo sapiens) cretins. J Hum Evol 62(2):201–224

Brown P, Sutikna T, Morwood MJ, Soejono RP et al (2004) A new small-bodied hominin from the Late Pleistocene of Flores, Indonesia. Nature 431(7012):1055–1061

Brusse C (2017) Making do without selection—review essay of "cultural evolution: conceptual challenges" by Tim Lewens. Biol Philos 32(2):307–319

Brusse C (2020) Signaling theories of religion: models and explanation. Relig Brain Behav 10(3):272–291. https://doi.org/10.1080/2153599X.2019.1678514

Bulbulia J (2004) Religious costs as adaptations that signal altruistic intention. Evol Cognit 10(1):19–38

Canguilhem G (1989) The normal and the pathological. Zone Books, New York

Carel H, Cooper R (eds) (2014) Health, illness and disease: philosophical essays. Routledge, New York

Caspari R, Lee S-H (2004) Older age becomes common late in human evolution. Proc Natl Acad Sci U S A 101(30):10895–10900

Caspari R, Lee S-H (2006) Is human longevity a consequence of cultural change or modern biology? Am J Phys Anthropol 129(4):512–517

Chin-Yee B, Upshur REG (2017) Re-evaluating concepts of biological function in clinical medicine: towards a new naturalistic theory of disease. Theor Med Bioeth 38(4):245–264

Cooper R (2017) Health and disease. In: Marcum J (ed) Bloomsbury companion to contemporary philosophy of medicine. Bloomsbury Academic, London, pp 275–296

Diamond JM (1998) Guns, germs and steel: a short history of everybody for the last 13,000 years. Random House, New York

Eckhardt RB, Henneberg M, Weller AS, Hsu KJ (2014) Rare events in earth history include the LB1 human skeleton from Flores, Indonesia, as a developmental singularity, not a unique taxon. Proc Natl Acad Sci 111(33):11961–11966

Grauer AL (ed) (2012) A companion to paleopathology. Wiley-Blackwell, Malden

Grauer AL (2018) A century of paleopathology. Am J Phys Anthropol 165(4):904–914

Griffiths PE (2006) Function, homology, and character individuation. Philos Sci 73(1):1–25

Griffiths PE, Matthewson J (2018) Evolution, dysfunction, and disease: a reappraisal. Br J Philos Sci 69(2):301–327

Handfield T (2020) The coevolution of sacred value and religion. Relig, Brain Behav 10(3):252–271. https://doi.org/10.1080/2153599X.2019.1678512

Hawkes K, Coxworth JE (2013) Grandmothers and the evolution of human longevity: a review of findings and future directions. Evol Anthropol Issues News Rev 22(6):294–302

Hawkes K, O'Connell JF (2005) How old is human longevity? J Hum Evol 49(5):650–653

Henneberg M, Thorne A (2004) Flores human may be pathological Homo sapiens. Before Farm 4(1):2–4

Henneberg M, Eckhardt RB, Chavanaves S, Hsu KJ (2014) Evolved developmental homeostasis disturbed in LB1 from Flores, Indonesia, denotes Down syndrome and not diagnostic traits of the invalid species Homo floresiensis. Proc Natl Acad Sci 111(33):11967–11972

Henrich J (2015) The secret of our success: how culture is driving human evolution, domesticating our species, and making us smarter. Princeton University Press, Princeton

Heyes C (2016) Tim Lewens: cultural evolution. Br J Philos Sci 67(4):1189–1193. https://doi.org/10.1093/bjps/axv05

Heyes C (2018) Cognitive gadgets: the cultural evolution of thinking. Harvard University Press, Cambridge, MA

Humber JM, Almeder RF (eds) (1997) What is disease? Humana Press, Totowa

Huneman P, Lambert G, Silberstein M (eds) (2015) Classification, disease and evidence: new essays in the philosophy of medicine. Springer, Dordrecht

Kaifu Y, Kono RT, Sutikna T, Saptomo EW, Jatmiko Awe RD, Baba H (2015) Descriptions of the dental remains of Homo floresiensis. Anthropol Sci. https://doi.org/10.1537/ase.150501

Kingma E (2007) What is it to be healthy? Analysis 67(294):128–133

Kingma E (2014) Health and disease: social constructivism as a combination of naturalism and normativism. In: Carel H, Cooper R (eds) Health, illness and disease: philosophical essays. Routledge, New York, pp 37–56

Klepinger LL (1983) Differential diagnosis in paleopathology and the concept of disease evolution. Med Anthropol 7(1):73–77

Laland KN (2017) Darwin's unfinished symphony: how culture made the human mind. Princeton University Press, Princeton

Larsen CS (2006) The agricultural revolution as environmental catastrophe: implications for health and lifestyle in the Holocene. Quat Int 150(1):12–20

Lee RB, DeVore I (eds) (1968) Man the hunter. Aldine de Gruyte, New York

Lewens T (2015) Cultural evolution: conceptual challenges. Oxford University Press, Oxford

Marcum JA (2016) The Bloomsbury companion to contemporary philosophy of medicine. Bloomsbury Academic, London

Matthewson J, Griffiths PE (2017) Biological criteria of disease: four ways of going wrong. J Med Philos 42(4):447–466

McBrearty S (2007) Down with the revolution. In: Mellars P, Boyle L, Bar-Yosef O, Stringer C (eds) Rethinking the human revolution: new behavioural and biological perspectives on the origin and dispersal of modern humans. McDonald Institute Archaeological Publications, Cambridge, pp 133–151

Mitchell SD (2000) Dimensions of scientific law. Philos Sci 67(2):242–265

Montgomery SH (2013) Primate brains, the 'island rule' and the evolution of Homo floresiensis. J Hum Evol 65(6):750–760

Morwood MJ, Soejono RP, Roberts RG, Sutikna T, Turney CSM, Westaway KE et al (2004) Archaeology and age of a new hominin from Flores in eastern Indonesia. Nature 431(7012):1087–1091

Murphy D (2020) Concepts of disease and health. In: Zalta EN (ed) The Stanford encyclopedia of philosophy (summer 2020 edition). https://plato.stanford.edu/archives/sum2020/entries/health-disease

Neander K (1995) Misrepresenting and malfunctioning. Philos Stud 79(2):109–141

Neander K (2016) Mental illness, concept of. In: Routledge encyclopedia of philosophy, 1st edn. Routledge, London. https://doi.org/10.4324/9780415249126-V021-1

Norenzayan A, Shariff AF, Gervais WM, Willard AK, McNamara RA, Slingerland E, Henrich J (2016) The cultural evolution of prosocial religions. Behav Brain Sci. https://doi.org/10.1017/S0140525X14001356

Nunn CL, Samson DR (2018) Sleep in a comparative context: investigating how human sleep differs from sleep in other primates. Am J Phys Anthropol 166(3):601–612

Obendorf PJ, Oxnard CE, Kefford BJ (2008) Are the small human-like fossils found on Flores human endemic cretins? Proc R Soc B Biol Sci 275(1640):1287–1296

Ogilvie MD, Curran BK, Trinkaus E (1989) Incidence and patterning of dental enamel hypoplasia among the Neandertals. Am J Phys Anthropol 79(1):25–41

Pearson OM, Cordero RM, Busby AM (2006) How different were Neanderthals' habitual activities? A comparative analysis with diverse groups of recent humans. In: Hublin J-J, Harvati K, Harrison T (eds) Neanderthals revisited: new approaches and perspectives. Springer, Dordrecht, pp 135–156

Perreault C (2012) The pace of cultural evolution. PLoS One. https://doi.org/10.1371/journal.pone.0045150

Ramsey G, De Block A (2017) Is cultural fitness hopelessly confused? Br J Philos Sci 68(2):305–328

Robson SL, Wood B (2008) Hominin life history: reconstruction and evolution. J Anat 212(4):394–425

Saborido C, Moreno A (2015) Biological pathology from an organizational perspective. Theor Med Bioeth 36:83–95. https://doi.org/10.1007/s11017-015-9318-8

Samson DR, Nunn CL (2015) Sleep intensity and the evolution of human cognition. Evol Anthropol 24(6):225–237

Schramme T, Edwards S (eds) (2017) Handbook of the philosophy of medicine. Springer, Dordrecht

Solomon M, Simon JR, Kincaid H (eds) (2017) The Routledge companion to philosophy of medicine. Routledge, New York

Spikins P, Needham A, Wright B, Dytham C, Gatta M, Hitchens G (2019) Living to fight another day: the ecological and evolutionary significance of Neanderthal healthcare. Quat Sci Rev 217:98–118

Sterelny K (2010) Minds: extended or scaffolded? Phenomenol Cogn Sci 9(4):465–481

Sterelny K (2012) The evolved apprentice: how evolution made humans unique. MIT Press, Cambridge, MA

Sterelny K (2015) Optimizing engines: rational choice in the Neolithic? Philos Sci 82(3):402–423

Tucci S, Vohr SH, McCoy RC, Vernot B, Robinson MR, Barbieri C et al (2018) Evolutionary history and adaptation of a human pygmy population of Flores Island, Indonesia. Science 361(6401):511–516

Ungar P (2004) Dental topography and diets of Australopithecus afarensis and early Homo. J Hum Evol 46(5):605–622

Ungar P (2017) Evolution's bite: a story of teeth, diet, and human origins. Princeton University Press, Princeton

Wakefield JC (1992) The concept of mental disorder: on the boundary between biological facts and social values. Am Psychol 47(3):373–388

Wakefield JC (2007) The concept of mental disorder: diagnostic implications of the harmful dysfunction analysis. World Psychiatry 6(3):149–156

Wrangham R (2010) Catching fire: how cooking made us human. Profile, London

Part III
Cognition, Language and Normativity

Chapter 8
Embodied and Extended Numerical Cognition

Marilynn Johnson and Caleb Everett

Abstract In this chapter we consider the theories of embodied cognition and extended mind with respect to the human ability to engage in numerical cognition. Such an enquiry requires first distinguishing between our innate number sense and the sort of numerical reasoning that is unique to humans. We provide anthropological and linguistic research to defend the thesis that places the body at the center of our development of numerical reasoning. We then draw on archaeological research to suggest a rough date for when ancient humans first were able to represent numerical information beyond the body and in enduring material artifacts. We conclude by briefly describing how these capacities for embodied and extended numerical cognition shaped our world.

Keywords Number words · Counting · Embodiment · Extended mind · Numerical cognition · Calendars

8.1 Introduction

What is the relationship of embodiment to our capacity to think numerically? It might seem, at first, that the capacity for numerical thought would be a paradigm case for the computational theory of mind. However, as we argue, numerical thought is embodied.

M. Johnson (✉)
Department of Philosophy, University of San Diego, San Diego, CA, USA
e-mail: marilynnjohnson@sandiego.edu

C. Everett
Department of Anthropology, University of Miami, Miami, FL, USA
e-mail: caleb@miami.edu

© Springer Nature Switzerland AG 2021
A. Killin, S. Allen-Hermanson (eds.), *Explorations in Archaeology and Philosophy*, Synthese Library 433, https://doi.org/10.1007/978-3-030-61052-4_8

Consideration of the relationship of embodiment to our capacity to think numerically requires first distinguishing quantical from numerical cognition. Quantical
reasoning, QR, is present during infancy, shared with other species, and allows us to
discriminate quantities up to three, automatically, without counting. By contrast,
numerical reasoning, NR, the ability to discriminate quantities greater than three,
requires us to have acquired number words. As we will defend, the genesis of number words stems from the special relationship humans have with our fingers and
toes. In this way, NR is necessarily embodied. By contrast, QR is not.

The development of number words was the first important shift enabling the
advent of NR; a second important shift took place when we developed the ability to
represent quantities outside the body. We discuss the first evidence of numerical
cognition as represented in our digits and in non-digital artifacts, and consider how
the latter development of non-anatomical numerical representation shaped our
capacities to think. We see the evidence we present here as providing support for the
criticality of the body and external representations to NR.

We present this discussion in terms of philosophical theories of embodied and
extended mind. The application of such theories will be broad. Concomitantly, we
hope to steer clear of ongoing philosophical debates regarding the boundaries of the
mind, and of associated debates regarding the usage of terms like 'belief', 'consciousness', and 'cognitive' (Chalmers and Clark 1998; Prinz 2008; Adams and
Aizawa 2008).

8.1.1 Numbers in Computational Theory of Mind

Numerical reasoning has historically been understood to be a paradigm case of the
sort of processing that can be done by computers (Rescorla 2017). Although scientists and programmers continue to develop more and more lifelike AI, mathematical
problem-solving is one area where computers have long excelled. Although Siri
might be bad at understanding some of your requests, no one doubts her ability to
do math, or the fact that a graphing calculator can create a graph with more speed
and accuracy than anyone can by hand. Computers have calculated sums that would
be impossible without technology. For example, with computers working 24 h a day
for 105 days we have recently been able to calculate 22,459,157,718,361 digits of
pi (Revell 2017). More practical and equally impressive applications of the mathematical capacities of increasingly rapid silicon (and non-silicon) processors abound
in more commonplace devices and programs—from spreadsheets to statistical analysis platforms, to the algorithms at the core of many common apps for social network analysis, direction-finding, and internet searches. Binary-based computational
math pervades modernity.

Because of such facts about computers, theories of embodiment may not seem to
have their most natural home with numbers. But a different question—and the one
we will consider in this chapter—is whether *our* numerical capacity supports the
computational theory of mind, or if NR is evidence of the theories of embodied and

extended mind. We will argue in the following sections that the body and artifacts outside the body played an essential causal role in the development of numbers, and continue to bear an important connection to how we conceive of and use numbers today.

8.1.2   Embodied Cognition

Before we continue with our discussion of numbers and forms of numerical cognition, some philosophical background is needed. The view of the mind as something distinct from the body was most famously presented in Descartes' *Meditations* in 1641. As highlighted in his contemporaneous letters with Princess Elizabeth of Bohemia, Descartes draws a fundamental and problematic distinction between the self as a thinking thing and the self as a body made of matter (Atherton 1994).

If there is nothing special about our selves as situated in or constituted by our bodies then computers could be an apt metaphor for the mind. This view had its heyday in the 1960s and 1970s but in the past 20 years there has been pushback against such theories, and many philosophers working on cognitive science and consciousness today have rejected the computer as a metaphor for the mind (Rescorla 2017; Churchland 2017; MacFarquar 2018). Many philosophers focus instead on what is special about ourselves as embodied beings, and, to further extend the realm of the self, on how the mind may extend from the body and to the tools we use (Chalmers and Clark 1998; Prinz 2008; Adams and Aizawa 2008; Menary 2015; MacFarquar 2018).

In a well-known philosophical thought experiment, we are asked to consider having our brains removed and put into a vat (Harman 1973). We are to imagine that all our sensory experience would be simulated by computer programs connected to our brains. According to proponents of the embodied cognition thesis, this thought experiment could never be fully realized, regardless of the advancements of science—because being me, and experiencing the world as I experience it, is inseparable from my experience in a physical body (Prinz 2008). According to the embodiment theorist, any agent's experience could never be accurately simulated by the brain alone, for that experience is inextricably tied to bodily states (Prinz 2008).

Some critics caution that the claims of the embodiment thesis may have been overblown, overhyped, or plain wrong (Prinz 2008; Adams and Aizawa 2008). Jesse Prinz writes that although embodied cognition is trendy, "the philosophical equivalent of a blockbuster", "excitement is not always correlated with truth" (Prinz 2008). Prinz argues that the theory of embodied cognition does not "hold the basic key to explaining consciousness" and argues for the more minimal thesis that "certain aspects of consciousness may depend on systems involved in perceiving and controlling the body" (Prinz 2008, pp. 1–2). Although we make use of the embodied cognition thesis here, we do not weigh in on its precise relationship to consciousness. Our argument is consistent with Prinz's more minimal take on the theory, as well as with a construal of the embodiment thesis that places consciousness at its center.

8.1.3   Extended Mind

The extended mind thesis expands the notion of the self beyond the mind and the body—as in the embodied mind thesis—and into the external world. As Dave Chalmers and Andy Clark argue in their canonical 1998 paper, our relationship to tools in the world, such as reminder notes, bears certain similarities to our relationship to our memory. They present us with the case of Otto, who has Alzheimer's and relies on his notebook to remind him of things, such as that the Museum of Modern Art is on 53rd St. In other words, as Chalmers and Clark put it, "his notebook plays the role usually played by biological memory" (1998, p. 12). They argue that because of examples such as this, we ought to say that Otto has *beliefs* that are in his notebook, rather than in his memory.

Chalmers and Clark's extended mind thesis highlights the question of whether belief is necessarily an internal mental phenomenon, or if a belief can be understood as an externally accessible, action-guiding proposition. If the former, then Otto does not have the belief that the MoMa is on 53rd St, if the latter, he does. Chalmers and Clark argue that even if we do not ordinarily use the word 'belief' in this way, we *ought* to, because such a notion of belief is more explanatorily useful (1998, p. 14).

In describing the extended mind thesis Fred Adams and Ken Aizawa put the theory as follows: "This is the view that when a student takes notes in class, the student literally commits information to memory. When someone uses pencil and paper to compute large sums, cognitive processes extend to the pencil and paper themselves" (Adams and Aizawa 2008, p. 79). Adams and Aizawa go on to argue that, contra the extended mind thesis as they see it, tool use is a case of "cognitive processes interacting with portions of the noncognitive environment", rather than "a matter of cognitive processing throughout" (Adams and Aizawa 2008, p. 80). They recognize that there is a potential for such discussions to devolve into terminological debates about words such as 'cognitive'. However, what they argue is that the cognitive should be understood as particular to the processes of the brain (Adams and Aizawa 2008).

Whether it is more explanatorily useful to speak of beliefs as things that can be in notebooks, as Chalmers and Clark argue, or as mental states that must be 'in the cranium', as Adams and Aizawa argue, remains to be seen, and we do not wish to weigh in on these ongoing debates. However, the broader point that Chalmers and Clark make is relevant to our discussion. There are certain external tools in the world that we use in a way that is similar in important respects to how we use memory. Through the creation of symbols that can create occurrent belief states we are able to offload certain tasks beyond the brain itself. At some point in our history this happened with numbers. In later sections we will propose when this transition occurred.

8.1.4 Our Aims

Before we proceed with presenting linguistic evidence that supports our view, let us say a word about the framing of this chapter. Our discussion here is presented in terms of philosophical theories on embodiment and the extended mind thesis. These are philosophical ideas that, although enjoying more prominence in the past 20 years, are not without their nuances and critics (see above and Prinz 2008; Adams and Aizawa 2008; Menary 2015; Rescorla 2017). In this discussion we characterize one range of viewpoints as the theory of embodiment and another range of viewpoints as the extended mind hypothesis; we attempt to stay out of some of the debates within the literature on embodiment and the extended mind. This is a necessary step to being able to say something about how these theories relate to the linguistic and archaeological evidence we present, without getting mired in the philosophical details.

The theories of embodiment and the extended mind are the framework on which we present and frame our ideas and findings about number words and counting. Some other theorists who have discussed numbers and theories of embodiment or extended mind have done so with the aim of defending some *particular* position within the debates on embodiment or the extended mind thesis—or some other position within this space such as 'radical enactivism' (Zahidi and Myin 2016) or 'cognitive integration' (Menary 2015)—with numerical cognition as their specific evidence in favor of their preferred view of the mind. This is *not* our aim. Our aim is not to defend any particular theory of the mind, but to sketch out a more general version of the embodied mind thesis and extended mind thesis and demonstrate how linguistic and archaeological evidence fits within this picture. When attempting to connect disparate threads across disciplines it is necessary to paint with a broad brush; this makes it possible to situate research from linguistics and archaeology within a simplified philosophical framework.

8.2 Cross-Cultural and Cross-Linguistic Evidence Highlights the Embodied Bases of *Numerical* Cognition

Now that we have presented the philosophical framing, we will now present empirical research that supports the thesis that humans' acquisition of number words was embodied. We also offer evidence that this acquisition was essential to the advent of NR, which is unique to our species, in contrast to QR. In the subsequent section we will return to the extended mind thesis.

Recent work in the field of cognitive science, based in large measure on cross-linguistic and cross-cultural studies, has aimed to draw a distinction between two kinds of cognition: numerical cognition and 'quantical' cognition. Two similar efforts to elucidate this distinction were made in recent publications: Núñez (2017) and Everett (2017). While Everett (2017) draws a distinction between quantitative (as opposed to 'quantical', a term coined by Núñez) and numerical thought, the relevant claims made in these works are remarkably similar. Both authors contend that there is a pressing and heretofore unnoticed need to dissociate those facets of quantitative cognition that are innate and generally inexact, henceforth *quantical* cognition, from the culturally and linguistically contingent *numerical* cognition that allows humans to count and, more broadly, to precisely distinguish all quantities. Núñez and Everett suggest, independently, that numerical and quantical cognition are inappropriately conflated in research on how humans think with and about quantities. They highlight clear drawbacks associated with this conflation, principally the muddying of the nature of humans' native capacities for discriminating quantities and the associated muddying of the role that these native capacities have in generating more elaborate forms of numerical thought. It is worth briefly outlining and disentangling numerical and quantical thought, and the empirical bases on which Núñez (2017) and Everett (2017) rest their claims. We do so next, prior to outlining the ways in which embodied processes were (and are) essential to the development of uniquely human *numerical* thought.

Humans, like a variety of other species, appear to possess a native and abstract capacity for distinguishing quantities. Shortly after birth we are capable of comparing and discriminating the quantities of given sets of stimuli, if the ratio between the sets is sufficiently large. As one example of this abstract capacity, recent experiments with day-old infants demonstrated that babies are generally capable of distinguishing 18 colored dots from, say, 6 dots, even after confounding variables (such as absolute size of stimuli, stimuli movement, etc.) are controlled. In such a case, the ratio between the two sets of stimuli is pronounced at 3:1, facilitating discrimination. Yet prelinguistic infants are also capable of consistently discriminating sets of items if the ratio describing their discrepancy is as low as 2:1. With ratios lower than 2:1, infants struggle with differentiating the relevant quantities. This native capacity for discriminating sets, assuming the ratio describing their respective amounts is sufficiently large, is often referred to as the 'approximate number sense' (Dehaene 1997). The abstract nature of this apparently native sense, which is typically housed in the intraparietal sulcus judging from a host of cortical imaging studies (Everett 2017), is evident in the cross-modal nature of some of the sets of stimuli discriminated by infants. Gaze tasks suggest that infants recognize, approximately, the quantitative similarity of a given set of audio stimuli, like a series of beeps, and an equinumerous set of visual stimuli, like dots (Xu and Spelke 2000; Izard et al. 2009).

Humans are also natively equipped with the capacity for exactly discriminating quantities less than four. This is evident in work with prelinguistic children (Wynn 1992), but also with adults in numberless cultures (Spaepen et al. 2011; Everett and Madora 2012, *inter alia*). Some debate exists as to whether or not this exact discrimination is truly distinct from the approximate 'number' sense, or whether our ability to discriminate smaller quantities follows from, e.g., the fact that the ratio between small quantities like 1 and 2 is sufficiently large to allow for their consistent subitization (Piantadosi 2016).

Setting aside such concerns, the abilities to subitize small quantities and to approximately discriminate large quantities are often referred to, together, as humans' innate 'number' sense and as a key part of our 'numerical' cognition. It is generally agreed, and certainly not contested here, that these innate capacities buttress the edifice of other more exact forms of numerical cognition, including arithmetic and the like. What is contestable is whether our imprecise native abilities should be referred to as 'numerical'.

Furthermore, while it is generally agreed that humans' native capacities for discriminating quantities is critical to the scaffolding of more robust forms of numerical thought, intense debate persists as to how this scaffolding occurs ontogenetically, or as to how it occurred diachronically (Overmann 2015; Everett 2015). Much of the motivation for this debate also underscores a key reason it is problematic to refer to humans' native quantitative reasoning as 'numerical cognition': these native capacities are largely similar to the capacities observed in many other species. Species with quantity discrimination skills that are at least roughly similar to those observed in prelinguistic infants and numberless adults include a variety of other primates, as well as a host of phylogenetically distant vertebrates (Brannon and Park 2015; Agrillo 2015).

It is unclear whether the quantitative skills of some of these species are homologous or analogous features, but it is increasingly clear that some of these skills are pervasive in nature. This pervasiveness makes troublesome the common terminological choice, evident throughout the cross-disciplinary literature on this topic, to refer to native human quantitative capacities as 'numerical' or as evidence for a human 'number sense.' After all, there is a gross discrepancy between these pervasive abilities, shared by so many species, and the ability to, say, distinguish 6 from 7 items consistently—an ability only observed in human populations that have acquired words and other symbols for exact quantities. As Núñez elegantly notes:

> The adjective 'numerical' in 'numerical cognition', however, is crucially over-inclusive: any cognition or behavior relating to quantity in babies, monkeys, rats, or fish—whether exact or inexact, symbolic or non-symbolic, operational or not—is labeled as being 'numerical'. This loose over-inclusiveness licenses stating—teleologically—that thousands of species, from fish to humans, by virtue of being able to discriminate quantities, de facto have 'number representations' as a result of biological evolution (Núñez 2017, p. 419).

The common and unfortunate terminological choice licenses implicit assessments of the nature of elaborate human *numerical* cognition, which is unique and only evident in those members of the species who are also members of numerate cultures. This sort of cognition clearly does not owe itself simply to basic neurobiology, given that it is not a cross-cultural universal, and is not simply a natural byproduct, nor even a straightforward cultural refinement, of our innate 'number sense' since that 'number sense' did not evolve for actually numerical purposes.

The relationship between truly numerical cognition and the coarse quantitative distinctions enabled by our neurobiology is hardly direct—as evidenced by the lack of precise numerical cognition in other species that seem to share many of our native abilities for quantitative reasoning. (Excepting certain members of other species that have been laboriously trained with number words and symbols, who also share some of our numerical cognition but concomitantly highlight the essential role that numbers play in enabling truly numerical thought—see discussion in Everett 2017.) For such reasons, Núñez (2017) suggests that our relevant biologically endowed capacities be referred to as 'quantical', rather than 'numerical'.

Numerical cognition, while relying on a phylogenetically primitive system for quantical cognition, is culturally and linguistically dependent. A fair amount of cross-cultural and developmental data now suggests that humans lack the capacity for consistently and exactly discriminating most quantities prior to their acquisition of numbers—words and other symbols for precise quantities (Everett 2017). From a diachronic perspective, the cultural acquisition of numbers is somewhat haphazard and contingent on a host of idiosyncratic factors including patterns of contact between diverse linguistic communities.

Nevertheless, at its core the process of the cultural acquisition of truly numerical concepts (rather than quantical ones) has an embodied etiology. While scholars have long recognized that number systems are often motivated by human anatomical characteristics, the extent of these physical motivations has still been underappreciated. Next we offer a brief overview of the cross-linguistic evidence that demonstrates the extent to which human anatomy motivates both the germination and florescence of linguistic systems of numbers, and thereby ultimately allows for truly numerical cognition across the bulk of the world's cultures.

As background to this overview, though, consider how quantical information surfaces in the world's languages: the vast majority of the world's languages make grammatical, rather than lexical, distinctions between the quantities distinguishable with our innate capacity for quantical reasoning. Most notably, the bulk of the world's languages distinguish grammatically between one entity and more than one entity via a singular vs. plural distinction. This basic distinction surfaces in nominal plurality/singularity, verb-subject agreement patterns, and sundry other grammatical phenomena. In rarer cases grammars also distinguish between singular and dual categories of entities, or, rarer still, singular, dual, and trial. Other potential categories, principally the paucal and plural categories, are imprecise. In other words, what the world's grammars discriminate correlates neatly with what native quantical reasoning allows speakers to discriminate: 1, 2, 3, and other larger quantities. This seems unlikely to be a coincidence, a point advanced recently in Everett (2017) and Franzon et al. (2019). (Though it should be acknowledged that most languages do not utilize a grammatical dual or trial, meaning that there is hardly an exact correspondence between quantical reasoning and grammar across all the world's languages.)

In contrast, truly numerical concepts, related to higher quantities that are not precisely discriminated with our native hardware, are not encoded morphologically or syntactically, but lexically, in the words themselves. Precise words for higher quantities exist in nearly all, but critically not some, of the world's languages. Tellingly, these words for quantities have clear anatomical bases in the vast majority of the world's languages. This fact is evidenced in myriad ways, including the typological commonality of number words with decimal bases. While the commonality of decimal bases has been acknowledged for many years, though, its extent is still perhaps underacknowledged. So, it is worth considering what the cross-linguistic data say about the extent to which the human hands facilitate the creation of number words.

In many languages the words for five or ten are etymologically related to the word for hand. Consider some examples from two Amazonian languages, Jarawara and Karitiâna, on which one of us has done field research. In (1) and (2) we see the words for five and ten in Jarawara:

(1) yehe ohari
     hand one
     "five", literally "one hand"

(2) yehe ka-fama
     hand with-two
     "ten", literally, "with
     two hands"

In (3) the word for five in Karitiâna reveals the same manual source:

(3) yj-pyt
     our-hand
     "five", literally
     "our hand"

This manual basis is also evident for larger Karitiâna numbers like eleven, as we seen in (4):

(4) myhint   yj-py   ota   oot
     one   our-hand   another take
     "eleven", literally "take one and our
     other hand"

Many of the world's languages have such transparently manual (and ultimately digital) origins. The word for five and/or ten is the lexical base on which higher numbers are constructed in languages like Jarawara and Karitiana, with numbers like 'six' and 'eleven' taking the form of 'five plus one', 'ten plus a finger', and so on.

The word for five serves to kickstart, diachronically, the growth of number-term systems in the world's languages. This is evident in a recent survey of Australian languages, which are not nearly as anumeric as some presume (Bowern and Zentz 2012). Worldwide, anatomically motivated words for 'five' and 'ten' seem critical to the intra-linguistic growth of number systems, and serve as the base for most number words. These words then become critical to the transmission of precise numerical concepts cross-culturally and cross-generationally.

Surprisingly to some, humans do not seem to grasp the associated numerical concepts prior to learning the relevant number words. Framed differently: the manual basis of number systems is not simply the result of convenient labels of pre-existing numerical concepts acquired through our 'number sense', the manual basis is what allows for the acquisition of numerical concepts. One wonders, then, how exactly the fingers/hands enable(d) some people to transcend quantical reasoning and create precise numbers for higher quantities, entering the world of the truly numerical. We address this issue below.

While languages like Karitiâna and Jarawara have number words with transparently digital origins, numbers are manually-sourced in most cultures, though the manual sources are evident more subtly. In most languages, numbers are decimally based. This is true in English, as evidenced in words like 'twenty-one' and 'thirty-one', wherein we begin counting at 'one' again with the addition of each ten units—words that are derived historically from a multiplicative and additive strategy that is now somewhat opaque (e.g., 'twenty-one' derives from 'two-ten-one').

Nevertheless the manual basis of English is somewhat obscured by the lack of a discernible etymological relationship between five and/or ten and 'hand' or 'hands', and by the fact that some number words like eleven and twelve are not as transparently decimal in their structure as their corresponding number words are in many languages (like Mandarin). Furthermore, English-speaking cultures generally use other bases, like the sexagesimal and duodecimal, for time-telling and navigation, further obscuring the manual historical underpinnings of all our numbers. We do, however, use the English word 'digit' to refer both to numerals and to our phalanges.

The manual basis of English numbers stretches back millennia, prior to the advent of written numerals in Mesopotamia and other regions. Both historically and ontogenetically, verbal numbers predate other kinds of numbers. (The written numerals we use today, with their own decimal basis of Indic origins, are a particularly recent innovation.) Work in historical linguistics has demonstrated that Proto-Indo-European used a decimal system, and this ancestral tongue was likely spoken over 6000 years ago. Most of the world's people speak a language that has a decimal base. Both Proto-Indo-European and Proto-Sino-Tibetan had decimal bases, and speakers of Indo-European and Sino-Tibetan languages, which have inherited these decimal bases from millennia ago, represent over half of the world's population. The ancestral tongues from which the other most pervasive language families descended, including Proto-Afro-Asiatic, Proto-Austronesian, and Proto-Niger-Congo, also had decimal number systems (Everett 2017).

Yet the current pervasiveness of decimal systems is not simply the result of the success of a few language families over the last few thousand years or so. Comrie (2013) conducted a worldwide survey of the kinds of number systems evident in the world's languages. His sample consists of 196 languages. While there are about 7000 mutually unintelligible languages in the world today, Comrie's sample represents all major families and regions, and is a reasonable indicator of patterns in all languages.

Of 196 languages, Comrie observes that 20 have limited number systems. These include 'one-two-many' systems that are found primarily in Australia in Amazonia. Of the remaining 176 languages that have robust systems of number, 125 (71%) are decimally based for numbers beyond ten. (Often smaller numbers are quinary based, even in languages in which higher numbers are decimally based.) Twenty-two of the languages (12.5%) have hybrid decimal/vigesimal systems, while 20 (11.4%) have pure vigesimal systems. Framed slightly differently, about 95% of the languages with robust number systems have numbers that are digitally based—oriented according to the numbers of fingers and/or toes on the human body. Furthermore, of the remaining nine languages, five of these have numbers that are based on an 'extended body-part system'. According to Comrie's survey, then, less than 3% of languages with robust number systems structure their numbers around something other than features of the human body.

With exceedingly few exceptions, languages with number words greater than 'five' create numbers via the body. It should be noted that the 'exceptions', while not cases of limited number in the strictest sense, also do not include number systems that are open-ended, with limitless numerical referents. For example, the senary (base-6) systems of some languages in New Guinea, which owe themselves at least partially to the manner in which yams are stored in groups of six, are not generally used in elaborate counting and arithmetic. In short, the body has clearly been critical, more critical than is often realized, to the development of number words.

The claim that the body is critical to the historical development of number words, across the vast majority of the world's cultures, suggests that the body is critical to the entrance of cultures into the world of numerical, rather than quantical, concepts. A surfeit of experimental data has demonstrated that number words are essential to children acquiring basic numerical concepts like the one-to-one correspondence between large sets. Work with anumeric adults has converged on the same conclusion (Spaepen et al. 2011; Everett and Madora 2012).

Rather than simply serving as labels for concepts that humans are natively predisposed to acquire during development, number words work as placeholders for concepts that children realize they must acquire. Learning numbers is not a matter of labeling concepts but of 'concepting labels' (Everett 2017; Carey 2009). Relatively early on during language acquisition, kids learn that number words come in a sequence, but do not appreciate that this sequence represents a growing magnitude. At a critical stage they learn the *successor principle*, realizing that each number word represents 'one more' than the quantity that precedes it. Native quantical reasoning appears to facilitate the acquisition of this principle, since kids can naturally recognize that two is greater than one, and that three is greater than two. Nevertheless, debate remains as to the role that native quantical reasoning plays in the acquisition of numerical concepts like 'five'. This is true, at least in part, because other species possess quantical reasoning yet do not enter the world of the numerical unless they are trained with numbers of human origins. Interestingly, most kids also rely on their hands and finger-counting as they learn the successor principle. So the structure of the human body remains critical to the ontogenetic acquisition of number words like 'five' that were once created by innovators who also benefited from the structure of the human hands during the relevant innovation.

Ultimately, then, precise numerical concepts owe themselves either directly or indirectly to humans' engagement with their fingers and, to a lesser extent, their toes. Why are our bodies so critical to transcending the quantical? Why have humans been able to rely on their manual digits to transcend the quantical while other primates have not? The facile answer is that we are the only linguistic species, capable of naming quantities via patterns in our hands. Or perhaps—to make the point in a more sophisticated way—the mechanism or 'module' that underpins our acquisition of grammar also underpins our capacity to recognize cardinality, a requirement for counting (Hauser et al. 2002; De Cruz 2008). Yet language is not a sufficient criterion for the acquisition of numerical concepts, as evidenced by linguistic yet anumeric people (Everett 2015).

And if acquiring numbers, even numbers as small as 'five', is not simply a matter of acquiring labels for things that the world carves up (since most anumeric people cannot consistently discriminate 5 from 6 of the same item[1]), how exactly do the hands and fingers really enable numerical concepts? As noted above, a key numerical concept is the appreciation of one-to-one correspondence between sets larger than four. Critically, humans and other species without number words or other number symbols are not able to consistently appreciate one-to-one correspondence for such sets. Human hands may facilitate the innovation of numbers because they are symmetrical and expose us continually to visual and tactile (or even proprioceptive) one-to-one correspondence for two naturally occurring sets of five items. Five is—critically—larger than the precise quantities we can distinguish with quantical reasoning.

While other primates have quantical reasoning and even symmetrical hands, no other species has these characteristics *and* bipedalism, which at least partially explains humans' unparalleled manual focus. This manual fixation apparently allows for the occasional and haphazard innovation of number words like 'five'/'hand' by number inventors who concretize the otherwise ephemeral realization that the quantity of fingers on one hand is equal to the quantity of fingers on another, or perhaps the realization that the quantity of fingers on one hand is equal to a number of small valuable items organized on palm of the hand, alongside those fingers. Of course there may be other characteristics of our species, including neurophysiological ones, that also draw humans into the world of the numerical. But the characteristics of our hands, and our continuous engagement with our non-locomotive appendages, were critical to the genesis of NC judging from the extant cross-linguistic, cross-cultural, and cross-species data. Our capacity for numerical thought and use of number words is causally tied to our existence as embodied beings with ten fingers and toes.

---

[1] This does not mean, of course, that a parent would not, say, be able to tell that one of her six children is missing from a lineup. Similarly, if we saw a map of the United States with (say) Florida missing, we would surely notice. This does not require us to count to 49 or suggest we are using numerical concepts. In neither of these cases does the recognition that something is missing require numerical reasoning but visual recognition that something is off from the norm. Visual recognition only goes so far, however, and very large families in our society may resort to counting to make sure the entire family is present on certain occasions. Thanks to Sean Allen-Hermanson for posing the question with the first example.

8.3 Archaeological Evidence

8.3.1 The Extended Mind and Archaeological Evidence of Numerical Representation

A significant further development in the capacity for numerical thought occurred when ancient humans moved beyond the ability to represent numbers linguistically, and were able to represent them in lasting symbols. It was this step that allowed our relationship with numbers to move beyond embodied cognition and into the territory of the extended mind thesis.

Recall from the introductory sections that the extended mind thesis, as proposed by Chalmers and Clark in their 1998 paper, concerns the ways our relationship to external tools can mirror our relationship to non-occurrent, but accessible, mental states. Chalmers and Clark ask us to consider Otto, the man with Alzheimer's, who uses a notebook to remind him that the MoMA is on 53rd St. A parallel is drawn between the sentence 'The MoMA is on 53rd Street' as written in the notebook and an internal mental state that Otto could have had with this same content.

As noted earlier, we hope to use this philosophical machinery without weighing in on its most controversial potential implications. That is, we do not wish to weigh in on the debate surrounding whether or not Otto's notebook is an extension of his mind and if the contents therein constitute Otto's mental states. However, drawing on the more modest construal, what the extended mind thesis does is highlight the ways certain external representations can change cognition and make accessible mental states that would be impossible without the use of tools. What we will consider now is the ways external representations of numbers changed our capacities to think numerically, and what archaeological remains constitutes evidence of numerical representation.

8.3.2 Numerical Representations

Archaeological evidence is such that it is necessarily a record of *externalized* mental states, that led to behaviors which modified the world in such a way that an enduring material record was created. Empirical facts about decay sharply reduce the quality of this evidence. Despite these limitations—which should be kept in mind throughout this discussion—a number of archaeologists have attempted to find the point at which humans first developed number tools that extended beyond the body, a change that extended our capacity for numerical cognition.

Today mathematicians do calculations by manipulating the Arabic numerals that we have adopted around the world to represent numbers, or by having computers do this for them. One reason that the Arabic numeral system overtook the system of Roman numerals that was previously used in the West is because mathematical calculations are extremely difficult to do with Roman numerals. Multiplication of the quantities twenty-seven by twenty-seven is much easier to do when it is depicted as 27 × 27 rather than as XXVII by XXVII.

In the previous sections we detailed a number of the ways that number words in societies across the globe bear evidence of the connection to the body at their genesis. This connection to the body is also seen with the numerical representation of Roman numerals. Unlike the Arabic numerals which provide an edge in calculations, the Roman numerals can be thought of as bearing a resemblance to what they represent. A single I in Roman numerals resembles one finger; a V resembles a hand, or five fingers, and X resembles two hands together, or ten fingers total. With Roman numerals we again see the primacy of the body, and this time not just in the spoken word itself but in the way that number is externalized and represented on material objects.

8.3.3   6000-Year-Old Clay Tokens and the Mind

Our current practice of depicting quantities with the Arabic numeral system is a result of a process of humankind developing numerical representations and keeping those that best suit our purposes. It is far from given that such representations would resemble Arabic or Roman numerals, and archaeologists must develop some account of what sorts of items from the archaeological record were used to represent quantities. Lambros Malafouris, in his 2013 book *How Things Shape the Mind*, engages with some of the same philosophical topics as we have here, including the extended mind thesis as applied specifically to numerical cognition, and argues that clay tokens found in 4000 BCE are the first archaeological evidence of numerical thought (Malafouris 2013, p. 113).

Rather than steering clear of the most controversial philosophical implications of the extended mind theory, as we attempt to do here, Malafouris embraces a radical interpretation of the extended cognition view wholeheartedly in his work. As Colin Renfrew writes in his laudatory introduction, Malafouris examines how "the human mental capacities that have their primary location in the brain (within the skull) are not separable in any serious consideration from their expression in action" (p. ix) and argues that "the mind is to be understood as embodied, indeed as extended beyond the body" (p. xi). In presenting his account, Malafouris makes the point in an even stronger way, writing that in the "gray zone of material engagement", "brains, bodies, and things conflate, mutually catalyzing and constituting one another" (p. 5). The claim that brains, bodies, and things conflate is stronger than the idea proposed by philosophers who are proponents of the extended mind thesis.

The claim Malafouris makes in this discussion is to parse, and on at least one interpretation seems clearly false and open to a number of obvious objections. Does Malafouris mean to claim that bodies, brains, and things *literally* conflate? Clearly it is not the case because the 6000 year old clay tokens he discusses are still here and can be held in the hand of an archaeologist, while the bodies of whoever it was that made and used those clay tokens have been reduced to bones or less, and the brains have long since decayed to nothing. Bodies, brains, and things persist over different timescales. It is this very fact that makes it worthwhile to take some thought, such as 'MoMA is on 53rd Street' and to put it down in a notebook. It is one thing to say that the notebook should count as a part of the mind and quite another to say that bodies, brains, and things conflate.

Malafouris does not attempt to get around such objections by clarifying precisely what he means by the claim that "brains, bodies, and things conflate". He writes, "too much clarity and too great an emphasis on definitions could be misleading in a context where transgressing the common wisdom about minds and things is often a precondition for success" (p. 9). His argument certainly does not suffer from an overabundance of clarity. Malafouris says he is transgressing the common wisdom about minds and things but does not tell us how. On its face this makes his position seem implausible. Without clarifying what else he could have meant it remains so.

Perhaps in an attempt to be as charitable as possible to Malafouris we could characterize his view as close to the Chalmers and Clark position and take it to be that minds (not brains) are sometimes outside the skull, and even outside the body. Such a construal might make his view appear to be a straightforward application of the extended mind thesis to the archaeological record.

However, Malafouris himself makes it clear that this is not what his view amounts to. Instead, he writes that his conclusions go beyond what the philosophers he draws on were willing to commit to. On this point he writes:

Most philosophical treatments remain epistemically agnostic about material culture's properties and about its active role in human life and evolution. Even embodied cognitive science (Anderson 2003; Wheeler 2005; Chemero 2009; Clark 1997, 2008), which explicitly
recognizes the intrinsic relationship between brain/body and environment, often seems
oblivious to the phenomenal properties of the material medium that envelops and shapes
our lives. Although the material world is recognized as a 'causal influence' rather than a
'mere stimulus', it is rarely seen as playing a 'constitutive' role (Malafouris 2013, p. 10).

We see here Malafouris commits himself further to the position that the mind is constituted by things.[2] He argues in the quote above that philosophers' failure to reach this conclusion is a result of their being 'oblivious' to materiality. Malafouris then expresses disappointment that the philosophers who developed the theories he draws on have not been led to his conclusion, writing, "at the present stage of research, philosophy of mind remains skeptical and undecided about entering the treacherous territory of the extended mind proper" (pp. 10–11). It is this purported failing of philosophy to take material culture seriously that Malafouris aims to rec-tify in his book.

---

[2] It is not clear if Malafouris mistakenly believes that Chalmers and Clark (1998) hold the view the Otto's notebook plays a 'mere' causal role rather than a constitutive one, or if he believes this paper (which, as of publication, has over 5000 citations) is one of the 'rare' exceptions to philosophers allegedly overlooking this possibility.

8.3.4   Malafouris on Numbers in the Archaeological Record

What does it look like when Malafouris's 'extended mind proper', as he sees it, is applied to the archaeological record? A fruitful place to look is his discussion of numbers, which, as we noted above, leads us to consideration of which artifacts in the archaeological record are evidence of numerical cognition. Our position is that the capacity to represent numbers in external tools that could then create occurrent mental states is an advancement that changed our capacity for different types of numerical cognition.

Malafouris adopts a more radical position with respect to number symbols. He writes:

> ...meaningful engagement of material signs is the precondition for the emergence of symbolism. These physical relations and interactions between the body and cultural artifacts should not be taken as mere 'indications' of 'internal' and invisible mental processes; they should, rather, be taken as an important form of thinking (Malafouris 2013, p. 105)

A few things are remarkable about the claims made here. First, this suggestion that material signs are a precondition for the emergence of symbolism calls into question what it was that motivated the creation of the material sign in the first place. If there was not first symbolism in the form of a mental representation, why would the material sign have been created? Malafouris frames his discussion in terms of the mental capacities we detailed in the previous section, and asks "Could *Homo sapiens* alone—that is, in the absence of external material support—have ever have [sic] moved beyond approximation" (p. 106)? By approximation, he means the sort of number approximation that we have called 'quantical'.

Most researchers believe that number words are what enable humans to move beyond approximation to exact number sense (p. 109), as we have defended here. Malafouris adopts a different position: that number words are not necessary to have number concepts (p. 110). On this point, he cites work by Daniel Everett with the Piraha and writes: "Another interesting possibility is that it isn't the lack of number names but the lack of a 'counting routine' or a 'technology for counting' that keeps the Piraha from developing exact numerical thinking" (p. 110). He asks us to consider how "humans conceive or grasp the quantity of 10 when no linguistic quantifier, and no symbol to express it, is yet available" (p. 110). Malafouris concludes this discussion with the statement that language "is not sufficient" to account for humans' development of the concept of number (p. 111).

Is it true that some *Homo sapiens* do not have a 'technology for counting'? If what we have defended above is correct, then, contra Malafouris, fingers represent a 'technology for counting' that is possessed by all humans. We need only look to the hands for a tool to represent the quantity of ten. As we have argued, the evidence that the hands have a longstanding connection to the quantity of 10 is found cross-linguistically. We do not need to look beyond the bounds of our bodies to understand how we utilized fingers—something visible, something symmetrical, something we can manipulate at will, something outside the brain itself, something with proprioceptive qualities, something all humans have—to develop number words up to 5, 10, 20, and number systems that are grounded in these quantities. Indeed it is difficult to conceive of better 'material support' for such a task.

However, Malafouris proposes instead that numerical thinking arises with the creation of forms of external material support, such as clay tokens. He writes: "the emergence of symbolic numerical thinking, in the particular context I am discussing, begins with the invention of the clay-token system" (p. 113). In other words, this is an argument that we had number tools before we had number concepts. Malafouris makes this explicitly clear when he writes that "at this early stage of concrete counting the concept of number had not yet emerged" (p. 114). According to Malafouris, we got numbers after we began to represent numbers in the enduring material record. In the context he considers, around 4000 BCE. This conclusion goes against what we have defended here and, as Malafouris notes, against the received view. The idea that *Homo sapiens* had concrete counting, as with the clay tokens, before we had the concept of number strains credulity. The burden lies with Malafouris to explain how something so implausible, and that goes against the established view, could be true.

8.3.5 Hands as Technology for Counting

When seeking to find evidence of a 'technology for counting' in the archaeological record, we need not look to clay tokens from 4000 BCE but to the fingers of skeletal remains. Of course, the presence of this 'technology' is not evidence of counting or number concepts. Non-human primates also have ten fingers and do not have number concepts. But we are all endowed with this 'permanent tool' that has numerous advantages as a means of counting. Among other advantages, research has also shown that the physical manipulation of fingers aids in number acquisition in children, and being able to use the body in mathematical tasks improves performance (Nathan 2014). The body does serve as an always-accessible technology, but beyond this, our proprioceptive relationship with our hands means that there are additional advantages beyond merely the body as 'tool'.

At the same time, there are certain limitations to using the fingers and toes as your technology for counting. Numbers counted on the fingers can only go up so high, and numbers cannot be 'held' and accessed later beyond when the counting is performed. The capacity for storing or recording numbers using only the hands is limited. There are advantages to creating additional 'technology for counting' beyond what we are naturally endowed with. An ideal tool could be created or modified when some mental state is active, ignored, and then returned to again to activate such a mental state. A tool that, for example, tells me that it has been 5 days since the last full moon would be helpful if I am in a culture that needs to track the tides. As with Otto's notebook that reminds him the MoMA is on 53rd Street, such a tool could help to activate mental states that would be too much of a drain to hold available in memory.

8.3.6   Artificial Memory Systems as Technology for Counting

In papers spanning 20 years, archaeologist Francesco d'Errico has proposed that we understand certain artifacts to be what he calls 'artificial memory systems' or AMSs, objects "conceived and produced to store, process and/or transmit numerical information" (d'Errico 1998; d'Errico et al. 2003; d'Errico et al. 2018). Such artifacts fit the bill for what we have been describing as the role of tools in the extended mind thesis. In presenting such artifacts d'Errico writes, "A fundamental turning point in the evolution of human cognitive abilities and cultural transmission was when humans were first able to store concepts with the aid of material symbols and to anchor or even locate memory outside the individual brain" (d'Errico et al. 2003, p. 31). Notice that contra Malafouris, d'Errico talks of storing concepts in material symbols, not of the material symbols constituting mental states.

An example of what d'Errico understands as an AMS is the notched bone first discussed by Alexander Marshack. Marshack argued that such markings track lunar phases (Marshack 1991; Dehaene 1997, pp. 95–96; d'Errico et al. 2003, p. 32). In his book *The Roots of Civilization: The Cognitive Beginnings of Man's First Art, Symbol and Notation* Marshack develops his hypothesis with respect to an Ishango bone from around 25,000 kya (Marshack 1991, p. 32). This bone has 59 notches and Marshack proposes that this is a two-month calendar (Marshack 1991, p. 30). Marshack himself notes that he has not established this hypothesis in any definitive way, writing "This first crude test of counts, worked out from the photographs and drawings of the Ishango bone, presents us with the *possibility*, then, that we may have a lunar phrasing and notation. It gives us no certainty, one way or another, but also it does not eliminate the lunar possibility" (Marshack 1991, p. 31). If we follow Marshack and d'Errico in their hypothesis that these notches were used to "store, process and/or transmit numerical information" in a way that allowed hominins to "locate memory outside the individual brain" that still leaves a good deal of room for interpretation about the specific numerical information contained therein. As d'Errico notes, "archaeologists have proposed a number of hypotheses to explain these markings. They have been interpreted as *marques de chasse* (marks recording the number of prey killed), devices to keep track of songs, or the number of people attending a ceremony, or other notational/calculation systems" (d'Errico et al. 2003, p. 32). Marshack's proposal is perhaps the most well-known, but his results are far from definitive, as he himself saw.

In recently published work on a notched hyena femur from approximately 72–60,000 kya, found in Les Pradelles, France, d'Errico argues that the notches are number symbols, but stops short of hypothesizing precisely what these numbers tracked. d'Errico argues that this artifact is the farthest back in history that we have evidence of AMSs (d'Errico et al. 2018). Because *Homo sapiens* were not yet in Europe at this time, this means that these incisions were made by Neanderthals. d'Errico uses a number of techniques, including microscopic and morphometric study, as well as experimental reconstruction to support his hypothesis. If number notches are evidence of the existence of number words, as has been proposed

(Dehaene 1997, p. 95), then this means that Neanderthals had number words. They had the same ten fingers and toes as we did and perhaps their body played a similar role in the development of those number words as they did in ours. We also cannot rule out the possibility that the origin of number concepts lies with a shared common ancestor,[3] perhaps *Homo heidelbergensis,* or another hominin, if the divergence between the *Homo sapiens* and *Homo neanderthalensis* is to be found farther back in history (Gomez-Robles 2019).

Whether or not we would want to say that the tally marks created by *Homo sapiens* and Neanderthals represent an extension of the human mind would require weighing in on details of the extended mind thesis that we have attempted to stay agnostic to. At some point this specific question does become a mere terminological debate rather than a metaphysical one. However, to make the point in the most neutral terms, what such bones with tally marks clearly do represent is a new tool that allows us to call to mind information in a way that we were previously unable to. Adopting the term, AMS, or artificial memory system, allows us to identify artifacts that capture the spirit of the extended mind thesis, without committing ourselves to—or even going beyond—its most controversial construals, as Malafouris does. Such AMSs were technology outside of the body itself—external, enduring technology that was built on number words that were initially developed using our hands as the first 'technology for counting'.

With recognition of the complex relationships between numbers and what goes on in the brain, what goes on in the body, and what goes on in the world comes a recognition of the different selection pressures that act on both. Regardless of where one stakes out territory for 'the mind', 'consciousness', 'cognitive', and so on (Chalmers and Clark 1998; Adams and Aizawa 2008; Prinz 2008; Malafouris 2013) it is clear that different processes lead to change in things that are bodily and things that are not. This is especially important to note when considering the developmental story, as we do here. Natural selection led to us having ten fingers and ten toes, as do many other mammals. The cognitive ability to develop number words using our fingers and toes was a later developmental step for our species, that perhaps, as d'Errico's research seems to show, was shared with Neanderthals.

Language itself is culturally transmitted, as are tools. This means that language and tools can develop at a faster rate than we can change genetically (Tomasello 1999). As d'Errico notes in the conclusion to his 2018 paper:

> …the invention of number symbols appeared very recently and has required no biological change. Our brain has not undergone specific adaptations in order to be able to use number symbols. This suggests that is it quite possible, and this is what we would argue, that these cultural exaptations have not required concomitant significant inheritable biological changes (d'Errico et al. 2018, p. 8)

d'Errico's conclusion highlights the benefits of being able to recognize the different selection pressures and thus different timescale for change that are at play with genetic factors in the brain and body versus cultural factors in tools.

---

[3] Thanks to Anton Killin for raising this point.

8.4 Summary and Future Research Questions

Our hands are both something we have a proprioceptive relationship with, and something that we can perceive visually. Because the prehistorical notches made on bone left a physical indentation that could be felt and tracked with the fingers (see the images in d'Errico 2018), these tools may have been perceived through touch as well as visually. When we say that fingers and toes have played a special role in the invention of numbers, we have not specified if this is a claim that there is something about the proprioception of the body that is essential, or if hands are simply playing a role as a 'walking abacus' that we perceive visually and all happen to be endowed with. If it is the former this says something more about the *embodied* nature of this relationship.

This is, at least in part, an empirical question that could be studied. One potential place to look would be to number word acquisition in the blind. With words in general, language acquisition in blind children occurs at the same rate as in sighted children (Gleitman and Newport 1995). Assuming this includes number words as well, this suggests that it is the proprioception of our fingers, and not the visual perception of them that is essential to gaining number concepts ontogenetically, at least for children in numerate cultures. Whether proprioception is critical to the introduction of numerical concepts in a culture, and whether it was essential to the invention of number words, is another matter entirely. We suspect that it was at least beneficial, given the tactile symmetry of the fingers that seems to facilitate the appreciation of one-to-one correspondence for quantities greater than 3. Yet it is also worth noting that vision alone is sufficient to allow for the transmission of numerical concepts. After all, children with amelia, and digit-lacking individuals more generally, can also learn numbers. But, again, this is for individuals in cultures that already have number words; it is another matter whether or not number words could come about in a society where everyone had amelia or was digit-lacking.

Our discussion of numerical cognition is focused on number words and counting. In this scope of focus it differs from some of the recent literature on these topics, which focuses instead on the computational or number manipulation side of numerical cognition (Menary 2015). Numerical computation requires number words and counting to be in place prior to the development of these capacities and would follow concomitantly (Zahidi and Myin 2016; Flegg 2002). Richard Menary focuses on the ways symbols are manipulated in a proceduralized way, which, as he notes, is a recent phenomenon (Menary 2015, pp. 11–14). Numerical computation is a rarefied practice in the course of human history and a discussion of this capacity must be grounded in the fairly recent history of the Greek, Chinese, Arab, and Maya worlds (Renfrew and Bahn 2012; Ansary 2009). Perhaps it is best to describe such a focus as mathematics, where our focus here has been its precursor numbers and counting, and on the cross-linguistic and cross-cultural data that supports this more general story.

And lastly, our focus was in ways narrower than those that consider calendrical systems, construed broadly, as their focus. In the previous section, we considered calendars as tools of externalizing number systems, but not all forms of calendars are necessarily numerical. If we understand calendars as an external feature that "helps to recognize and record temporal events" (De Smedt and De Cruz 2011, p. 66) then features of the natural world, would be included, and do not require numerosity to play this role. Such natural features would include the phases of the moon, the flowering of certain plants, the location of constellations, and features as basic as the changing of the seasons, such as the changing of the leaves or the first frost. Johan De Smedt and Helen De Cruz (2011) have argued that Palaeolithic rock art that depicts animals with identifiably seasonal fur patterns or behavior is a sort of calendar because these depictions played the role of "storing ecologically relevant information about the seasons" (De Smedt and De Cruz 2011, p. 70). Ideally one would want more evidence that these depictions in fact played this role, but nonetheless this discussion of calendars does not make use of numbers, and thus is different from our focus here.

To put it mildly, the expansion of numerical thought has had pervasive effects on the human experience. The effects of this particular sort of embodied thought have been radical and transformative, and are obviously wide-ranging. Consider for a moment some of the cultural and material practices that are associated with or a direct result of the availability of numerical cognition—practices that would not be possible were we to rely only on quantical cognition as humans have done for the bulk of their existence, and as some still do. The discrimination of time in discrete units that can be enumerated, and the general division of time, is the result of numerical cognition. The manner in which most of us demarcate the progression of time, governed as it is by an esoteric and vestigial Mesopotamian base 60 mathematical system, is possible only with numerical cognition and with particular number bases. Seconds, minutes, and hours are some of many non-material numerical constructs that help to govern our experience. More fundamentally, the tracking of days and lunar cycles, natural as opposed to cultural phenomena, requires numerical cognition. It is unlikely a coincidence that societies with very infrequent references to time and temporal progression, like the Tupi-Kawahib or Pirahã of Amazonia, are societies with few if any numbers.

At the material level, examples of the pervasive influence of numerical cognition also abound. We suspect that, as you read this, few if any of the human-made items in your surroundings—from smooth walls, to regular flooring, to your clothing or even the fabric of that clothing—would be possible without the precise measurement that is itself reliant on numerical, rather than quantical, cognition.

One major socio-cultural shift that is at least partially contingent on numerical cognition is urbanization. Urbanization, of course, was a byproduct of dense settlement patterns that were made possible by agriculture, since the latter allowed for the food stores requisite of densely structured populations. In contrast, hunting, gathering, and horticulture require less densely structured groups, and cannot sustain large groups of people in packed configurations. Agriculture, in turn, relies heavily on numerical cognition in ways that hunting and gathering do not. Inter alia, much of agriculture relies on the precise discrimination of astronomical patterns, and on the precise measurement of seeds, tilled rows, and so forth. Mathematics evolved only after the advent of agriculture that enabled urban settlements with diverse vocations, including some that did not contribute directly to food production. These and other factors suggest that agriculture and mathematics coevolved (like much of human culture), benefiting each other in direct and indirect ways. One of many critical results of this coevolution was the advent of literacy, which is arguably a byproduct of symbolic notations developed in Mesopotamia and elsewhere to track quantities of grain and other agricultural products.

The pervasive cognitive, material, and socio-cultural effects of truly numerical thought may obscure the fact that these ubiquitous effects are culturally and linguistically contingent and certainly not native characteristics of our species. It was a multi-stage process for our species to invent number words and develop systems of notion that suit our purposes and externalize information in a way that lightens our cognitive load. These linguistic and symbolic inventions, if you will, are the direct and indirect products of the outgrowth of numerical cognition from quantical cognition—an outgrowth made possible by the structure of our bodies and only then externalized in the material record.

Acknowledgements We would like to thank Sean Allen-Hermanson, Francesco d'Errico, Peter Godfrey-Smith, Anton Killin, and Ronald Planer for their helpful comments on the paper.

References

Adams F, Aizawa K (2008) Why the mind is still in the head. In: Robbins P, Aydede M (eds) Cambridge handbook of situated cognition. Cambridge University Press, Cambridge, pp 78–95

Agrillo C (2015) Numerical and arithmetic abilities in non-primate species. In: Kadosh RC, Dowker A (eds) Oxford handbook of numerical cognition. Oxford University Press, Oxford, pp 214–236

Anderson ML (2003) Embodied cognition: a field guide. Artificial Intelligence 149:91–130

Ansary T (2009) Destiny disrupted: a history of the world through Islamic eyes. Public Affairs, New York

Atherton M (1994) Women philosophers of the early modern period. Hackett, Cambridge

Bowern C, Zentz J (2012) Diversity in the numeral systems of Australian languages. Anthropological Linguistics 54:133–160

Brannon E, Park J (2015) Phylogeny and ontogeny of mathematical and numerical understanding. In: Kadosh RC, Dowker A (eds) Oxford handbook of numerical cognition. Oxford University Press, Oxford, pp 203–213

Carey S (2009) The origin of concepts. Oxford University Press, Oxford

Chalmers D, Clark A (1998) The extended mind. Analysis 58(1):7–19

Chemero A (2009) Radical embodied cognitive science. MIT Press, Cambridge, MA

Churchland P (2017) Neurophilosophy. In: Smith DL (ed) How biology shapes philosophy. Cambridge University Press, Cambridge, pp 72–94

Clark A (1997) Being there: putting brain, body, and world together again. MIT Press, Cambridge, MA

Clark A (2008) Supersizing the mind. Oxford University Press, Oxford

Comrie B (2013) Numeral bases.. The World Atlas of Language Structures. http://wals.info/chapter/131

De Cruz H (2008) An extended mind perspective on natural number representation. Philos Psychol 21(4):475–490

De Smedt J, De Cruz H (2011) The role of material culture in human time representation: calendrical systems as extensions of mental time travel. Adapt Behav 19:63–76

D'Errico F (1998) Paleolithic origins of artificial memory systems: an evolutionary perspective. In: Renfrew C, Scarre C (eds) Cognition and material culture: the archaeology of symbolic storage. McDonald Institute, Cambridge, pp 19–50

D'Errico F, Henshilwood C, Lawson G, Vanhaeren M et al (2003) Archaeological evidence for the emergence of language, symbolism, and music—an alternative multidisciplinary perspective. J World Prehist 17(1):1–70

D'Errico F, Doyon L, Colagé I, Queffelec A et al (2018) From number sense to number symbols: an archaeological perspective. Philos Trans R Soc B. https://doi.org/10.1098/rstb.2016.0518

Dehaene S (1997) The number sense: how the mind creates mathematics. Oxford University Press, Oxford

Everett C (2015) Lexical and grammatical number are cognitively and historically dissociable. Curr Anthropol 57:351

Everett C (2017) Numbers and the making of us. Harvard University Press, Cambridge, MA

Everett C, Madora K (2012) Quantity recognition among speakers of an anumeric language. Cogn Sci 36:130–141

Flegg G (2002) Numbers: their history and meaning. Dover, New York

Franzon F, Zanini C, Rugani R (2019) Do non-verbal systems shape grammar? Numerical cognition and number morphology compared. Mind and Language 34:37–58

Gleitman L, Newport E (1995) The invention of language by children: environmental and biological influences on the acquisition of language. In: Gleitman L, Liberman M (eds) Language: an invitation to cognitive science, 2nd edn. MIT Press, Cambridge, MA

Gomez-Robles A (2019) Dental evolutionary rates and its implications for the neanderthal-modern human divergence. Sci Adv. https://doi.org/10.1126/sciadv.aaw1268

Harman G (1973) Thought. Princeton University Press, Princeton

Hauser M, Chomsky N, Fitch W (2002) The faculty of language: what is it, who has it, and how did it evolve? Science 298(5598):1569–1579

Izard V, Sann C, Spelke E, Streri A (2009) Newborn infants perceive abstract numbers. Proc Natl Acad Sci 106:10382–10385

MacFarquhar L (2018) Mind expander: a philosopher asks where we begin and where we end. The New Yorker. April 2, 2018, pp 62–73.

Malafouris L (2013) How things shape the mind. MIT Press, Cambridge, MA

Marshack A (1991) The roots of civilization: the cognitive beginnings of man's first art, symbol and notation. Moyer Bell Limited, Rhode Island

Menary R (2015) Mathematical cognition: a case of enculturation. In: Metzinger T, Windt JM (eds) Open mind: 25. Frankfurt am Main. https://doi.org/10.15502/9783958570818

Nathan N (2014) Grounded mathematical reasoning. In: Shapiro L (ed) Routledge handbook of embodied cognition. Routledge, Oxon/New York, pp 171–183

Núñez R (2017) Is there really an evolved capacity for number? Trends Cogn Sci 21(6):409–424

Overmann K (2015) Numerosity structures the expression of quantity in lexical numbers and grammatical number. Curr Anthropol 56:638–653

Piantadosi S (2016) A rational analysis of the approximate number system. Psychon Bull Rev 23:877–886

Prinz J (2008) Is consciousness embodied? In: Robbins P, Aydede M (eds) Cambridge handbook of situated cognition. Cambridge University Press, Cambridge, pp 419–436

Renfrew C, Bahn P (2012) Archaeology: theories, methods, practice. Thames & Hudson, London

Rescorla M (2017) The computational theory of mind. In: Zalta EN (ed) The Stanford encyclopedia of philosophy, spring 2017 edn. https://plato.stanford.edu/archives/spr2017/entries/computational-mind

Revell T (2017) Celebrate pi day with 9 trillion more digits than ever before. New Scientist. March 14, 2017

Spaepen E, Coppola M, Spelke E, Carey S, Goldin-Meadow S (2011) Number without a language model. Proc Natl Acad Sci 108:3163–3168

Tomasello M (1999) The cultural origins of human cognition. Harvard University Press, Cambridge, MA

Wheeler M (2005) Reconstructing the cognitive world. MIT Press, Cambridge, MA

Wynn K (1992) Addition and subtraction by human infants. Nature 358:749–750

Xu F, Spelke E (2000) Large number discrimination in 6-month-old infants. Cognition 74:B1–B11

Zahidi K, Myin E (2016) Radically enactive numerical cognition. In: Etzelmuller G, Tewes C (eds) Embodiment in evolution and culture. Mohr-Siebrek, Tübingen, pp 57–71

Chapter 9
Late Pleistocene Dual Process Minds

Murray Clarke

Abstract The global dispersal of prehistoric ancient humans from Africa to North America, and the existence of artistic innovation evidenced in the Late Pleistocene are, by now, parts of a familiar and fascinating story. But the explanation of how our human career was possible cries out for clarification. In this chapter, I argue that dual process theory can provide the needed explanation. My claim will be that the advent of System-2 reasoning running offline, aided by executive cognitive control and language, and facilitated by neural plasticity, made possible the remarkable human dispersal from Africa to North America by way of the Middle East and Asia. System-1 modular adaptations, together with System-2 reasoning, gave rise to the flexible, culturally-informed, mental operations that were essential to shape the psychologically modern mind. The result was, inter alia, the surprising Late Pleistocene dispersal of ancient humans to North America.

Keywords Late Pleistocene hominins · Dual process theory · Evolutionary Psychology · Psychologically modern minds · Reasoning

9.1 Introduction

In the Late Pleistocene, ancient humans became highly encephalized (the average brain sizes of *H. sapiens* increased to 1370 cc—a 17% increase over *H. heidelbergensis*: Dunbar 2014) and underwent powerful cumulative cultural evolution. The latest data suggests that modern humans dispersed from Africa starting around 70,000 years ago (Rito et al. 2019); some eventually arrived in Europe and their descendants are responsible for the famous and truly impressive Upper Palaeolithic cave paintings and figurines which appear there from around 40,000 years ago (Bahn 2012). Others settled elsewhere and eventually gave rise to the complex

M. Clarke (✉)
Department of Philosophy, Concordia University, Montreal, QC, Canada
e-mail: murray.clarke@concordia.ca

© Springer Nature Switzerland AG 2021
A. Killin, S. Allen-Hermanson (eds.), *Explorations in Archaeology and Philosophy*, Synthese Library 433, https://doi.org/10.1007/978-3-030-61052-4_9

cultural traditions we see all over the world today. Their ancient artistic activities left archaeological traces too, and some examples even pre-date the oldest known European cave art (e.g., Bednarik 2013; Aubert et al. 2014, Aubert et al. 2018).

The story of prehistoric human dispersal, artistic innovation, and so on—the 'human career'—is now a familiar one, much discussed by archaeologists and evolutionary theorists, but an explanation of how that career was possible cries out for clarification. In what follows, I avail myself of the resources of dual process theory to suggest that it was the advent of System-2 reasoning running offline (aided by executive cognitive control and language, and facilitated by neural plasticity), which together with System-1 modular adaptations, gave rise to the flexible, culturally-informed mental operations that were necessary to shape the psychologically modern mind. Thus in Sect. 9.2 I frame my discussion by appeal to dual process theories that posit two multi-purpose reasoning systems, System-1 and System-2, respectively: an evolutionarily old mind and an evolutionarily new (or modern) mind. Following that I situate Late Pleistocene minds within the context of recent debates between 'capital EP' Evolutionary Psychologists (such as Pinker, Buss, Cosmides and Tooby, Symons, Daly, among others) and 'lower case' evolutionary psychologists (such as Buller and Hardcastle, Sterelny, Griffiths, Richerson and Boyd, among others) in Sect. 9.4. Whereas 'capital EP' theorists (Sect. 9.3) propose a massively modular conception of the mind, a mind which evolved due to environmental pressures in the Pleistocene, 'lower case' theorists (Sect. 9.4) reject the massive modularity and externalism of 'capital EP' Evolutionary Psychology. Combining aspects of both 'upper case' and 'lower case' evolutionary psychology, I argue for Late Pleistocene dual process minds. In Sect. 9.5 I consider whether Cecilia Heyes' *Cognitive Gadgets* view offers a superior account of mechanisms and suggest that the verdict is not yet out. In Sect. 9.6 I offer concluding remarks.

9.2 Dual Process Theory

It was William James (1890) who first offered a dual process model of the mind in his book, *The Principles of Psychology*. But, the 'two minds' literature has its more contemporary roots in the cognitive revolution of the 1960s and 1970s. For instance, we find this view in work by Evans in the 1970s and it was prominent in the work of Reber (1993) and others in the 1990s. Dual process theorists argue that there are two minds in each cranium. These two minds employ two distinct processing mechanisms and employ different procedures to deal with deductive reasoning, decision-making, and social judgement. As Frankish and Evans put it: "Typically, one of the processes is characterized as fast, effortless, automatic, non-conscious, inflexible, heavily contextualized, and undemanding of working memory, and the other as slow, effortful, controlled, conscious, flexible, de-contextualized, and demanding of working memory" (Frankish and Evans 2009, p. 1). And dual process accounts of learning and memory have also been developed, "...typically positing a non-conscious implicit system, which is slow learning but fast access, and a conscious explicit one, which is fast learning but slow access" (Evans and Frankish 2009, p. 1). Human cognition is then seen as involving two multi-purpose reasoning systems, System-1 and System-2. The former has the fast characteristics and the latter, the slow characteristics. Of course, there are a variety of differences among the positions held in this debate. Table 9.1 lists the typical properties of System-1 and System-2 theories of cognition, drawing on Evans (2009, see p. 34).

It is often claimed that System-1 is early evolving, shared with other animals and includes implicit learning and modular cognition. In contrast, System-2 is recent, uniquely human, and is related to working memory and general intelligence. Reber (1993, ch 3) argued for the primacy of the 'implicit' and proposed that consciousness, as executive conscious thought, was a late arrival in evolutionary terms. Implicit learning of complex stimulus domains is acquired largely without the involvement of any top-down, conscious control. The processes of socialization and language acquisition are two examples of this. On this account, unconscious cognition is the default and dominant system, while conscious cognition is a uniquely human and recently acquired plug-in that does considerably less than one might have expected in the sense that consciousness is not the primary cognitive system and often does not exert executive functions (Evans and Frankish 2009, p. 15). Reber did much to build Table 9.1 by including age of evolution as a general claim, implicit and explicit knowledge, and the idea that implicit but not explicit cognition is shared with other animals (Evans and Frankish 2009, p. 16). Moreover, Reber "…also argued that implicit function had low variability across individuals and was independent of general intelligence" (Evans and Frankish 2009, p. 16). But it was Stanovich who introduced the System-1 and System-2 terminology. That said, System-1 is much more like Cosmides' and Tooby's massive modularity position (e.g., Cosmides and Tooby 2010): a mind that largely, though not entirely, consists of domain-specific inputs and domain-specific computational processors or mini-computers. Whereas Fodor's peripheral modularity combined with a nonmodular, general intelligence capacity at the center of the mind is more like the hybrid System-1/System-2 combination (Fodor 1983). Fodor famously proposed that input

Table 9.1 Typical properties of System-1 and System-2

System-1	System-2
Evolutionarily old	Evolutionarily new
Shared with animals	Distinctively human
Unconscious/preconscious	Conscious
Automatic	Controlled/volitional
Fast/parallel	Slow/sequential
Independent of language	Associated with language
Associative	Rule-based
Belief-based/pragmatic reasoning	Abstract/logical reasoning
Implicit knowledge	Explicit knowledge
Independent of cognitive capacity	Dependent on cognitive capacity
Sub-personal	Personal

systems, like perception, language, and audition, are modular. In contrast, for Fodor, central systems, responsible for belief fixation and practical reasoning, are non-modular. The peripheral modular systems take raw sensory data from its inputs, and transforms them, via sensory transducers, into hypotheses about the layout of objects in the world. This information is then sent to the central system for the purpose of belief fixation. At that point abduction (or inference to the best explanation) generates thoughts that guide action-guiding, modular, output systems that are responsible for behavior.

Perhaps, Richard Samuels's nonmodular, domain-general processors combined with domain-specific information (or the 'Library Model') would constitute an example of a System-2 account of reasoning (Samuels 1998). On this view, Samuels argued that instead of there being many computational mechanisms that are composed of innate, domain-specific cognitive structure, there might be innate, domain-specific bodies of knowledge such as knowledge of language, knowledge of physical objects, and knowledge of number. As he notes:

> Each system of knowledge applies to a distinct set of entities and phenomena. For example, knowledge of language applies to sentences and their constituents, knowledge of physical objects applies to macroscopic material bodies and their behavior, and knowledge of number applies to sets and to mathematical operations such as addition (Samuels 1998, p. 380).

Returning to Evans, his idea is that System-1 and System-2 are responsible for type 1 and type 2 processing, respectively. Type 1 processes are fast, automatic, have high-processing capacity and require only low effort, while type 2 processes are slow, controlled, of limited capacity, and require high effort to utilize. Finally, Evans added a System-3 processor to deal with conflict and control issues concerning the interaction of System-1 and System-2 (Evans 2009). But this seems gratuitous since System-2 can do everything that System-3 does. For example, suppose I am driving a car while talking to someone. An accident occurs ahead of us and I need to respond. This is a case where the automatic System-1 driving capacity may conflict with the System-2 intentional, conscious conversation. Evans thinks one needs an additional processor to resolve such conflicts. This is because working memory, since it is involved with System-2, cannot be called upon to monitor such hazards. In effect, working memory would have to be able to monitor a system using working memory. As he claims: "The system responsible for monitoring hazards and recruiting working memory cannot be working memory itself. Nor should we confuse the 'executive function' property of working memory with the kind of control I am talking about here" (Evans 2009, p. 48). But this overlooks the fact that System-2 is flexible enough to monitor its own workings with the aid of working memory. If that were not true, we would not be able to cut a conversation short in order to, for instance, run to teach a lecture. But we do things like this all the time. Surely, that is what we mean when we say that someone is multi-tasking. At any rate, for present purposes, I propose to drop the idea of a third processor.

An important point that Samuels (2009) makes is that each system involves clusters of co-instantiating properties. That is, processes that exhibit one property typically possess the other properties. This matters since the idea that clusters exist suggests an underlying suite of mechanisms sub-serving such co-instantiation. This inference then paves the way to posit a "bipartite division between cognitive mechanisms" (Samuels 2009, p. 131). In short, the idea is that there are natural kinds that underwrite cognition (where Samuels construes 'natural kind' in Richard Boyd's sense as homeostatic property clusters (Boyd 1991, 1999)). According to Samuels, dual process theorists endorse two claims:

1. Dual-Cluster Thesis: cognitive processes tend to exhibit either the S1 or S2 clusters.
2. Dual-Systems Thesis: there is a division in our cognitive architecture—a division between cognitive systems—that explains this clustering effect (Samuels 2009, p. 132)

Samuels thinks that there are two ways of developing these generic claims: these are the Token and Type Theses. The Token Thesis maintains that there are just two *particular* cognitive mechanisms or systems. The System-1 mechanism sub-serves cognitive processes that exhibit the S1-property cluster. The System-2 mechanism sub-serves cognitive processes that exhibit the S2-property cluster. On this thesis, each human mind exhibits a fundamental, bipartite division into *just* these two *particular* systems. This view is not very promising; Samuels provides no examples of defenders of such a simplified Token Thesis. But he does cite Sloman as someone who defends a restricted Token Thesis concerning just reasoning (Sloman 1996). In contrast, according to the Type Thesis, each mind is constructed out of two types or *kinds* of cognitive system: systems of the first kind sub-serve processes that tend to exhibit the S1-cluster (Type 1 Systems); systems of the second kind sub-serve processes that tend to exhibit the S2-cluster (Type 2 Systems). But there is no old/new mind dichotomy or 'Token Thesis' at play, for these types are multiply realizable. The Token Thesis implies the Type Thesis but not vice-versa. As such, the Type Thesis is logically weaker than the Token Thesis. Samuels defends the Type Thesis as more plausible than the Token Thesis principally because there seem to be *many* System-1 and System-2 devices in the mind. This is a claim that Evans also endorses. For instance, the human visual system involves many subsystems for depth perception, color identification, and categorization (Palmer 1999). And, these subsystems themselves decompose into smaller units, and so on. As Samuels suggests:

> If anything like this story is correct—and virtually all vision scientists assume it is—then there are obviously going to be more than two systems. Indeed, if we focus on lower levels in the hierarchy of decomposition, we should expect to find loads of them—still more if we are permitted to sum mechanisms across all different levels in the decomposition hierarchy (Samuels 2009, pp. 133–134).

Clearly, this story is incompatible with the idea that there are just two particular systems that compose human cognition, an old mind and a new mind. For instance, the visual system and the reasoning system are distinct and so do not decompose together as part of a single System-1 and part of a single System-2. Suppose that Samuels is correct, and the Type Thesis is true. That is, there are many System-1 and System-2 devices that make up the mind. I take this claim to be largely noncontroversial, but not without critics, in both psychology and cognitive science (and will only assume what it requires in what follows). Critics have questioned the evidence on which dual-process accounts are based and pointed to some inconsistencies among the various models (Kruglanski and Gigerenzer 2011). Evans and Stanovich have produced a very plausible response to these objections (Evans and Stanovich 2013). However, I will not engage with this debate in this chapter. Rather, I ask, how ought we fit this account into a picture of hominin evolution? As the name would suggest, the 'old mind' is the one forged in the deep past of the hominin lineage. I now turn to the work from the 1980s and 1990s by Leda Cosmides in order to determine what the nature of this old mind might have been. Note that I will use the terms 'old mind' and 'new mind' in what follows in Samuel's sense as involving multiple old and new, i.e., System-1 and System-2, devices.

9.3 'Capital EP' Evolutionary Psychology

In the evolutionary-psychological literature, there have been two waves of theorizing. These are typically referred to as 'capital EP' Evolutionary Psychology and 'lower case' evolutionary psychology. As mentioned at the outset of this chapter, the first wave was spearheaded by Leda Cosmides and John Tooby in the 1980s, and popularized by Stephen Pinker in the 1990s. The second wave of theorizing began to take shape in the mid-to-late 1990s and gained steam in the new millennium with work by Buller, Sterelny, Griffiths, and others.

According to advocates of the first wave, i.e., Evolutionary Psychology, the mind comprises 'modules' or domain-specific programs: context- and content-specific computational/informational processors. According to Cosmides and Tooby:

> Our minds consist of a large number of circuits that are functionally specialized. For example, we have some neural circuits whose design is specialized for vision. All they help you do is see. The design of other neural circuits is specialized for hearing. All they do is detect changes in air pressure, and extract information from it (Cosmides and Tooby 2013 pp. 89–90).

To be sure, Cosmides and Tooby are referring to our current domain specific computational processors or circuits. These would include the social-exchange/cheater detection module, the spatial reorientation module, and a variety of other modules. As Tooby and Cosmides wrote in their foreword to Baron-Cohen (1995):

> …our cognitive architecture resembles a confederation of hundreds or thousands of functionally dedicated computers (often called modules) designed to solve adaptive problems endemic to our hunter-gatherer ancestors. Each of these devices has its own agenda and imposes its own exotic organization on different fragments of the world. There are specialized systems for grammar induction, for face recognition, for dead reckoning, for construing objects and for recognizing emotions from the face. There are mechanisms to detect animacy, eye direction, and cheating. There is a 'theory of mind' module…a variety of social inference modules…and a multitude of other elegant machines (Tooby and Cosmides [in Baron-Cohen 1995, pp. xiii–xiv]).

Famously, Stephen Pinker suggested that the mind can be considered to be a collection of instincts adapted for solving evolutionarily significant problems: the mind as a 'Swiss Army knife', consisting of many task-specific tools as modules, rather than a general-purpose computer (Pinker 1994). As such, tasks traditionally classified as central processes could well be modular, such as intuitive mechanics: that is, knowledge of the motions, forces, and deformations that objects undergo. What Tooby and Cosmides, on the one hand, and Pinker, on the other hand, did not have much to say about was the evolutionary timeline concerning when such modules came on board. In fact, Cosmides reported to me that she had trouble getting the journal, *Cognition*, to accept even one footnote referring to evolution into her 1989 National Science Award winning paper on the social-exchange module. At any rate, her focus, given her training, was on running psychology experiments, not establishing evolutionary timelines as a physical anthropologist might do. Instead, she appealed to the methodological notion of 'reverse-engineering' to stand in for a detailed examination of the archaeological and paleoanthropological record. Reverse engineering, in turn, amounted to the idea that one could cook up a story about the environmental demands on our Pleistocene ancestors and then construct hypotheses concerning what sort of domain-specific program would have been crucial for solving such problems. At that point, experiments would be completed to determine if such processors exist in contemporary humans. And many of those experiments seemed to show that they did exist (Cosmides and Tooby 1994, 1996, 1997a, 1997b; Barkow et al. 1992). But it is an important question to ask, at what point in evolutionary history did the complex functional adaptations that she was searching for, many of which were as complex as the human eye, come on board? Cosmides' answer to this was both decidedly vague and gradualist in spirit: over a *very* long period of time during the Pleistocene. The advent of farming and agriculture after 10,000 kya would, for instance, be within much too short a period of time to explain the evolution of complex functional adaptations. As Cosmides notes:

> Complex, functionally integrated designs like the vertebrate eye are built up slowly, change by change, subject to the constraint that each new design feature must solve a problem that affects reproduction better than the previous design. The few thousand years since the scattered appearance of agriculture is only a small stretch in evolutionary terms, less than 1% of the two million years our ancestors spent as Pleistocene hunter-gatherers. For this reason, it is unlikely that new complex designs—ones requiring the coordinated assembly of many novel, functionally integrated features—could evolve in so few generations (Barkow et al. 1992, p. 5).

No doubt she was correct about this. However, some philosophers, such as Stephen Downes, have taken her to task for restricting the relevant period where most adaptations of this sort took place to just the Pleistocene (Downes 2013). But, to be fair to Cosmides, she didn't say it all took place in a week! We are talking about 2.6 million years. And, she never said that evolution did not take place both before and after the Pleistocene. And, she never said that there was just one 'environment of evolutionary adaptation' or EEA. The notion of an EEA is a relative, contextual, and tensed notion (see Clarke 2004, pp. 57–69). At any rate, Cosmides allows that complex functional adaptations took place throughout the Pleistocene and included computational processors, like the cheater detection module, that operated in social exchanges as exhibited in her experiments with the Wason selection task (Cosmides 1989; Cosmides and Tooby 1996).[1] In contrast, Downes argues, with Richerson and Boyd, that the evolution of lactose tolerance occurred in the last 10,000 years since the domestication of animals via gene-culture coevolution. Since this occurred after the Pleistocene Epoch, it shows that human evolution of key human mechanisms did occur in the Holocene Period. Of course, Cosmides and Tooby can argue that they are only making a claim about human psychological mechanisms: complex functional mental adaptations. Hence, this example does not refute their claim, but it does show that evolutionary change in humans can occur more rapidly than Cosmides and Tooby allow for. Beyond time-line issues, it is really concerns about causality that are at the heart of Downes', and Richerson and Boyd's, and Sterelny's rejection of Cosmides and Tooby's approach. They think that Cosmides and Tooby adopt a top-down causal picture where universal complex functional mental adaptations get triggered by environmental cues that lead to cultural traits such as forms of cooperation and moral norms. In contrast, as Downes points out (about Richerson and Boyd): "they maintain '[C]ulturally evolved traits affect the relative fitness of different genotypes in many ways', and this leads them to oppose the view that 'cultural evolution is molded by our evolved psychology, but not the reverse" (Downes 2013, p. 97). For instance, niche construction provides an example of how a bottom-up approach works. Nest building, dam building, and the building of colony dwelling structures provides an environment within which one's genotype will be altered rather quickly in evolutionary terms. The developmental point here is that the processes of niche construction and gene-culture coevolution cause one's genotype to be altered. That is, as Downes points out, "In Kim Sterelny's hands, human downstream niche construction provides numerous ways in which to impact upon evolutionary change in the human lineage, specifically with respect to the production of novel behavioral repertoires" (Downes 2013, p. 98). But do Cosmides and Tooby actually deny the developmental point that the processes of niche construction and gene-culture coevolution cause one's genotype to be altered? The answer is decidedly 'No'. In their 1992 article, "The Psychological Foundations of Culture", they are at pains to deny this and explicitly argue that genes and the environment are equal partners in the production of behavior and the expression of culture. As they note:

> …every feature of every phenotype is fully and equally codetermined by the interaction of the organism's genes (embedded in its initial package of zygotic cellular machinery) and its ontogenetic environments—meaning everything else that impinges on it. By changing either the genes or the environment any outcome can be changed, so the interaction of the two is always part of every complete explanation of any human phenomenon. As with all interactions, the product simply cannot be sensibly analyzed into separate genetically deter-

---

[1] See Clarke (2004, ch 4–5), for a detailed summary of this literature. The Wason Selection Task is the most intensively studied reasoning task in the history of psychology. It involves determining whether subjects reason in accordance with modus tollens, a standard deductive reasoning rule, in a variety of experimental set-ups.

mined and environmentally determined components or degrees of influence. For this reason, everything, from the most delicate nuance of Richard Strauss's last performance of Beethoven's Fifth Symphony to the presence of calcium salts in his bones at birth, is totally and to exactly the same extent genetically and environmentally codetermined. 'Biology' cannot be segregated off into some traits and not others (Tooby and Cosmides 1992, pp. 83–84).

Clearly, Tooby and Cosmides reject the top down model and, instead, accept what might be called the 'Codetermination Model'. That is, they think that nature and nurture play equal causal roles in the production of behavior and the adaptations that are selected for. I conclude that their critics have misread their position. That said, I think that Cosmides and Tooby ignored some of the complex ways in which cultural products and niche construction can impact one's genotype. But such examples do not refute Evolutionary Psychology, but rather, provide amendments to, and an enrichment of, that picture.

But what accounts for the genesis of the new mind (System-2)? I suggest that this type of mind is just the sort of thing that might help to explain the cumulative cultural evolution of the Late Pleistocene. If such a mind slowly developed from, say, 200,000 kya, and was in place by the time humans dispersed from Africa, and ran offline, then the reasoning flexibility needed for adaptation to many new environments would be possible. Lots of planning would be an essential feature of life both with respect to migration and determining how to hunt in various environments. That a new kind of mind evolved during this period and functioned, in a discrete area of the brain, alongside old minds is theoretically and empirically possible. It must also be emphasized that the variety of new environments encountered in this period placed novel constraints on how evolution would occur. The very nature of this ever changing environment might well speed up the rate of evolutionary change well beyond previous, more incremental, evolutionary changes. Of course, another approach and the one favored by Carruthers, has it that virtual System-2 reasoning emerges from System-1 via language in the sense that a virtual feedback loop occurs that is produced by our capacity for the mental rehearsal of action schemata (Carruthers 2006). On Carruthers' view, there is no actual System-2 hardware but only mental capacities that produce effects that mimic such a System-2 reasoning processor. As Frankish and Evans note: "In the case of utterances, Carruthers argues, such rehearsal generates auditory feedback (inner speech) that is processed by the speech comprehension subsystem and tends to produce effects at the modular level appropriate to the thoughts the utterances express" (Evans and Frankish 2009, p. 22). The result is that a massively modular architecture can solve the major problem of explaining how something approximating inference to the best explanation can occur without positing a distinct domain-general thinking device. But Carruthers' view is highly speculative and has not received any empirical support to date. In contrast, in my view, the idea of multiple System-2 processors has become the default assumption in psychology. At any rate, all of the features of System-2 cognition appear to be evidenced in spades in, for example, the representational art preserved in the caves of Upper Palaeolithic Europe. So let us turn to 'lower case' evolutionary psychology to see if progress can be made by adopting aspects of that perspective.

9.4 'Lower Case' Evolutionary Psychology

Is there an evolutionary story about how System-2 reasoning came on board? In my view: yes; a plausible empirical, neurological account of this sort of cognitive functioning is possible. It is here where the new, 'lower case' evolutionary psychologists, Buller, Sterelny, among others, can help explain how a modern mind came on board by way of neural plasticity. Buller (2005) provides a powerful critique of 'capital EP' Evolutionary Psychology. Instead of a long list of unconscious, modular adaptations coming to fixation during the Pleistocene, producing a mental 'Swiss Army knife', Buller argued that the mind is not written in stone, or the Stone Age! There is no Swiss Army knife. Instead, the mind is enormously flexible, largely the product of learning and is forged by a brain that is constantly changing due to a 'proliferate and prune' developmental process. Jeffrey Elman's 'proliferate and prune' account of brain development has the resources to explain how a System-2, domain-general reasoning processor came to be (Elman et al. 1996). The philosopher Stephen Davies (2012) captures the basic approach that more recent evolutionary psychologists have adopted when he says:

> ...contemporary evolutionary psychologists are now more likely to stress humans' neural and behavioral plasticity in response to contingency, variety, and change. They emphasize how highly developed we are socially and culturally. They stress our cognitive differences as well as commonalities. And they are more liable to hold that evolution works in tandem with culture rather than in opposition to it (Davies 2012, p. 41).

Buller and Hardcastle (2005) work in this more recent tradition and they have appealed heavily to the work by Jeffrey Elman and his colleagues on the brain. On Elman's view the brain is not genetically specified, there is no set of special-purpose brain circuits constructed by a genetic program that specifies its design. Instead, Buller and Hardcastle via Elman claim that:

> From the time the human brain begins to develop in utero, at about twenty-five days after conception, it increases by a remarkable 250,000 cells per minute, and this rate of cell production continues until birth. The production of these cells takes place in two different 'zones', the ventricular and the subventricular. The cells that make up the evolutionarily oldest parts of the brain are produced in the ventricular zone. These are the cells that make up the midbrain and the limbic system, which are regions of the brain controlling motor coordination, sexual response, and emotion (such as the fear response). In contrast, the cells produced in the subventricular zone make up the evolutionarily most recent addition to the brain, the neocortex, which carries out the 'higher' cognitive functions (Buller and Hardcastle 2005, p. 131).

Evolutionary Psychologists like Cosmides and Tooby want to say that modules are complex information-processing mechanisms which execute sophisticated 'Darwinian algorithms' in solving adaptive problems (see, e.g., Buller and

Hardcastle 2005, p. 130). As such, these modules are likely to be found in the neo-cortex. The neocortex is made up of cells in the subventricular zone. The major structures of the neocortex and their primary subdivisions are under strict genetic control; however, the fine-grained structures that perform special cognitive functions are not under such strict genetic control. These structures are not added on from birth to adulthood. Instead, the adult brain contains fewer cells and connections than the infant brain. The mechanism of subtraction is cell competition and cell death (Buller and Hardcastle 2005, p. 132). In short, a process of 'pruning' occurs. The process of pruning the excess connections and cells forms the brain circuits that carry out a variety of specialized cognitive functions. Crucial to the process of pruning is the environmental inputs that result in the development of the brain. Likewise, spontaneous neural firings internal to the brain are important to the development of specialized circuits such as the retina, without which normal vision will not develop. According to Elman, cortical development depends both on *additive events*, which overproduce neurons and connections, and *subtractive events*, which selectively eliminate neurons and connections. So additive events provide a mass of clay, which subtractive events 'sculpt' into functional form. The upshot, says Buller and Hardcastle is that:

> The process of 'proliferate and prune' can produce relatively stable brain circuits that specialize primarily in particular information-processing tasks. In other words, the process of proliferate-and-prune can produce brain circuits that closely resemble Evolutionary Psychology's postulated modules. Some of these circuits even function to solve adaptive problems, and they can even be produced with some regularity across populations and down lineages, even more closely resembling Evolutionary Psychology's postulated adaptive-problem-solving modules. However, the degree to which different brains develop similar cortical circuitry is due more to their encountering similar environmental inputs during development than to a 'genetic program' that 'specifies' recurrent developmental outcomes. Thus, although the adult brain can be characterized by 'modular' information-processing structures, these are environmentally shaped, not 'genetically specified,' outcomes of development. For it is primarily environmental inputs to the brain that determine the subtractive events that shape its cognitive-processing structures (Buller and Hardcastle 2005, p. 134).

This sort of neural plasticity leads to phenotypic plasticity; the brain literally remakes itself, as they note:

> This kind of flexibility entails not only the possibility of multiple developmental outcomes, which are contingent on the environment, but also the possibility of change or reorganization of structure in response to changes in the environment. In other words, the concept of neural plasticity refers not to a genotype's ability to produce different adaptive phenotypes, but to the brain's ability to remake itself in response to changing environmental demands. This makes neural plasticity an instance of phenotypic plasticity (Buller and Hardcastle 2005, p. 137).

For instance, Evolutionary Psychologists argue that there is a face recognition module that develops from birth onward (McKone 2009; McKone and Palermo 2010; Buller and Hardcastle 2005, p. 150). It is a fixed, developmental mini-computer that allows the young child, once the module is triggered by the sight of its mother, to hone in on those faces that it must rely on for survival. In contrast, on a neural plasticity view of face recognition, the child has a nativist pre-set that picks up on three blobs. Later, the child, given suitable experiences with the mother, begins to distinguish her from other faces in the crowd. However, that same pre-set might have been developed in a different environment to pick up on other sorts of faces and then it would have become very good at discerning those types of faces, say of bears. As such, there is no completely predetermined module that unfolds in developmental time. Of course, Machery and Barrett (2006) have challenged some of the claims of Buller and Hardcastle but I do not intend to arbitrate such disputes here.

Other researchers who would question Cosmides and Tooby include Sterelny (2013) and Richerson and Boyd (2005), who say that culture is a more important force in evolution than has traditionally been understood. The knowledge that a generation passes on to another generation is a much more potent and central force than has been acknowledged. Richerson and Boyd ask us to consider the impact of major temperature changes that have occurred over hominin evolution. Suppose a group of hominins living in Madrid at one time, move to Scandinavia 100 years later. One might expect that individual learning would trump imitation with respect to adapting to the new environment. But, as they say:

> Odd as it may seem, in many kinds of variable environments, the best strategy is to rely mostly on imitation, not your own individual learning. Some individuals may discover ways to cope with the new situation, and if the not-so-smart and not-so-lucky can imitate them, then the lucky or clever of the next generation can add other tricks. In this way the ability to imitate can generate the cumulative cultural evolution of new adaptations at blinding speed compared with organic evolution. …When lots of imitation is mixed with a little bit of individual learning, populations can adapt in ways that outreach the abilities of any individual genius (Richerson and Boyd 2005, p. 269).

And, the creation of a niche, such as a beaver dam, illustrates how organisms can create the environment within which their species evolves. Sterelny (2013) has argued that this can lead to an 'arms war' where conspecifics evolve in response to the response of others. The cumulative iteration of cultural responses to this arms war in species such as humans indicates that culture is itself a central driver of evolution; culture must be seen as central to evolution and not just its handmaiden. As Sterelny states: "Humans interact with their physical, biological, social and technical environment; we molded our world as well as responding to it. These feedback loops explain the rapidity of the hominin divergence from the chimp lineages" (Sterelny 2013, p. 276). He also rejects the 'key innovation' pictures of the rise of the hominins that suggest that it was one key change (such as language, reproductive cooperation or cooking) that explains behavioral modernity (see also Sterelny 2011). On his view, many factors were implicated. Sterelny offers the 'apprentice learning model' to explain part of the rise of hominins (Sterelny 2012). It differs from the key innovation pictures of the rise of hominins in three ways, because:

1. There is no key innovation that explains the rise of hominins; the australopithecines diverged from their last common ancestor with chimpanzees and bonobos in a number of "respects, and that divergence intensified and accelerated through positive feedback" (Sterelny 2013, p. 276). However, enhanced social learning via the apprentice learning model is very important. The younger members of

hominin society "learned skills, norms and factual information by doing," from the previous generation of experts and so "… the flow of information (and misinformation) across social networks and between generations is one of the most distinctive features of social life" (Sterelny 2013, p. 276).

2. The model situates agents in their local social and physical environment. As such, the focus is taken off hominin cognition and genetically canalized features. As he notes: "The hominin career depends on assembling and stabilizing environments (especially developmental environments) that support technical and social capacities, not just on the evolution of genes that help build the right wetware" (Sterelny 2013, p. 276).

3. The model is dynamic. Much of Evolutionary Psychology has focused on the end result, adaptations. Instead, Sterelny focuses on the incremental evolution of the distinctive forms of hominin social learning.

I don't doubt that Sterelny's picture of hominin history is right in many ways: learning skills from experts was, and is, crucial. However, I would want to put more emphasis on the role of genetic evolution, cranial size increase, and other nativist features of the hominin lineage than does Sterelny, Richerson and Boyd, and others. Hence, I largely agree with Sterelny but would emphasize the nativism aspect more than he does. That is why I think we need an account that explicitly appeals to the notion of a System-2 account of inference that is dependent on the neural plasticity of the brain. We need to explain how we arrived at a mind that is far more flexible than expected and a mind that is the result of neural plasticity informed by culture, understood broadly. The neural plasticity of the modern, System-2, mind is exactly the sort of mind needed to negotiate the variable environments and challenges that our ancestors faced. This does not mean that the System-1, inflexible old mind vanished. Rather, each type of mind had a role to play and eventually, they became intertwined in our contemporary mind, though, plausibly, without the aid of Evans' postulated System-3 processor. Hence, I think we can expect that a blend of 'upper case' and 'lower case' evolutionary psychology is needed to explain the data. My own view lies between these approaches, but is tilted towards the 'upper case' evolutionary psychology perspective.

9.5 Heyes' *Cognitive Gadgets*

A new position and approach to the larger debate about the nature of the mind has recently been defended by Cecilia Heyes in her book *Cognitive Gadgets* (Heyes 2018). Heyes calls her approach 'cultural evolutionary psychology'. Heyes wants to provide a largely cognitive science-based account of how natural selection operates on culture to produce cognitive gadgets rather than the cognitive instincts that, for instance, 'upper case' Evolutionary Psychology posited.

As such, her position is even more at odds with mine than is 'lower case' evolutionary psychology (à la Sterelny, Richerson and Boyd) and so provides a useful contrast. In her book, Heyes argues that genes provide only a starter kit (Heyes 2018, ch. 3) for the eventual development of 'Big Special' psychological attributes such as language, theory of mind, imitation, and selective social learning. As she says:

> ...this book suggests that, although the vast majority of adult humans have these Big Special cognitive mechanisms, we do not genetically inherit programs for their development. Rather, we genetically inherit 'Small Ordinary' psychological attributes: the propensity to develop relatively simple mechanisms that closely resemble those found in other animals, including chimpanzees. Genetic evolution has tweaked the human mind (Heyes 2018, p. 53).

What is genetically provided is a highly social temperament in humans, thus allowing humans to 'tolerate, seek, and thrive' on the company of other agents. This feature allowed humans to acquire knowledge, skills, and cognitive mechanisms from other agents. In addition, attentional biases served to focus the attention of human infants on other agents from birth. As she notes: "We are driven from very early infancy to look at biological motion and faces, and to listen to human voices" (Heyes 2018, p. 53). Thirdly, humans have powerful central processors: "...mechanisms of learning, memory, and control that extract, filter, store, and use information. Each of these processors is domain-general, crunching data from all input domains using the same set of computations, and taxon-general, present in a wide range of animal species" (p. 53). Given central processors with lots of speed and capacity, such processors are: "Shaped and fed throughout development by the torrents of culturally evolved information flowing in from other agents, domain-general central processors not only capture this information but use it to build new, domain-specific cognitive mechanisms—the Big Special mechanisms that make humans such peculiar animals" (p. 54). Thus, Big Special mechanisms are set up by cultural evolution assisted by "souped-up, genetically inherited mechanisms of learning and memory, using raw materials that are, from birth, channeled into infant minds by genetically inherited temperamental and attentional biases" (p. 54).

The idea that there are genetically acquired central processors that lead to the production of domain-specific culturally acquired *beliefs* and *methods* is not new. We find this idea in Dennett's arguments utilizing Dawkins' meme concept (e.g., Dennett 1991, 1995, 2017), and in Richerson and Boyd's notion of culturally-acquired methods for producing, for instance, canoes (Richerson and Boyd 2005). What is new with Heyes is the idea that it is not just beliefs and methods that are culturally acquired, but *mechanisms*. As Heyes puts this: not just 'grist' (beliefs, behaviors, and behavioral dispositions, etc.) but 'mills' are involved. Mills are domain-specific mechanisms handed down through cultural evolution. The key insight here is the idea that mechanisms can be handed down through cultural evolution. In contrast, the 'California School' of cultural evolution, found in the work of Richerson and Boyd and Henrich, holds that all mechanisms are genetically acquired. This is also true of Dennett's position. As Heyes notes with respect to the issue of how units of behavior and ideas are to be divided into discrete chunks of psychological grist or unitized such that they are explicable by Darwinian selection: "However, it is very hard to answer in a principled and rigorous way questions about the unitization of psychological grist—ideas and behavior…. Arguably, this is the reason why, after thirty years of conceptual development, memetics still has not been converted into an empirical research program, and its hypotheses still rarely inspire observation and experiment" (Heyes 2018, pp. 37–38). It would seem that absent some physical mechanism that would imitate what genes provided for natural selection on the standard evolutionary story of Evolutionary Psychology, there was only loose talk about cultural evolution to explain how the memetics story went. By focusing on mechanisms, Heyes hopes to win the day by providing a physical substrate that could be replicated in subsequent generations via cultural evolution. But wait, actually, the idea that there might be domain-general, content-neutral, capacities that lead to domain-specific mechanisms is not new either. One position that Cosmides (1989) contrasts with her social contract view concerning the Wason Selection Task was the pragmatic reasoning schema position of Cheng and Holyoak (1985). The Cheng and Holyoak view is that humans possess a content-neutral, domain-general, inductive reasoning capacity that leads to culturally acquired pragmatic reasoning schemas. That is, Cheng and Holyoak were defending exactly the sort of view that Heyes' champions: a basic learning kit that includes a domain-general, inductive, learning processor that leads to a Big Special domain-specific pragmatic reasoning mechanism. Hence, as I said in my 2004 book, Cheng and Holyoak:

> …push the same explanatory variables back one step by suggesting that humans reason using 'pragmatic reasoning schemas' that were induced through recurrent experience within goal-defined domains (Cosmides 1989, p. 191). The schemas are content-dependent, whereas, the inductive cognitive processes that give rise to such schemas are content independent. Differential experience is a key variable used to explain which schemas are built and which are not. For instance, the following is a permission schema: "Rule 1: If the action is to be taken, then the precondition must be satisfied." Cheng and Holyoak think that most of the thematic problems that subjects have done well on are permission rules, such as rule 1. All social contract rules involve permission, but not all permission rules involve a social contract. This, as Cosmides notes, is because the social contract statement: "If one is to take the benefit, then one must pay the cost," entails the permission rule: "If one is to take action A, then one must satisfy precondition P." But the reverse is not true. All benefits taken are actions taken, but not all actions taken are benefits taken. As Cosmides notes: "A permission rule is also a social contract rule only when the subjects interpreted the 'action to be taken' as a rationed benefit, and the 'precondition to be satisfied' as a cost requirement" (Ibid, p. 237). This makes the domain of the permission schema larger than that of the social contract algorithms. This difference has empirical consequences. In particular, permission rules that are not social contracts should result in content effects on the Wason selection task; according to social contract theory, this should not happen. In part 2 of her study, Cosmides establishes the falsity of the pragmatic reasoning theory's permission rules in that non-social contract permission rules fail to result in content effects on the Wason selection task (Clarke 2004, p. 96).

As I noted in the book:

> For instance, in experiment 5 subjects attempted to solve two Wason selection tasks. Each test booklet involved two rules, one occurring with a social contract surrounding story and the other rule occurring with a non-social contract permission rule story. The rules were: "If a student is to be assigned to Grover High School, then that student must live in Grover City," and "If a student is to be assigned to Milton Hugh School, then that student must live

in the town of Milton." The surrounding story for the social contract problem had it that being assigned to Grover High was a benefit (compared to be assigned to Hanover High), while living in Grover City is a cost (compared to living in Hanover). The surrounding story for the non-social contract problem gave the rule a social purpose: following the rule will allow the board of education to develop the statistics needed to assign teachers sensibly to each school. No mention was made of costs or benefits with respect to the rule in this second case; both places and schools are, therefore, portrayed as being of equal value (Clarke 2004, pp. 96–97).

In experiment 5, 75% of subjects selected the right answer concerning the social contract rule. In contrast, only 30% of subjects selected the correct answer where a non-social contract permission rule was in play. This is precisely what one would expect if Cosmides' social contract view was correct while Cheng and Holyoak's pragmatic reasoning schema view was incorrect. This suggests that the cost-benefit representations of the social contract theory have psychological reality, while the induction-based pragmatic reasoning schemas do not. On the Cheng-Holyoak view one would have expected good reasoning on a permission schema rule regardless of whether a social contract surround story was provided. But this did not happen. Instead, only permission rules offered in the context of a social contract surround story resulted in good reasoning. Without such a social contract surround story, the percentage of answers that conform to the propositional calculus was low, only 30%. So, in experiment five subjects would have to carry out a Wason Selection Task given a social contract surround story and the rule that says: "If a student is to be assigned to Grover High School (P), then that student must live in Grover City (Q)". The subjects would then be asked to determine which of four cards need to be turned over to see if any violations of this rule have been committed. The four cards would say on their face: Grover High School (P), Milton High School (not-P), Grover City (Q), and Milton (not-Q). One side of each card refers to a school and the other side refers to a location. Of course, the P's and Q's would not appear in the rule or on the cards, I add these to clarify the logic involved. To satisfy the propositional calculus, the correct result is that one must turn over the P and not-Q cards only in order to see if there is a not-Q on the flipside of the P card or a P on the flipside of the not-Q card. This is because a conditional is only false when the antecedent is true and the consequent is false. The result, as mentioned, was that 75% of students selected the correct cards when this rule occurred with the social contract surround story. In contrast, when the same rule occurred with a non-social contract surround story, only 30% of the subjects selected the correct cards. According to Cheng and Holyoak, given that both selections tasks involve the same pragmatic reasoning schema permission rule, both selection tasks should have produced the same very good results. But it was only when a social contract surround story was added to the permission rule that good results ensued. It was on this basis that it was determined that the pragmatic reasoning schema view was likely false. Of course, other formally identical experiments that Cosmides performed produced the same result consolidating her conclusion that the content effects, i.e., that different content produces different results, on the Wason Selection Task can only be explained by appeal to the social contract view. That is, social contract content produces better reasoning than non-social contract content on Wason Selection Tasks.

The upshot of part two of Cosmides' study is that the pragmatic reasoning schema position is not supported by the evidence. The claim that humans innately possess a domain-general, content-independent, inductive reasoning capacity that is employed to produce a domain-specific, content dependent, pragmatic reasoning schema mechanism that is learned through recurrent experience was not supported by the evidence. That is, people do not reason better on the Wason Selection Task when a pragmatic permission schema is built into the task but only when a social contract is. That is what explains the content effects on the Wason Selection Task; that is why some tasks produce reasoning that is in better conformity to the deductive norms of logic. The univocal, over-arching domain-general processors that are presupposed by the pragmatic reasoning schema view (and by Heyes) have no psychology reality. Cosmides also showed that the associationist position (of Wason 1968 and Johnson-Laird 1983), according to which it is familiarity that explains the content effects on the Wason Selection Task, fails as it is incompatible with her data. The reason that these two views fail, according to Cosmides, is that they share the false common assumption that the same cognitive processes govern reasoning about different domains. That is, they share the false assumption that domain-general mechanisms govern all reasoning. In contrast, her conclusion was that "…humans possess special-purpose, domain-specific, mental algorithms that help us solve important and recurrent adaptive problems" (Clarke 2004, p. 98). As Cosmides pointed out: "The more important the adaptive problem, the more intensely selection should have specialized and improved the performance of the mechanism for solving it" (Cosmides 1989, p. 193). Cosmides' defense of a domain-specific account of the content effects on the Wason Selection Task provided a powerful argument against the domain-general accounts of the associationist view and the pragmatic reasoning schema view.

At any rate, the pragmatic reasoning schema view, at least with respect to human reasoning, is exactly the kind of position that Heyes wants to defend by talking about cognitive gadgets. It would be an example of what Heyes calls an "explicitly metacognitive social learning strategy" (Heyes 2018, p. 94), which is a kind of selective social learning. As such, it is similar to examples that Heyes gives like 'copy the boat builder with the largest fleet' and 'copy digital natives' (people born since the advent of the Internet). The permission rule, "If one is to take action A, then one must satisfy precondition P", is just such a metacognitive social learning strategy that promotes cultural inheritance by enhancing the 'exclusivity, specificity, and accuracy of social learning'. On Heyes' view, such permission rules would be learned from other people and applied at the output stage. Such permission rules would be cognitive gadgets.

The difference is that Cosmides actually produced an award-winning article that tested a cognitive gadget account against a so-called 'High Church' Evolutionary Psychology, social contract view. Cosmides won this award exactly because the evidence that I just detailed forcefully supported the social contract view. Of course, it does not follow that there are no cognitive gadgets of the sort that Heyes defends concerning the other topics that she considers: language, imitation, theory of mind. Nevertheless, Cosmides did actually succeed in testing her theory against the familiarity theory of associationists like Wason (1968) and Johnson-Laird (1983), and against the pragmatic reasoning schema position of Cheng and Holyoak (1985) in a precise way with stunning success. Heyes has not tested her theory in a way that begins to match the precision of the study that Cosmides did in 1989. This is because Cosmides offered the scientific community something close to a crucial experiment of the sort that is rarely seen in science: a test that successfully pits one theory against others such that the outcome provided largely incontestable evidence for just one theory. This ought to give us pause to reflect. Often, Heyes claims that the cognitive gadget theory is compatible with the evidence, but she does not deny that the Chomsky nativist accounts of language or the High Church Evolutionary Psychology position on face recognition can accommodate the data too. This is certainly a reasonable foundation to base a theory on. But nowhere does Heyes offer a crucial experiment of the sort that Cosmides did and that is precisely why Cosmides' account of reasoning is superior to the Heyes-style cognitive gadget account of learning that is embedded in Cheng and Holyoak's pragmatic reasoning schema view. It should be noted, ironically, that Heyes also takes a shot at what she sarcastically calls 'High Church' Evolutionary Psychology when she says, while extolling the virtues of her approach, "We don't have to guess how cognitive mechanisms were put together by genetic evolution in the Pleistocene past. Through laboratory experiments and field studies, we can watch them being constructed in people alive today" (Heyes 2018, p. 222). But this criticism emphatically does not apply to Cosmides' work since the evolutionary psychology that forms the basis of Cosmides' account of reasoning is, in fact, based on a powerful empirical study and the discussion of evolution was, famously, restricted to a footnote (it was 'famous' because Cosmides had to beg the editor to be allowed any reference to evolutionary theory). That said, I think that it isn't clear whether the cognitive gadget story is correct about other topics but it cannot be denied that it has some plausibility. What I can say is that as things stand now, with respect to reasoning, the gadget story is not supported by the evidence. The gadget story is incompatible with the evidence that Cosmides provided concerning reasoning over 30 years ago. There is little doubt that the social contract theory is best seen as a System-1 theory though Cosmides does not present it as such. But she does not deny that some domain-general mechanisms may guide other parts of our reasoning and that they are standardly taken to be System-2 processors. As Cosmides and Tooby say: "We're not arguing that there are no general rules. We are just suggesting that psychologists should consider the hypothesis that a given performance is generated by domain-specific mechanisms on an equal basis with the hypothesis that it is generated by domain-general mechanisms, rather than either ruling it out a priori or accepting a lower standard of evidence for the domain-general hypothesis" (Cosmides and Tooby 1997a, b, p. 159). And, in fact, Cosmides and Tooby argued for a domain-general Bayesian processor (see Cosmides and Tooby 1996). Certainly, the domain-general accounts of the associationists and the pragmatic reasoning schema are System-2 interpretations of the Wason Selection Task. Given the evident virtues of a dual process theory of reasoning then one would need to demonstrate how each theory fits into the overall picture of reasoning.

9.6 Conclusion

Hominin cognitive and cumulative cultural evolution, on my account, was facilitated by a System-2 mind, operating offline in the sense that one can run this system without acting on its results. At the same time, a System-1 mind handled simpler standard tasks operating online. This System-2 mind would operate offline in that it could imagine, without acting on, problem-solutions to the day-to-day problems that presented themselves to ancient hominins without having to pay any immediate price for its initial models. Such a mind could determine how best to hunt large animals, how to create better tools with new materials, how to construct shelters and, later, more durable mudbrick dwellings. As the arms race developed, this mind could help the System-1 cheater detection module determine how best to exchange goods and services with conspecifics. At the same time, more primitive mating strategies afforded by other System-1 modules could resolve their own issues.

There are many open questions remaining. Clearly, what is needed for further progress is a more sophisticated treatment concerning exactly how System-1 and System-2 minds jointly resolved the adaptive problems posed by the environments within which they operated to produce a complete architectural portrait of this period. It must also be pointed out, as Kornblith has persuasively argued, that System-1 and System-2 are deeply intertwined in human reasoning, these systems interact in a multitude of ways (Kornblith 2012, ch 5). In the end, however, we won't need to choose exclusively between the approaches of the Evolutionary Psychologists and the evolutionary psychologists: aspects of each approach will explain different parts of the hominin puzzle. And, one might add, Heyes' cognitive gadget view may yet prove its worth as part of the larger picture of mind especially insofar as it embraces the dual process theory (understood in Heyes' distinctive cultural evolutionary psychology sense).

Acknowledgements I would like to thank Jesse Prinz and Edouard Machery for comments on the earlier version of this paper that was read at the Digging Deeper: Archaeological and Philosophical Perspectives conference (December 1-3, 2017). I would especially like to thank Anton Killin and Sean Allen-Hermanson for comments on several drafts of this paper. Those comments are deeply appreciated.

References

Aubert M, Brumm A, Ramli M, Sutikna T et al (2014) Pleistocene cave art from Sulawesi, Indonesia. Nature 514:223–227

Aubert M, Setiawan P, Oktaviana AA, Brum A et al (2018) Paleolithic cave art in Borneo. Nature 564:254–257

Bahn P (2012) Cave art: a guide to the decorated ice age caves of Europe. Frances Lincoln, London

Barkow J, Cosmides L, Tooby J (eds) (1992) The adapted mind: evolutionary psychology and the generation of culture. Oxford University Press, New York

Baron-Cohen S (1995) Mindblindness: an essay on autism and theory of mind. MIT Press, Cambridge, MA

Bednarik RG (2013) Pleistocene paleoart of Asia. Arts 2(2):46–76

Boyd R (1991) Realism, anti-foundationalism and the enthusiasm for natural kinds. Philos Stud 61:127–148

Boyd R (1999) Homeostasis, species, and higher taxa. In: Wilson RA (ed) Species: new interdisciplinary essays. MIT Press, Cambridge, MA

Buller D (2005) Adapting minds. MIT Press, Cambridge, MA

Buller D, Hardcastle V (2005) Modularity. In: Buller D (ed) Adapting minds. MIT Press, Cambridge, MA, pp 127–200

Carruthers P (2006) The architecture of the mind. Oxford University Press, New York

Cheng P, Holyoak K (1985) Pragmatic reasoning schemas. Cogn Psychol 17:391–416

Clarke M (2004) Reconstructing reason and representation. MIT Press, Cambridge, MA

Clarke M (2010) Concepts, intuitions, and epistemic norms. Logos Episteme 1(2):269–286

Coghlan A (2015) First humans to leave Africa went to China, not Europe New Scientist, 3043, October 17, 2015

Cosmides L (1989) The logic of social exchange: has natural selection shaped how humans reason? Cognition 31:187–276

Cosmides L, Tooby J (1992) Cognitive adaptions for social exchange. In: Barkow J, Cosmides L, Tooby J (eds) The adapted mind: evolutionary psychology and the generation of culture. Oxford University Press, New York, pp 163–228

Cosmides L, Tooby J (1994) Origins of domain-specificity: the evolution of functional organization. In: Hirschfield L, Gelman S (eds) Mapping the mind: domain specificity in cognition and culture. Cambridge University Press, New York, pp 85–116

Cosmides L, Tooby J (1996) Are humans good intuitive statisticians after all? Rethinking some conclusions from the literature on judgement under uncertainty. Cognition 58:1–73

Cosmides L, Tooby J (1997a) Dissecting the computational architecture of social inference mechanisms. In: Ciba Foundation Symposium (ed) Characterizing human psychological adaptations. Wiley, New York, pp 132–159

Cosmides L, Tooby J (1997b) The modular nature of human intelligence. In: Scheibel A, Schopf JW (eds) The origin and evolution of intelligence. Jones & Bartlett, Sudbury, pp 71–101

Cosmides L, Tooby J (2010) Universal minds: human nature and the science of evolutionary psychology. Orion, London

Cosmides L, Tooby J (2013) On the universality of human nature and the uniqueness of the individual: the role of genetics and adaptation. In: Downes S, Machery E (eds) Arguing about human nature. Routledge, New York, pp 217–244

Davies S (2012) The artful species. Oxford University Press, Oxford

Dennett DC (1991) Consciousness explained. Little, Brown, and Company, Boston

Dennett DC (1995) Darwin's dangerous idea. Simon and Shuster, New York

Dennett DC (2017) From bacteria to Bach and back: the evolution of minds. WW Norton, New York

Downes S (2013) The basic components of the human mind were not solidified during the Pleistocene epoch. In: Downes S, Machery E (eds) Arguing about human nature. Routledge, New York, pp 93–102

Downes S, Machery E (eds) (2013) Arguing about human nature. Routledge, New York

Dunbar R (2014) Human evolution. Pelican, London

Elman J, Bates E, Johnson M, Karmiloff-Smith A, Parisi D, Plunkett K (1996) Rethinking innateness: a connectionist perspective on development. MIT Press, Cambridge, MA

Evans J (2009) How many dual-process theories do we need? One, two, or many? In: Evans J, Frankish K (eds) In two minds: dual processes and beyond. Oxford University Press, Oxford, pp 33–54

Evans J, Frankish K (2009) The duality of mind: an historical perspective. In: Evans J, Frankish K (eds) In two minds: dual processes and beyond. Oxford University Press, Oxford, pp 1–32

Evans J, Stanovich KE (2013) Dual-process theories of higher cognition: advancing the debate. Perspect Psychol Sci 8(3):223–241

Fodor J (1983) The modularity of mind. MIT Press, Cambridge, MA

Heyes C (2018) Cognitive gadgets: the cultural evolution of thinking. Harvard University Press, Cambridge, MA

James W (1890) The principles of psychology. Holt, New York

Johnson-Laird P (1983) Mental models. Harvard University Press, Cambridge, MA

Kornblith, H (2012) On reflection. Oxford University Press, Oxford

Kruglanski AW, Gigerenzer G (2011) Intuitive and deliberate judgements are based on common principles. Psychol Rev 118(1):97–109

Machery E, Barrett C (2006) David J. Buller: evolutionary psychology and the persistent quest for human nature. Philos Sci 73(2):232–246

McKone E (2009) Holistic processing for faces operates over a wide range of sizes but is strongest at identification rather than conversational distances. Vis Res 49:268–283

McKone E, Palermo R (2010) A strong role for nature in face recognition. Proc Natl Acad Sci 107:4795–4796

Palmer S (1999) Vision science: photons to phenomenology. MIT Press, Cambridge, MA

Pinker S (1994) The language instinct. William Morrow, New York

Reber AS (1993) Implicit learning and tacit knowledge. Oxford University Press, Oxford

Richerson P, Boyd R (2005) Not by genes alone. University of Chicago Press, Chicago

Rito T, Vieira D, Conde-Sousa E, Pereira L, Mellars P, Richards MB, Soares P (2019) A dispersal of Homo sapiens from southern to eastern Africa immediately preceded the out-of-Africa migration. Sci Rep 9:472B

Samuels R (1998) Evolutionary psychology and the massive modularity hypothesis. Br J Philos Sci 49(4):575–602

Samuels R (2009) The magical number two, plus or minus: dual-process theory as a theory of cognitive kinds. In: Evans J, Frankish K (eds) In two minds: dual processes and beyond. Oxford University Press, Oxford, pp 129–148

Sloman SA (1996) The empirical case for two systems of reasoning. Psychol Bull 119:3–22

Sterelny K (2011) From hominins to humans: how sapiens became behaviorally human. Philos Trans R Soc B 366:809–822

Sterelny K (2012) The evolved apprentice. MIT Press, Cambridge, MA

Sterelny K (2013) The informational commonwealth. In: Downes S, Machery E (eds) Arguing about human nature. Routledge, New York, pp 274–288

Tooby J, Cosmides L (1992) The psychological foundations of culture. In: Barkow J, Cosmides L, Tooby J (eds) The adapted mind: evolutionary psychology and the generation of culture. Oxford University Press, New York, pp 19–79

Wason P (1968) Reasoning about a rule. Q J Exp Psychol 20:273–281

Chapter 10
Theory of Mind, System-2 Thinking, and the Origins of Language

Ronald J. Planer

Abstract There is growing acceptance among language evolution researchers that an increase in our ancestors' theory of mind capacities was critical to the origins of language. However, little attention has been paid to the question of how those capacities were in fact upgraded. This article develops a novel hypothesis, grounded in contemporary cognitive neuroscience, on which our theory of mind capacities improved as a result of an increase in our System-2 thinking capacities, in turn based in an increase in our working memory capacities. I contrast this hypothesis with what would appear to be the default position among language evolution researchers, namely, that our theory of mind became more powerful as a result of genetic change to a domain-specific mindreading system which we share with other great apes. While the latter hypothesis is not implausible, it arguably enjoys less empirical support at present than does the alternative hypothesis I develop.

Keywords Attention · Episodic memory · Evolution of language · Self-control · Theory of mind · Working memory

10.1 Introduction

The view that the evolution of language was to a significant extent caused by an upgrade in our theory of mind capacities has gained increasing acceptance in recent years (see, e.g., Tomasello 2008; Sperber and Orrigi 2012; Fitch 2010; Sterelny 2012; Bar-On 2013; Gamble et al. 2014; Scott-Phillips 2014; Bar-On and Moore 2017; Moore 2017a, b; Planer 2017a, b; Planer and Godfrey-Smith 2020; Planer and

R. J. Planer (✉)
School of Philosophy, Australian National University, Acton, ACT, Australia
e-mail: ronald.planer@anu.edu.au

© Springer Nature Switzerland AG 2021
A. Killin, S. Allen-Hermanson (eds.), *Explorations in Archaeology and Philosophy*, Synthese Library 433, https://doi.org/10.1007/978-3-030-61052-4_10

Sterelny forthcoming).[1] This is in part due to a growing appreciation of the role that mindreading plays *in* human linguistic communication (especially in pragmatics). But it is also due to many researchers moving away from a strong, nativistic picture of language and more towards a cultural-evolutionary one. For on the latter view, language was fundamentally something we had to create together, as opposed to something we merely had to externalize, and it is very hard to see how that might have gone unless our ancestors were increasingly 'tuned in' to each other's mental states.

Despite this growing consensus, there is no generally accepted account of how our theory of mind capacities were in fact upgraded. (In contrast, why such an upgrade would have evolved is uncontroversial; that was surely due to increasing social complexity in our line, of both a competitive and cooperative variety.) This chapter separates out and explains two (compatible) proposals. One of these is obvious, and is probably the default view among many language evolution researchers. This is that humans' innately-channeled[2] theory of mind capacities were upgraded. There was some genetic change (or more likely, some series of changes) linked specifically to those capacities that made them more powerful. Such a view is not implausible. However, there is an alternative view which I seek to outline and motivate in this chapter. On this view, our theory of mind capacities were upgraded as a result of a general upgrade to our working memory system. In particular, enhanced working memory led to enhanced System-2 thinking which in turn led to enhanced mindreading (among other things). This possibility has been neglected, likely because of the tendency to view System-2 mindreading, if not System-2 thinking in general, as language dependent. I will argue that, on at least one plausible conception of System-2 thinking, this is misguided.

Cognitive archaeologists Wynn and Coolidge (2011) have argued for a very different proposal about the role of working memory in the evolution of language and behavioral modernity more generally. Aiming to interpret the archaeological record through the lens of a cognitive model of working memory, these authors connect an upgrade in working memory to differences in *sapiens* and Neanderthal cognition and behavior. However, no such claim is made here. Moreover, Wynn and Coolidge do not discuss mindreading, the central focus of this chapter. Cole (2015), on the other hand, discusses the kind of skilled action required to produce symmetrical stone technologies (such as Acheulean handaxes) and argues that theory of mind capacities such as third-order perspective taking are implicated. The picture developed in this chapter is compatible if not complementary (see also Planer 2017b), though again, I do not commit to any particular reading of the archaeological record here. That said, in future work, I do plan to use the framework advanced below to

---

[1] Within this camp, there is, however, a good deal of heterogeneity regarding the specific type of theory of mind capacities that were relevant, as well as the specific role those capacities played in language origins. For a nice overview of this issue, see Moore (2018).

[2] This is a nice term Peter Carruthers has coined. As I shall use it, an 'innately-channeled' trait is one that is in part, and possibly in large part, genetically specified.

interpret key archaeological traces and suspect that doing so will produce novel and important insights.

I will proceed as follows. First, I give a brief overview of evidence pertaining to great ape mindreading and the development of mindreading in humans. Second, I sketch the 'two-systems' approach to handling the totality of this evidence. Third, I set out an account of working memory and System-2 thinking. The ideas at this point owe to a very impressive synthesis carried out by Peter Carruthers in recent years. Fourth, I explain how System-2 mindreading is possible in the absence of language. This allows us to see how, independently of language, an upgrade in System-2 thinking might have upgraded our theory of mind capacities. Finally, I evaluate the two hypotheses, highlighting the empirical strength of the System-2 mindreading hypothesis.

10.2 Mindreading in Great Apes and Young Humans

It is now widely agreed that humans have innately-channeled theory of mind capacities. This idea is well-supported by both comparative and developmental evidence. For our purposes, the most important set of comparative results come from great apes. Most of this work has looked at chimpanzee theory of mind capacities. There is very good evidence that great apes are proficient at recognizing others' goals and intentions. For example, chimpanzees and orangutans can distinguish between intentional and accidental behavior on the part of an experimenter, even when the two behaviors closely resemble one another (Tomasello et al. 1998; Tomasello and Carpenter 2005). In addition, chimpanzees can extract the intended purpose of a behavior of an experimenter even when that behavior fails to achieve its goal (Tomasello et al. 1987; Horner and Whiten 2005). Moreover, chimpanzees treat an experimenter who is attempting to give them food but failing differently from one who is simply withholding food from them (Call et al. 2004).

Further, there is good evidence that great apes understand others' attentional and perceptual states to some extent. For example, they will go out of their way to see what a conspecific is looking at (e.g., moving around a barrier to do so) (Rosati and Hare 2009). Chimpanzees and bonobos can use knowledge of whether an experimenter has previously interacted with an object to disambiguate what they are attending to (MacLean and Hare 2012).[3] Chimpanzees also use information about whether a competitor's line of sight to a food item is obstructed in deciding whether to go for that food (Hare et al. 2000; Hare et al. 2006; Braeuer et al. 2007). There is similar (though less) evidence that chimpanzees are also sensitive to what a competitor can hear (Melis et al. 2006).

Moreover, great apes appear to have some understanding of others' thoughts and beliefs. Chimpanzees can infer how a competitor is likely to have behaved based on

---

[3] Chimpanzees do a bit better on this task.

whether that individual *previously* had an unobstructed line of sight to where some food item was hidden (Kaminski et al. 2008). In other words, they apparently understand that seeing leads to knowing. Originally, it was thought that chimpanzees were nonetheless incapable of representing others' false beliefs (mainly because they failed to capitalize on those beliefs). But recent work adapting non-verbal tests used to probe infants' false-belief understanding (see below) suggest otherwise. These studies have found that not just chimpanzees but all great apes can pass versions of the false-belief test (Krupenye et al. 2016; Buttlemann et al. 2017). (I will return to the significance of this point later.)

Ironically, research focused on children's theory of mind capacities was spawned by the question of whether chimpanzees had such capacities (Premack and Woodruff 1978). The false-belief test[4] was suggested as a way to definitively answer the question of whether any agent was genuinely representing another's mental states (as opposed to simply relying on their own knowledge to predict others' behavior) (Bennett 1978; Dennett 1978; Pylyshyn 1978). In the decades that followed, research into children's theory of mind capacities was dominated by an interest in false-belief understanding. These experiments required children to make verbal responses. Initially, it was found that children do not reliably pass such tests until they are between the ages of 4 and 5 (Wimmer and Perner 1983). Later work suggested a somewhat earlier onset time for this ability.[5] Children were found to pass verbal tests involving the attribution of other mental states considerably earlier (e.g., they can pass verbal tests involving desire attribution by as early as 2 years of age (Wellman and Woolley 1990)).[6]

The field has since been radically reshaped by the use of non-verbal or 'implicit' mindreading tests. These tests include anticipatory looking, violation of expectations (as judged by how long an infant looks at a stimulus), and helping behavior. These results are widely (though not universally) taken to show that the ability to attribute goals and intentions to others emerges by or before 12 months of age (Johnson 2000; Csibra et al. 2003; Luo and Baillargeon 2005; Schlottmann and Ray 2010), as does the ability to attribute knowledge and ignorance (Liszkowski et al. 2006; Luo and Baillargeon 2007; Luo and Johnson 2009). This is followed by the emergence of the ability to attribute false beliefs to others, and to predict that they will behave in accordance with those beliefs, by 18 months (Onishi and Baillargeon 2005; Southgate et al. 2007, 2010; Surian et al. 2007; Scott and Baillareon 2009; Scott et al. 2010).

---

[4] I assume my readership is well-versed in the logic of this test. If not: a participant watches as another agent places an item in a particular place. The agent then leaves, and the item is moved to a new location in their absence. The question the participant must answer is where the agent will look for the object. The participant must attribute a false belief to the agent that the object is at its original location if they are to reliably answer correctly.

[5] See Wellman et al. (2001) for a meta-review of this research.

[6] See also Bartsch and Wellman (1995). They show that, from 2 years old, children regularly explain others' behavior in terms of desires and emotions.

Taken together, the comparative and developmental evidence strongly suggests that humans and other great apes share a mindreading system, and that this system is genetically specified to a significant extent. It is very hard to see how the capacities of either infants (given their very young age) or other great apes can be explained in terms of individual or social learning. Moreover, that explanation would have to explain the degree of convergence between infant and great ape theory of mind capacities. Why should those capacities turn out so similar, given the very significant differences in learning environments? The other main option is to deny that infants and other great apes are actually mindreading at all; one might claim, for example, that their performance reflects mere 'behavior rules' (Penn and Povinelli 2007, 2013) or associationist learning (Heyes 2018). Both of these suggestions are quite problematic in my view, however, though for reasons I shall not enter into here. Suffice it to say that this chapter sets aside such skepticism about infant and great ape mindreading.

10.3 A 'Two-Systems' Account of Mindreading

One attractive way of handling the totality of this comparative and developmental evidence is to give a 'two-systems' account (see, especially, Apperly 2010; Butterfill and Apperly 2013). This kind of account accepts the above line of argument. It adds that humans can also reason about mental states using domain-general cognitive processes. This type of mindreading is conscious and flexible. It draws upon local cultural information, such as narratives and norms, and the schemas they give rise to. Accordingly, it is to a significant extent culturally acquired, and hence culturally variable. It does not supplant one's innately-channeled mindreading system, but rather exists alongside it.

This treatment of human mindreading slots into a more general picture of human cognitive architecture. On this picture, the human mind features two systems, System-1 and System-2 (Evans and Over 1996; Sloman 1996; Stanovich 1999; Kahneman 2002; Frankish 2012). When we consciously solve a logic puzzle, for example, that is System-2 thinking. In contrast, when we see a car hurdling towards us and quickly jump out of the way, that is System-1 thinking.

More fully, System-1 is conceived of as a large collection of sub-systems dedicated to belief and desire formation, as well as decision making. Despite their varied functions, these systems are held to instantiate a set of common properties. Researchers differ on exactly which properties these are, however, and the extent to which they are shared. On one common view (associated with evolutionary psychology), System-1 systems are automatic, operate swiftly (by comparison with System-2), and run in parallel. They are shared by all typical humans, and most are evolutionarily old, and hence shared with other species. They are characterized by a degree of operational fixedness; to the extent that their operation is modifiable, it may be modifiable only during some critical period or as a result of protracted learning. Their operation cannot be modified *at will*. System-1 systems instantiate domain-specific algorithms which work well in their proprietary domain but may be useless (or worse) outside it. Often these algorithms are 'quick and dirty' rather than logically or mathematically sound. They have been designed to work well enough given stringent time constraints, rather than to respect principles of rationality. Finally, while the outputs of System-1 systems may be conscious (often they are not), their operation is unconscious, and more generally, opaque to introspection.

In contrast to this 'grab bag' of specialist systems, System-2 is conceived of as a single domain-general system. Its operation is slow, effortful, and serial. We are consciously aware of its operation, and can verbally report on it. There are significant individual differences in how, and how often, people deploy System-2, which account (at least in part) for differences in general intelligence. System-2 is readily modifiable, both at a time and over time; its operation is sensitive to instruction, as when we adopt the advice of an expert as to how to approach some problem. More generally, System-2 can be trained to follow complex rules, including logical and mathematical ones. System-2 makes extensive use of natural language in the form of mentally rehearsed sentences and other constructions—inner speech. We often use System-2 to engage in extended dialogue with ourselves as we attempt to make up our mind on some topic or plan a course of action. This activity makes up most of what we have in mind by 'thinking' or 'reasoning' in the everyday sense of those terms. It has been frequently assumed by researchers that only humans possess System-2.

In terms of mindreading, then: the idea would be that humans' System-1 contains a specialist mindreading system. This system develops early and generates the behavior we see in non-verbal mindreading tests. The ability to reason about mental states using System-2 emerges later, and is shaped by the child's social and cultural environment (among other things). Explicit or verbal measures of mindreading reflect the coming online of System-2 mindreading.

I will assume that something like this account of human mindreading is on the right track. If so, then there are two ways mindreading might have been upgraded over the course of human evolution: we may have become better at System-1 mindreading or better at System-2 mindreading. However, in order to figure as part of an explanation for the origins of language, we cannot presuppose language. It is easy to see how that might go in the case of the first option, but less easy in the case of the second. What would System-2 thinking about mental states even look like in the absence of language to describe mental states? Let me explain.

10.4 System-2 Thinking and Working Memory

Peter Carruthers (2015) offers a distinctive take on the nature of System-2 thinking. His account is couched within a very impressive synthesis of research on the nature of working memory. In this section, I provide an overview of this synthesis (adding a few details that are key given the aims of this chapter) and of his account of

System-2 thinking, respectively.[7] I then turn to the issue of System-2 mindreading in the next section.

10.4.1 The Nature of Working Memory

For Carruthers, working memory is a set of domain-general cognitive capacities supporting the selection, retention, rehearsal, and manipulation of mental representations. He points out that attention is fundamental to these processes.[8] When some representation is targeted with attention, it is globally broadcast. A globally broadcast representation is available for processing by a large number and range of cognitive and affective systems in the brain, and is experienced as a conscious mental state (Baars 1997, 2005).

Mechanistically, this is thought to work as follows. The brain attends to some information by boosting the activity of the group of neurons carrying that information, and typically, also by inhibiting the activity of other, close by groups of neurons. There is a threshold level of activity which, once reached, results in the information realized by the relevant group of neurons being distributed throughout the brain (Gazzaley et al. 2005). This is the neural basis of global broadcasting.

The work of widely propagating this information is primarily carried out by a neural network connecting the dorsolateral prefrontal cortex, the pre-motor and motor cortex (in particular, regions dedicated to our eye fields, which control eye movements), and the intraparietal sulcus (in both hemispheres of the brain). The intraparietal sulcus, in turn, projects to various mid-level sensory areas. Such areas compute more abstract representations than primary sensory areas (e.g., V1), but less abstract than high-level association and cognitive areas. These representations are imagistic or sensory-like, but can contain conceptual information (e.g., an image of a dog, together with a 'tag' or 'label' DOG[9]). (See Fig. 10.1 for the rough location of these brain areas and others that are discussed below.)

This network is under intentional control. It is directed by decisions made in light of a wide range of abstract cognitive states—an agent's beliefs, desires, goals, and values. For this reason, it is often referred to as the 'top-down attentional network', and is contrasted with another network referred to as the 'bottom-up attentional network'. The latter is typically asymmetrically-based in the right hemisphere. It connects the right ventrolateral prefrontal cortex and right medial frontal gyrus with the right ventral parietal cortex (located at the temporal-parietal juncture). This network also receives projections from the limbic system. Based on information about the individual's goals and values, the right ventral parietal cortex monitors

---

[7] I refer the reader to Carruthers (2015) for citations, though I include the most crucial ones here.

[8] See Carruthers (2015) for a discussion of how working memory, so conceived, relates to the traditional model originally proposed in Baddeley and Hitch (1974).

[9] As per the usual convention in philosophy, I use small caps when referring to mental content.

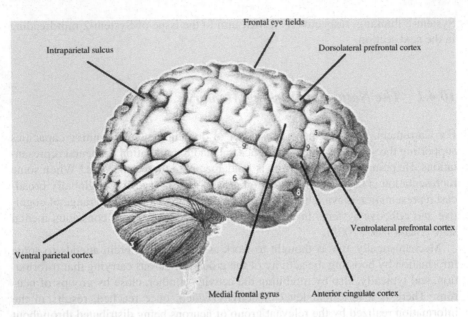

Frontal eye fields

Intraparietal sulcus

Dorsolateral prefrontal cortex

Ventrolateral prefrontal cortex

Ventral parietal cortex

Medial frontal gyrus   Anterior cingulate cortex

Fig. 10.1 Right profile of the cerebral cortex with the location of key areas involved in attention shown [Brain drawing: Popular Science/Public domain]

unattended-to mid-level sensory representations for ones that may be relevant to the agent.

The bottom-up network is not capable of directly allocating attention to mental representations, however. Rather, it must coopt the long-range projections linking the prefrontal cortex with the intraparietal sulcus. When the right ventral parietal cortex detects a representation that matches its search criteria, it relays signals to the right ventral prefrontal cortex requesting that attention be directed to the representation in question. Whether this request is granted depends on competitive processing that takes place in the right ventral prefrontal cortex, the anterior cingulate cortex, the right medial frontal gyrus, and the dorsolateral prefrontal cortex. So long as attention is not shifted, the content of the representations which the right ventral parietal cortex has highlighted as relevant remain *distractors*.[10]

An interim summary. Selecting a representation with attention serves to load that representation into working memory. It remains there by way of attention to that representation being sustained, which is effortful. The neurobiological evidence

---

[10] In addition to these two networks, there are well-known subcortical networks that influence attention. These networks that are responsible for the reflexive reorientation of attention to high-intensity stimuli (e.g., a sudden loud noise). Unlike the bottom-up attentional network, these networks are capable of directly seizing control of the long-range connections that are part of the top-down attentional network; they do not have to 'make their case' to the top-down system in order for attention to be reallocated, as does the bottom-up system. When attention has been shifted by a subcortical network, the prefrontal cortex evaluates whether the novel attentional target is indeed highly relevant, and if not, attention is typically shifted back to the original target.

provides an answer to which representations it is that can figure into working memory. These are mid-level sensory representations which are imagistic, but which typically feature some conceptual information. We now turn to mental rehearsal and the manipulation of representations in working memory, respectively, which build on the foregoing processes.

As Carruthers points out, a number of authors have plausibly connected mental rehearsal with forward models of action in the brain (see, e.g., Wolpert and Flanagan 2001; Wolpert et al. 2003; Jeannerod 2006). The defining feature of such models is the production of an efference copy (think a xerox) of a motor command that has been sent to the muscles for execution. The brain makes a variety of uses of such copies. In order to accurately control bodily movements, the brain needs to know the location and velocity of specific body parts. Information about these variables can be sourced via the senses, but only after much processing of perceptual information. This takes time and opens the door to noise. To the extent that an efference copy of a recently launched motor command accurately predicts the immediate future location and velocity of specific body parts, one can more or less make immediate use of that copy to estimate the value of these variables. This allows for correction of bodily movements even before they are initiated.

But efference copies are also used to make sensory predictions: the brain generates mental images of how the act will look, sound, or feel as it unfolds. These predictions can influence how the brain handles incoming perceptual information. For example, they may be used to attenuate sensory effects that are expected on the basis of our motor behavior and (consequently) enhance those that are not. Thus, the same tactile stimulation is experienced as less ticklish when we deliver it to ourselves, than when it is generated exogenously (Blakemore et al. 1999). More generally, the brain uses correspondence (or lack thereof) between predicted and observed sensory effects to distinguish self- and other-generated actions. The comparison of predicted and observed sensory effects can also be used to reduce uncertainty about the properties of stimuli (or correct misimpressions about them), allowing the brain to select the forward model most appropriate to the context. For example, one model might predict that pressing on some object will result in an indentation in that object,[11] whereas another model might predict that the object's shape will remain unchanged. A match between the first of these predictions and the observed effect of the action tells the brain that the object is soft (Wolpert and Flanagan 2001).[12]

Mental rehearsal can be analyzed as follows: An action schema is activated, leading to a motor command and an efference copy of that command, but the motor command to the muscles is cancelled so that overt action does not occur. This requires self-control in the form of motor inhibition. Motor inhibition has been the most exhaustively studied form of self-control, and consequently is very well

---

[11] Or more strictly speaking, that it will result in percepts consistent with that object's having been indented.

[12] There is an obvious connection with predictive processing models of cognition here. I take no stand in this chapter on the plausibility of the latter as an overarching theory of cognitive architecture.

understood. Both the dorsolateral and right ventral prefrontal cortex are implicated in this form of control, the latter being especially active. More generally, it would appear that there is a common neural basis to self-control in the brain, involving many of the same prefrontal areas that are recruited by the top-down and bottom-up attentional networks: the two just mentioned, together with the anterior cingulate cortex and medial prefrontal gyrus (as part of the medial prefrontal cortex). As in the case of motor inhibition, the right ventral prefrontal cortex appears to play a special role in other forms of self-control, being the most consistently and highly activated (Cohen and Lieberman 2010). The efference copy of the motor command remains behind, however, and goes on to drive mental imagery of the predicted effects of the action per usual (i.e., as if the action had in fact been performed). If and when attention is brought to this imagery, it enters working memory, and is globally broadcast. This imagery enters consciousness, and we have the experience of imagining ourselves performing the corresponding act and of observing (some of) its consequences.

Finally, turn to the manipulation of working memory representations. Carruthers plausibly understands manipulation as the transformation of conscious mental imagery. To take a simple example: suppose you were asked to recall how to cook an omelet. You start by imagining an empty bowl. Then, you imagine cracking some eggs on the lip of the bowl and emptying them into it. Then, you imagine whipping the eggs, and so on. What you are doing is mentally rehearsing a *series* of actions and attending to the predicted sensory effects at each stage. This mental activity can be exploratory or done with a particular goal or end state in mind. We experience manipulation as either a gradual change to a single mental image over time or as a series of intuitively related mental images (the line here is blurry).

This treatment of mental rehearsal and manipulation is supported by the famous work on mental rotation carried out by Roger Shepard and colleagues (see, e.g., Shepard and Metzler 1971; Shepard and Feng 1972; Shepard 1978). In the classical setup, participants are presented with a two-dimensional image of a complex and irregular three-dimensional shape. They are then shown a second image of a similar object and asked if it is the same object rotated a certain number of degrees. What has been found (time and again) is that there is a linear relationship between partici-pants' response time and the number of degrees the object would have to be rotated if it were to look like the second image. This result is well-explained (and not easily explained otherwise) by the hypothesis that participants mentally rehearse the action of manually rotating the object some number of degrees while monitoring the resul-tant visual imagery for a match with the target stimulus. More recent fMRI research has shown that the intraparietal sulcus—the hub from which attention spreads out to mid-level sensory areas in the brain—is especially active during tasks involving mental rotation (Prather and Sathien 2002; Gogos et al. 2010).

Pulling all this together, then: attention and its direction forms the heart of work-ing memory. Representations gain entry to working memory by way of being tar-geted by attention. They remain there if and only as long as attention can be held on them. This yields two types of executive work: strategic or selective control, that is, the selection of one representation[13] (or set of representations) with attention as opposed to another; and cognitive inhibition, that is, the suppression of signals attempting to reorient the focus of attention to other representations. (Ignoring the distractors generated by the bottom-up attentional system is a paradigm case of cognitive inhibition.) Mental rehearsal and manipulation likewise involve targeting specific representations with attention and sustaining that attention, but there is more going on in these cases. Mental rehearsal (at least of the intentional variety) involves an executive command to activate a particular action schema. The resulting motor command is cancelled, but the efference copy goes on to drive mental imagery of the action, with attention focused on this imagery. Mental rehearsal thus additionally involves motor inhibition. And manipulation is more complex still insofar as it prototypically involves executing a series of mentally rehearsed actions, often until some (mentally represented) condition is satisfied. Manipulation may further or instead involve the direction of imagination in a way that is not directly tied to any action the agent can him- or herself perform (as when we imagine the trajectory of a falling object). The contents of working memory are neurally realized in the mid-level sensory areas where they are (perhaps only ephemerally) stored, while the various forms of control implemented by working memory appear to make use of a largely overlapping set of prefrontal cortical and subcortical areas. (The significance of these neural details will become clearer below.)

10.4.2   System-2 Thinking

On Carruthers' account of System-2 thinking, such thinking is realized by *cycles* of System-1 activity made possible by working memory. In virtue of being in working memory, information is globally broadcast and hence available for processing by a large number and range of System-1 systems. Each cycle of System-1 activity modifies the cognitive and affective environment against which the next cycle of activity operates.[14] The account has numerous benefits (not least its economy), though I will not rehearse those benefits here. Suffice it to say that it answers several outstanding questions for the standard picture of System-1/System-2 architecture.

More fully, as I read Carruthers, an episode of System-2 thinking can commence in one of two ways. The first is for an intuitive response generated by System-1 to be suppressed. That response will be an activated action schema of some kind (often a speech-action schema—that is, a schema corresponding to the production of a certain linguistic utterance—in the case of modern humans). The response is suppressed in the sense that the motor command issuing from it to the muscles is

---

[13] Construed broadly. This representation could be a rule, a strategy, etc.

[14] As such, System-2 thinking is inherently a mix of conscious and unconscious states and processes. One can think of it like an iceberg (to use an old metaphor): the 'tip' consists in the contents of working memory, which are conscious, the 'submerged portion' is all the processing that takes place in System-1, which is unconscious.

cancelled. An efference copy is nevertheless generated and goes on to drive mid-level sensory imagery in the usual way. This imagery is attended to and hence enters working memory at which point it is globally broadcast. The other way is for the agent to make an executive decision to load a particular mid-level sensory image to working memory, for example, an episodic memory (more on episodic memory later).[15]

System-1 processing of this content can produce a variety of cognitive and affective consequences. Recall from above that forward models of action already generate certain predictions about the effects of action (beyond how the act itself will look, sound, and/or feel). The operation of other System-1 systems can be expected to compute a wider range of effects in both space and time.[16] In addition, System-1 processing of act imagery can lead to new desires and/or modulate standing ones. In discussing this possibility, Carruthers draws upon influential work by Antonio Damasio (1994; 2003) (though the idea can be developed in other ways too). As observed earlier, we routinely engage in mental rehearsal of action (with attention to its associated imagery) to determine how to behave in some situation. On Damasio's model, this activity produces a cascade of physiological effects which, upon being perceived by the brain, result in *feelings*. The valence of these feelings function to adjust the strength of our desire to perform the simulated act.[17]

Crucially, System-1 processing of mid-level sensory representations can also issue in executive commands. For executive control is realized by a System-1 system, or more plausibly, by a subset of such systems. In particular, attention may be directed to other (offline or online) representations, entering them into working memory. These representations might work together with the content already in working memory, forming a more complex mental representation, and this representation may well elicit a different System-1 response than the original content on its own. Alternatively, attention may be directed to an entirely new set of relevant representations. In any case, the ability of System-1 processing to direct attention

[15] There are passages in Carruthers (2015) that suggest he thinks all System-2 thinking begins in the former way. This is especially true in some of his earlier work (e.g., Carruthers 2009). Other passages are consistent with the interpretation provided here. This view seems more extensionally adequate to me.

[16] Here it is crucial that the brain somehow manages to keep separate its representations of the actual state of the world from its representations of the state of the world *conditional upon the action being performed*. The latter have to be 'tagged' as hypothetical. As Carruthers himself admits, how the brain might achieve this is not obvious. However, we can, I think, take some consolation in the fact that the brain already does something like this in distinguishing between predicted sensory effects and observed ones. Moreover, the problem is not just Carruthers'; all researchers in this area owe an explanation of hypothetical reasoning at the end of the day.

[17] There is an interesting connection here with the phenomenological character of episodic memory and prospection. By definition, these states involve projecting the self through time. This may be necessary in order for these states to produce feelings which modulate our desires. We may have to imagine *ourselves* doing or experiencing various things. I leave this issue for another day, however.

enables the whole process to repeat, with each cycle yielding its own set of cognitive and affective consequences which form the backdrop for the next cycle.

As will by now be obvious, System-2 thinking is not language dependent on Carruthers' account. Such thinking can and often does (in modern humans) involve inner speech. This type of activity is a special case of the mental rehearsal of action, and can be analyzed accordingly (i.e., in terms of forward models of action). To engage in inner speech is to mentally rehearse a sentence or sentence fragment, attending to the predicted effects of that action, most obviously how it would sound if uttered. But System-2 thinking need not involve language: it can involve a series of visual mental images, for example. Recall the steps required for making an omelet, as discussed above. Importantly, this means System-2 thinking can figure in an explanation for the origins of human language. Moreover, while it is indeed plausible that language significantly upgrades System-2 thinking,[18] there is no reason to think that the flexibility, creativity, and reflectiveness of such thinking depends on having a natural language. Rather, those features are plausibly understood as inhering in the ability to repeatedly, and in a controlled, targeted, and cumulative fashion, probe the expert knowledge of System-1 systems until a satisfactory result is obtained.

10.5   System-2 Mindreading Without Language

No doubt much of our System-2 mindreading makes essential use of mental state vocabulary. This is not surprising, as mental states are paradigmatically hidden, abstract entities. I consciously think (in inner speech): *Where would she like to go to dinner tonight? She didn't seem to enjoy that Italian restaurant last week. What about the new Chinese restaurant downtown? She said that she heard that was very good.* But is this essential to System-2 mindreading? Must it take this form? I say 'no'.

Suppose I am due to meet a friend for coffee. I am tempted to respond to several more emails but will run late if I do so. I imagine my friend sitting alone at the café. I imagine him checking his watch and the two cups of coffee getting cold on the table. I imagine the disappointed look on his face and feel bad. I shut my laptop and head out the door. There is a sense here in which I did not think about my friend's mental states. What I did, rather, was imagine his disappointed look, etc. If, however, we accept the above account of System-1/System-2 architecture, then there is another sense in which I very likely *was* engaged in mindreading. By predicting these effects of my action and attending to them, those mental images were globally broadcast. Accordingly, they were made available to my System-1 mindreading system (among many others). To be sure, I was not conscious of the processing that

---

[18] Exactly how turns out to be a complex issue. Suffice it to say that Carruthers' own account underestimates some of the complexities. I plan to take up this issue in future work.

ensued, but it impacted on me in such way so as to dampen my desire to keep answering emails and to strengthen my desire to leave for the café.

Let us unpack this further. Suppose, as seems very plausible, that our System-1 mindreading system evolved to be activated by representations of agents. And suppose in addition that these representations are mid-level sensory ones.[19] If so, then by loading working memory with an imagistic representation depicting one or more agents, that imagery becomes globally broadcast, and will cause one's System-1 mindreading system to operate. This may result in (further) conceptual information being added to the representation. The content JAX SEES THE FOOD may be associated with an image of Jax having a clear line of sight to the food. This conceptual information can affect how the representation is processed by other System-1 systems (including ones that embody associatively learned information). So long as attention is held on this modified representation, it will remain available for processing by these other systems (and for *re*processing by the theory of mind system), potentially resulting in additional modifications. And so on. Alternatively, the System-1 mindreading system may output a purely amodal representation with certain downstream cognitive and affective consequences. For example, it might produce the belief JAX INTENDS TO GET THE FOOD. This (unconscious) belief might cause one to imagine being attacked by Jax when one subsequently rehearses the action of retrieving the food oneself. The details obviously depend on the computational specs of the System-1 mindreading system and its broader computational environment. But on just about any set of such specs, an individual who can entertain offline agent representations and make repeated use of the System-1 mindreading system will have a more advanced set of theory of mind capacities than one who cannot.

Moreover, and crucially for present purposes, it is not hard to imagine how mindreading of this sort might shape communicative behavior, and hence how communication might be enhanced by an upgrade to System-2 mindreading. Here is a simple example: imagine two prelinguistic hominins, Send and Receive. Send wants Receive to see the animal up ahead. Receive is currently facing in the opposite direction. Send rehearses various actions, predicting their effects. He imagines tapping Receive on the shoulder and predicts (or more strictly speaking: some subpersonal system in him predicts) that Receive will orient towards him. Send then imagines extending his arm in the direction of the animal and predicts that Receive will orient in that direction. Perhaps this prediction is justified by Send's past experience; he has observed that others tend to pay attention to one's hands. This imagery leads Send's System-1 mindreading system[20] to output the representation

---

[19] The main argument here can be developed without this assumption. I make this assumption mainly to simplify the exposition (though I also think it is plausible). If the representations accepted by the System-1 mindreading system were in fact high-level conceptual ones, then there would have to be some system that maps imagistic representations involving agents to amodal ones about agents.

[20] The System-1 mindreading system may have also been involved at early steps in this process.

RECEIVE SEES THE ANIMAL. This information ratchets up Send's desire to perform that very act sequence, which he subsequently executes.

Here, signal creation is being facilitated by System-2 mindreading on the sender's (Send's) part. Absent this thinking, it would be unlikely, or at least much less likely, that Send would act in such a way. His System-2 mindreading results both in an action plan that is likely to solve his communicative problem as well as the motivation to perform it. In the view of some, the major challenge for language evolution theorists is to explain an increase in signal creation (see, e.g., Fitch 2010).[21] Support for this idea comes from the work of Cheney and Seyfarth with baboons.[22] Cheney and Seyfarth have shown that baboons are capable of appropriately semantically interpreting a massive array of calls, including wholly novel call sequences. Moreover, the interpretations baboons provide are shot through with complex social knowledge about group members and recent interactions among them. Also relevant here is the very impressive levels of interpretative competence enculturated apes like Kanzi can attain with natural languages.[23] What we do not find in the wild, nor really even in enculturated apes, is much innovativeness. This places an obvious brake on the evolution of a rich, expandable communication system, and may be due to cognitive limitations, motivational limitations, or (likely) both.

I suggest that enhanced System-2 mindreading is a plausible explanation for how signal production and hence creation was ramped up in our line. At the same time, System-2 mindreading can be important on the receiver's side too, and further facilitate signal creation. Put abstractly, such mindreading can increase the rate at which cues turn into signals. Cues are acts (or traits) that carry (useful) information for a receiver, but which do not have the function of doing so.[24] The more receivers are able to engage in deliberate, reflective thinking about others' behavior,[25] the more likely they are to discern the motivation behind that behavior. If interests sufficiently align, those receivers may behave in ways that are conducive to some actor achieving his or her (non-communicative) goals. In turn, this can lead to the act coming to be performed for that very reason, to elicit this effect in receivers, with the act often becoming exaggerated or stylized.

The greatest boon to signal creation, however, is provided by the joint exercise of System-2 mindreading in both sender and receiver. Senders are better able to improvise communicative acts, tailoring them to the specific circumstances (including receivers' mental states), while receivers are better able to recognize such acts *as* attempts to communicate, and to infer their meanings. Such mutual sensitivity to others' mental states is the natural home of on-the-fly signal creation.

---

[21] For a Gricean-inspired skeptical reply, see Bar-On and Moore (2017). I myself have more sympathy for the view than they do.

[22] See Cheney and Seyfarth (2008) for an excellent summary of this work.

[23] Kanzi's abilities are detailed in Savage-Rumbaugh et al. (1993).

[24] A classic example is the carbon dioxide produced by animals that mosquitoes use to locate them.

[25] Whether by sustaining a concurrent perceptual representation, allowing for it to be processed more thoroughly, or by recalling a past episode.

I have in other works fleshed out these processes and their relationship to deliberate, flexible mindreading in much greater detail (see, especially, Planer 2017a).[26] What I have not done, or done sufficiently in my view, is explain just how our ancestors came to possess such mindreading capacities in advance of language. I see the present discussion as going some way towards filling in this gap.

10.6 Two Hypotheses About Mindreading and Language Origins

Given the basis of System-2 thinking in working memory, any upgrade in working memory capacities will upgrade System-2 thinking. That includes System-2 mindreading. Hence, one way mindreading might have been upgraded over the course of human evolution is for working memory to have been upgraded. Moreover, mindreading might have been benefited independently of language. This means that the upgrade in mindreading can be called on to help explain even the earliest stages of language evolution.

That is one hypothesis. Alternatively, it might have been that our System-1 mindreading system became more powerful, and that this fueled language evolution. There was some genetic change (or series of changes) that directly upgraded that system. This upgrade might have taken a variety of forms. Most obviously, the set of attitudes (construed broadly) it represents might have expanded. (Note that, on the present account of System-1/System-2 cognitive architecture, any upgrade to the System-1 mindreading system would also upgrade System-2 mindreading. This is obviously a very different System-2 mindreading hypothesis than the one I have developed here, however.)

For a long time the evidence suggested just that. Chimpanzees were proficient mindreaders of goals and intentions. They understood (at least to some extent) perception and attention. But they could not represent what others believe. The implicit mindreading abilities of very young human infants, in contrast, showed that their possession of the concept of belief was innate or at least that it was genetically-primed to a very considerable extent. It appeared that there had been genetic change in our line which had served to enhance our innately-channeled mindreading abilities.

As reviewed above, however, the evidential situation has changed. With chimpanzees and other great apes now passing implicit false belief tests, the experimental evidence for a clear difference between great apes and humans' System-1 mindreading system has been eroded. This is not to say that there are no differences. In my view, that there might be such differences is not implausible, given the much

---

[26] Including whether or not we must conceive of communication involving the joint exercise of flexible mindreading as 'Gricean' or 'ostensive-inferential' communication. I have argued we need not (e.g., Planer 2017a; Planer and Godfrey-Smith 2020; Planer and Sterelny forthcoming).

greater social complexity in our line over the course of the Pleistocene. But these recent results do complicate what was once a very simple and intuitive contrast.

Is there empirical evidence to suggest that working memory capacities have been upgraded in our line? Great ape working memory has not been the target of intense study in the way that their theory of mind has. Moreover, some of the studies purporting to shed light on great ape working memory are more accurately interpreted as involving short term sensory memory or long-term memory (Carruthers 2015). On the other hand, there are studies targeting other great ape cognitive and behavioral capacities which plausibly carry information about their working memory. To me, the most telling evidence for *differences* in human and great ape working memory at present fall into this category. I briefly discuss two such forms of evidence.

The first concerns great ape cognitive and behavioral control. Great apes and children have been compared along this dimension using a variety of tasks. Great apes and young children perform comparably well on A-not-B tasks (Barth and Call 2006). In these tasks, a reward is repeatedly placed at one location (A) but is then switched to another location (B) during the test trial. The participant must inhibit his or her prepotent response to reach to location A and instead reach to B. Great apes and young children also perform similarly on detour-reaching tasks (Vlamings et al. 2010). Here, a participant must inhibit a direct reach for a reward and instead reach around a barrier.

Inhibitory control continues to develop in human children, however (Kochanska et al. 2000; Knopp and Neufeld 2003). In a study comparing the performance of human 3 year olds, 6 year olds, and chimpanzees, 6 year olds outperformed the other groups across a wide range of inhibitory tasks (Herrmann and Tomasello 2015). Most of these tasks required a more sophisticated form of control than A-not-B and detour-reaching tasks. For example, in one task, children and apes learned how to obtain a reward with a rake from inside a box. The experimental setup was then suddenly changed so that raking the reward in the usual way would result in the reward dropping through a trap door that had opened up. Both 3 year olds and chimpanzees found it very difficult to inhibit the prepotent raking response and adopt a modified strategy that avoided the trap, whereas 6 year olds adapted to the novel situation quite readily. This pattern is consistent with the results of earlier research using the reverse contingency paradigm, in which participants must point to the smaller of two rewards in order to obtain the larger reward (Boysen and Berntson 1995; Boyson et al. 1996). Other work in this area suggests that, even where great apes and older children can perform similarly on inhibitory tasks in terms of overall success rate, great apes have to expend more effort to do so (Tecwyn et al. 2013; Voelter and Call 2014).

Bringing this back to working memory: recall that working memory makes constitutive use of inhibitory control. Accordingly, these tests provide a valuable window on the working memory capacities of humans and great apes. Moreover, to the extent that inhibitory control and selective control (i.e., which representations to target and when) share a common neural basis, as it would appear they do, then the above tests carry even more information about working memory in these groups.

While great apes' capacity for self-control is no doubt quite advanced relative many other animals,[27] humans' is (perhaps unsurprisingly) more advanced still.[28]

Research on episodic memory is also relevant here. Episodic memories are representations of past particular events or 'episodes': for example, that of one's first martial arts lesson wherein one was shown various stances to imitate, or that of a friend's birthday dinner last year and the drinks which followed elsewhere. While episodic memories can and typically do involve conceptual information, they are primarily imagistic or sensory-based. When we entertain an episodic memory, we have a conscious experience of having been present at a particular place and time in the past. For this reason, episodic remembering is often described as mentally 'reliving' the past.

The neurobiological evidence strongly suggests episodic memory makes constitutive use of working memory. Neuroimaging studies show that episodic memory recruits many of the same brain areas that make up the bottom-up and top-down attentional networks described earlier (Wagner et al. 2005; Benoit and Schacter 2015; Spreng et al. 2009). The main difference is that episodic memory also makes central use of the medial temporal cortex, hippocampus, parahippocampus, and amygdala. The hippocampus in particular is believed to be crucial, providing the spatial and temporal context for the memory, which in turn serves to organize the memory's other elements into a cohesive whole (Buckner 2010). There is dense connectivity between these regions and the right ventral parietal cortex (Vincent et al. 2006), which (recall) is one of the main areas making up the bottom-up attentional network.[29]

---

[27] They outperform all monkeys on inhibitory control tests, for example (Amici et al. 2008; MacLean et al. 2014).

[28] It is a further question, and not one I will take up here, whether and to what extent the enhanced self-control of humans is genetically based. For the purposes of the present argument, it would not necessarily be a problem if that control was to a significant extent (or even *entirely*, though that is very unlikely) socially learned. There would only be a problem if one could not explain the evolution of enhanced self-control in humans without appeal to language. I see no reason to think that would be the case.

[29] Moreover, as Carruthers (2015) points out, understanding episodic memory as based in working memory is highly explanatory. It is now widely accepted that episodic memory is constructive in nature. What this means, basically, is that episodic remembering is always in part a process of *inferring* what happened at some past time. We do not directly call forth a record of that event which was written to our brain at that time. Rather, the brain generates a version of that event using various bits of information (some particular to that event, some more general), along with its current desires, feelings, values, and so on. Relatedly, it is also widely accepted that there is a tendency for episodic memories to be enriched over time. These features of episodic memory can be neatly explained if we assume such memories make constitutive use of working memory, and hence are globally broadcast. For recall that, globally broadcast representations are available for processing by a large number and range of System-1 systems. That includes systems responsible for categorization and inference, but also affective systems. The constructive nature of episodic memory reflects the operation of System-1 systems which serve to fill in additional detail about the event. (Some of this may simply involve making explicit information that is implicit in the memory.) These alterations to the content of the memory come to be associated with it, and are recalled with the memory in the future.

Researchers have employed various means of testing great ape episodic memory.[30] An early result in this area comes from a study performed by Menzel (1999). This study looked at the ability of chimpanzees to form enduring cognitive maps of an artificial environment. They first learned the whereabouts of various food items in the environment. In a memory test administered 16 h later, it was found that these chimps were able to communicate both the *type* of food and *where* in the environment it was located (they were language trained).[31] A similar experiment was carried out by Mendes and Call (2014) more recently. Food items were stored at various locations within an indoor enclosure, and the chimpanzees were allowed to discover these items. The experimenters baited these same locations with food on following days. After one or two exposures, the chimpanzees' ability to recall these locations was tested. The experimenters found that recall accuracy was high as long as 3 months later.

Using quite a different paradigm, Swartz and colleagues (Swartz et al. 2002, 2004) have looked at the ability of an adult male western lowland Gorilla ('King') to form enduring memories of particular events. For example, in one set of experiments, they used the human 'eye-witness' paradigm, which involves exposure to an unusual event. In King's case, the event in question was either being given an unusual food by an experimenter, exposure to an unfamiliar person, observing an unusual act performed by a familiar person (e.g., swinging a tennis racket), or observing an unfamiliar object (e.g., a toy frog). In a test phase held 15 min later, King was presented with three cards, only one of which referred to an object or person in the event he had witnessed (King had previously learned the meaning of each card). King identified the relevant card significantly more often than chance would predict, though his performance was far from perfect. As in the case of Mendes and Call, it is significant that the learning phase was very short (one- or two-shot), as opposed to a string of reinforcement episodes, which serves to further increase the analogy with human episodic memory.

The problem with these studies as tests of episodic memory is that they fail to test whether great apes associate information about a past event with information about its time of occurrence. Martin-Ordas and colleagues have attempted to address this shortcoming with the use of food-decay experiments (Martin-Ordas et al. 2010; Martin-Ordas and Call 2013). Chimpanzees, bonobos, and orangutans watched as experimenters baited various compartments with either frozen juice (a preferred but perishable food item) or a grape. The apes had previously witnessed how these food items decay over time. If the apes were allowed to retrieve the food after a short time (5 min), they tended to go for the frozen juice. If, however, they were made to wait an hour before retrieving the food, they tended to go for the grape. To explain this behavior, it seems necessary to attribute to the apes representations integrating *what* (food type), *where* (which compartment), and *when* (how long ago) information.

---

[30] For a nice overview of this research (and much else), see Rosati (2017).

[31] This fact about communication is important, as it strongly suggests global broadcasting of the information. The information is available to the 'language' production system, among others.

However, there is reason to think this capacity is much attenuated compared to humans. Dekleva et al. (2011) presented chimpanzees with a similar but more complex test. This study involved three food types, apple sauce, yogurt, and red bell peppers, which the chimpanzees strongly preferred in that order. Over a period of 10–13 days, they experienced apple sauce lasting for 15 min, but not 1 h, and yoghurt lasting for 1 h, but not 5 h. Red bell peppers always remained. During the test phase, chimpanzees watched as two of four compartments were baited with two of the three possible types of food. They then selected a compartment at 15 min, 1 h, and 5 h intervals. Under these conditions, the chimpanzees proved completely unable to form integrated what, where, when memories. Their choosing of preferred, present foods over less preferred, present foods was at chance levels. Indeed, they randomly chose among baited and non-baited compartments. That was true even at the 15 min test interval. How the chimpanzees chose thus appeared to be wholly unguided by any feature of the baiting event.

Limits on great ape episodic memory may be due to differences in their hippocampus and/or surrounding structures compared to humans. However, those limits might also arise from differences in working memory capacity itself. The latter would cohere with the former set of results on self-control (or more strictly speaking, my proffered interpretation of them). It may be that apes are unable to select relevant information at will with the same degree of precision as humans (strategic control). Or it may be that they find it more difficult to sustain selected information in working memory over the course of a problem-solving task. Or both.

To precisely pin down the differences between human and great ape working memory will require a lot more research. However, based on current evidence, it seems likely that there are indeed some real and important differences. This provides support to the hypothesis linking working memory, System-2 thinking, and mindreading, developed in this chapter.

10.7   Conclusion

I have looked at two (compatible) ways theory of mind capacities may have been upgraded in advance of language. Both are thus available to help explain how it is that language first evolved in our line. There are probably others that are plausible. For example, one might think a general increase in our conceptual repertoire might have upgraded our theory of mind capacities. Exposed to the same data set, an individual with a larger conceptual repertoire in general has a better chance of finding more resilient, more explanatory generalizations than one with a smaller repertoire. That includes generalizations about others' mental states and behavior.[32] This

---

[32] Suppose my friend often goes hiking. And suppose he also often goes fishing. These facts support my beliefs that he likes to hike and that he likes to fish, respectively. However, now suppose I acquire the concept of an outdoor sport. Then I might infer that my friend likes outdoor sports. This belief makes different predictions about how my friend will behave under various counterfactual

hypothesis strikes me as plausible, but it also makes substantive assumptions of its own about the cognitive architecture of mindreading. I plan to explore it in future work.

For now, suffice it to say that we have seen there is a plausible alternative to the hypothesis that our theory of mind capacities were enhanced via genetic change to a domain-specific mindreading system which we share with other great apes. This is reassuring, as the evidence in support of that hypothesis is not what it once was.

References

Amici F, Aureli F, Call J (2008) Fission-fusion dynamics, behavioral flexibility, and inhibitory control in primates. Curr Biol 18(18):1415–1419

Apperly I (2010) Mindreaders: the cognitive basis of 'theory of mind'. Psychology Press, New York

Baars BJ (1997) In the theater of consciousness: the workspace of the mind. Oxford University Press, New York

Baars BJ (2005) Global workspace theory of consciousness: toward a cognitive neuroscience of human experience. Prog Brain Res 150:45–53

Baddeley AD, Hitch G (1974) Working memory. In: Bower G (ed) Psychology of learning and motivation. Academic, New York, pp 47–89

Bar-On D (2013) Origins of meaning: must we 'go Gricean'? Mind and Language 28(3):342–375

Bar-On D, Moore R (2017) Pragmatic interpretation and signaler-receiver asymmetries in animal communication. In: Andrews K, Beck J (eds) The Routledge handbook of philosophy of animal minds. Routledge, New York, pp 291–300

Barth J, Call J (2006) Tracking the displacement of objects: a series of tasks with great apes (Pan troglodytes, Pan paniscus, Gorilla gorilla, and Pongo pygmaeus) and young children (Homo sapiens). J Exp Psychol Anim Behav Process 32(3):239

Bartsch K, Wellman HM (1995) Children talk about the mind. Oxford University Press, New York

Bennett J (1978) Some remarks about concepts. Behav Brain Sci 1(4):557–560

Benoit RG, Schacter DL (2015) Specifying the core network supporting episodic simulation and episodic memory by activation likelihood estimation. Neuropsychologia 75:450–457

Blakemore SJ, Frith CD, Wolpert DM (1999) Spatio-temporal prediction modulates the perception of self-produced stimuli. J Cogn Neurosci 11(5):551–559

Boysen ST, Berntson GG (1995) Responses to quantity: perceptual versus cognitive mechanisms in chimpanzees (Pan troglodytes). J Exp Psychol Anim Behav Process 21(1):82

Boysen ST, Berntson GG, Hannan MB, Cacioppo JT (1996) Quantity-based interference and symbolic representations in chimpanzees (Pan troglodytes). J Exp Psychol Anim Behav Process 22(1):76

Bräuer J, Call J, Tomasello M (2007) Chimpanzees really know what others can see in a competitive situation. Anim Cogn 10(4):439–448

Buckner RL (2010) The role of the hippocampus in prediction and imagination. Annu Rev Psychol 61:27–48

Buttelmann D et al (2017) Great apes distinguish true from false beliefs in an interactive helping task. PLoS One 12:4

Butterfill SA, Apperly IA (2013) How to construct a minimal theory of mind. Mind Lang 28(5):606–637

---

circumstances than either of the foregoing beliefs taken alone or together. For example, it predicts that were I to propose that we go mountain biking next weekend, my friend may well agree.

Call J, Hare B, Carpenter M, Tomasello M (2004) 'Unwilling' versus 'unable': chimpanzees' understanding of human intentional action. Dev Sci 7(4):488–498

Carruthers P (2009) An architecture for dual reasoning. In: Evans J, Frankish J (eds) In two minds: dual processes and beyond. Oxford University Press, Oxford, pp 109–127

Carruthers P (2015) The centered mind: what the science of working memory shows us about the nature of human thought. Oxford University Press, Oxford

Cheney DL, Seyfarth RM (2008) Baboon metaphysics: The evolution of a social mind. University of Chicago Press, Chicago, Illinois

Csibra G, Bíró S, Koós O, Gergely G (2003) One-year-old infants use teleological representations of actions productively. Cogn Sci 27(1):111–133

Cohen JR, Lieberman MD (2010) The common neural basis of exerting self-control in multiple domains. Self Control in Society, Mind, and Brain 9:141–162

Cole J (2015) Handaxe symmetry in the Lower and Middle Palaeolithic: implications for the Acheulean gaze. In: Coward F, Hosfield R, Pope M, Wenban-Smith F (eds) Settlement, society and cognition in human evolution: landscapes in mind. Cambridge University Press, Cambridge, pp 234–257

Damasio AR (1994) Descartes' error. Random House, London

Damasio AR (2003) Looking for Spinoza: joy, sorrow, and the feeling brain. Harcourt, New York

Dekleva M, Dufour V, De Vries H, Spruijt BM, Sterck EH (2011) Chimpanzees (Pan troglodytes) fail a what-where-when task but find rewards by using a location-based association strategy. PLoS One 6(2):e16593

Dennett DC (1978) Beliefs about beliefs [P&W, SR&B]. Behav Brain Sci 1(4):568–570

Evans JSB, Over DE (1996) Rationality in the selection task: epistemic utility versus uncertainty reduction. Psychol Rev 103(2):356–363

Fitch WT (2010) The evolution of language. Cambridge University Press, Cambridge

Frankish K (2012) Dual systems and dual attitudes. Mind Soc 11(1):41–51

Gamble CS, Gowlett JA, Dunbar R (2014) Thinking big: the archaeology of the social brain. Thames & Hudson, London

Gazzaley A, Cooney JW, McEvoy K, Knight RT, D'esposito M (2005) Top-down enhancement and suppression of the magnitude and speed of neural activity. J Cogn Neurosci 17(3):507–517

Gogos A, Gavrilescu M, Davison S, Searle K, Adams J, Rossell SL, ..., Egan GF (2010) Greater superior than inferior parietal lobule activation with increasing rotation angle during mental rotation: an fMRI study. Neuropsychologia 48(2):529–535

Hare B, Call J, Agnetta B, Tomasello M (2000) Chimpanzees know what conspecifics do and do not see. Anim Behav 59(4):771–785

Hare B, Call J, Tomasello M (2006) Chimpanzees deceive a human competitor by hiding. Cognition 101(3):495–514

Herrmann E, Tomasello M (2015) Focusing and shifting attention in human children (Homo sapiens) and chimpanzees (Pan troglodytes). J Comp Psychol 129(3):268

Heyes C (2018) Cognitive gadgets: the cultural evolution of thinking. Harvard University Press, Cambridge, MA

Horner V, Whiten A (2005) Causal knowledge and imitation/emulation switching in chimpanzees (Pan troglodytes) and children (Homo sapiens). Anim Cogn 8(3):164–181

Jeannerod M (2006) Motor cognition: what actions tell the self. Oxford University Press, Oxford

Johnson SC (2000) The recognition of mentalistic agents in infancy. Trends Cogn Sci 4(1):22–28

Kahneman D (2002) Maps of bounded rationality: a perspective on intuitive judgment and choice. Nobel Prize Lect 8:351–401

Kaminski J, Call J, Tomasello M (2008) Chimpanzees know what others know, but not what they believe. Cognition 109(2):224–234

Kochanska G, Murray KT, Harlan ET (2000) Effortful control in early childhood: continuity and change, antecedents, and implications for social development. Dev Psychol 36(2):220

Kopp CB, Neufeld SJ (2003) Emotional development during infancy. In: Davidson RJ, Scherer KR, Goldsmith HH (eds) Series in affective science: handbook of affective sciences. Oxford University Press, Oxford, pp 347–374

Krupenye C, Kano F, Hirata S, Call J, Tomasello M (2016) Great apes anticipate that other individuals will act according to false beliefs. Science 354(6308):110–114

Liszkowski U, Carpenter M, Striano T, Tomasello M (2006) 12-and 18-month-olds point to provide information for others. J Cogn Dev 7(2):173–187

Luo Y, Baillargeon R (2005) Can a self-propelled box have a goal? Psychological reasoning in 5-month-old infants. Psychol Sci 16(8):601–608

Luo Y, Baillargeon R (2007) Do 12.5-month-old infants consider what objects others can see when interpreting their actions? Cognition 105(3):489–512

Luo Y, Johnson SC (2009) Recognizing the role of perception in action at 6 months. Developmental Sci 12(1):142–149

Martin-Ordas G, Call J (2013) Episodic memory: a comparative approach. Front Behav Neurosci 7:63

Martin-Ordas G, Haun D, Colmenares F, Call J (2010) Keeping track of time: evidence for episodic-like memory in great apes. Anim Cogn 13(2):331–340

MacLean EL, Hare B (2012) Bonobos and chimpanzees infer the target of another's attention. Anim Behav 83(2):345–353

MacLean EL, Hare B, Nunn CL, Addessi E, Amici F, Anderson RC, ..., Boogert NJ (2014) The evolution of self-control. PNAS 111(20):E2140–E2148

Melis AP, Hare B, Tomasello M (2006) Engineering cooperation in chimpanzees: tolerance constraints on cooperation. Anim Behav 72(2):275–286

Mendes N, Call J (2014) Chimpanzees form long-term memories for food locations after limited exposure. Am J Primatol 76(5):485–495

Menzel CR (1999) Unprompted recall and reporting of hidden objects by a chimpanzee (Pan troglodytes) after extended delays. J Comp Psychol 113(4).426

Moore R (2017a) Social cognition, stag hunts, and the evolution of language. Biol Philos 32(6):797–818

Moore R (2017b) Pragmatics-first approaches to the evolution of language. Psychol Inq 28(2/3):206–210

Moore R (2018) Gricean communication, language development, and animal minds. Philos Compass 13(12):e12550

Onishi KH, Baillargeon R (2005) Do 15-month-old infants understand false beliefs? Science 308(5719):255–258

Penn DC, Povinelli DJ (2007) On the lack of evidence that non-human animals possess anything remotely resembling a 'theory of mind'. Philos Trans R Soc B 362(1480):731–744

Penn DC, Povinelli DJ (2013) The comparative delusion: beyond behavioristic and mentalistic explanations for nonhuman social cognition. In: Metcalfe J, Terrace HS (eds) Agency and joint attention. Oxford University Press, Oxford, pp 62–81

Planer RJ (2017a) Protolanguage might have evolved before ostensive communication. Biol Theory 12(2):72–84

Planer RJ (2017b) Talking about tools: did early Pleistocene hominins have a protolanguage? Biol Theory 12(4):211–221

Planer RJ, Godfrey-Smith P (2020) Communication and representation understood as sender-receiver coordination. Mind Lang. https://doi.org/10.1111/mila.12293

Planer RJ, Sterelny K (forthcoming) From signal to symbol: the evolution of language. MIT Press, Cambridge, MA

Prather SC, Sathian K (2002) Mental rotation of tactile stimuli. Cogn Brain Res 14(1):91–98

Premack D, Woodruff G (1978) Does the chimpanzee have a theory of mind? Behav Brain Sci 1(4):515–526

Pylyshyn ZW (1978) When is attribution of beliefs justified? [P&W]. Behav Brain Sci 1(4):592–593

Rosati AG (2017) Chimpanzee cognition and the roots of the human mind. In: Muller M, Wrangham R, Pilbeam D (eds) Chimpanzees and human evolution. Harvard University Press, Cambridge, MA, pp 703–745

Rosati AG, Hare B (2009) Looking past the model species: diversity in gaze-following skills across primates. Curr Opin Neurobiol 19(1):45–51

Savage-Rumbaugh ES, Murphy J, Sevcik RA, Brakke KE, Williams SL, Rumbaugh DM, Bates E (1993) Language comprehension in ape and child. Monogr Soc Res Child Dev 58(3/4):1–252

Schlottmann A, Ray E (2010) Goal attribution to schematic animals: do 6-month-olds perceive biological motion as animate? Dev Sci 13(1):1–10

Schwartz BL, Colon MR, Sanchez IC, Rodriguez I, Evans S (2002) Single-trial learning of 'what' and 'who' information in a gorilla (Gorilla gorilla gorilla): implications for episodic memory. Anim Cogn 5(2):85–90

Schwartz BL, Meissner CA, Hoffman M, Evans S, Frazier LD (2004) Event memory and misinformation effects in a gorilla (Gorilla gorilla gorilla). Anim Cogn 7(2):93–100

Scott RM, Baillargeon R (2009) Which penguin is this? Attributing false beliefs about object identity at 18 months. Child Dev 80(4):1172–1196

Scott RM, Baillargeon R, Song HJ, Leslie AM (2010) Attributing false beliefs about non-obvious properties at 18 months. Cogn Psychol 61(4):366–395

Scott-Phillips T (2014) Speaking our minds: why human communication is different, and how language evolved to make it special. Palgrave Macmillan, London

Shepard RN (1978) The mental image. Am Psychol 33(2):125

Shepard RN, Feng C (1972) A chronometric study of mental paper folding. Cogn Psychol 3(2):228–243

Shepard RN, Metzler J (1971) Mental rotation of three-dimensional objects. Science 171(3972):701–703

Sloman SA (1996) The empirical case for two systems of reasoning. Psychol Bull 119(1):3

Southgate V, Senju A, Csibra G (2007) Action anticipation through attribution of false belief by 2-year-olds. Psychol Sci 18(7):587–592

Southgate V, Chevallier C, Csibra G (2010) Seventeen-month-olds appeal to false beliefs to interpret others' referential communication. Dev Sci 13(6):907–912

Sperber D, Origgi G (2012) A pragmatic perspective on the evolution of language. In: Wilson D, Sperber D (eds) Meaning and relevance. Cambridge University Press, Cambridge, pp 331–338

Spreng RN, Mar RA, Kim AS (2009) The common neural basis of autobiographical memory, prospection, navigation, theory of mind, and the default mode: a quantitative meta-analysis. J Cogn Neurosci 21(3):489–510

Stanovich K (1999) Who is Rational? Studies of individual differences in reasoning. Lawrence Erlbaum.

Sterelny K (2012) Language, gesture, skill: the co-evolutionary foundations of language. Philos Trans R Soc B 367(1599):2141–2151

Surian L, Caldi S, Sperber D (2007) Attribution of beliefs by 13-month-old infants. Psychol Sci 18(7):580–586

Tecwyn EC, Thorpe SK, Chappell J (2013) A novel test of planning ability: great apes can plan step-by-step but not in advance of action. Behav Process 100:174–184

Tomasello M (2008) Origins of human communication. MIT Press, Cambridge, MA

Tomasello M, Carpenter M (2005) Intention reading and imitative learning. Perspectives on Imitation: From Neuroscience to Social Science 2:133–148

Tomasello M, Davis-Dasilva M, Camak L, Bard K (1987) Observational learning of tool-use by young chimpanzees. Hum Evol 2(2):175–183

Tomasello M, Call J, Hare B (1998) Five primate species follow the visual gaze of conspecifics. Anim Behav 55(4):1063–1069

Vincent JL, Snyder AZ, Fox MD, Shannon BJ, Andrews JR, Raichle ME, Buckner RL (2006) Coherent spontaneous activity identifies a hippocampal-parietal memory network. J Neurophysiol 96(6):3517–3531

Vlamings PH, Hare B, Call J (2010) Reaching around barriers: the performance of the great apes and 3-5-year-old children. Anim Cogn 13(2):273–285

Völter CJ, Call J (2014) Younger apes and human children plan their moves in a maze task. Cognition 130(2):186–203

Wagner AD, Shannon BJ, Kahn I, Buckner RL (2005) Parietal lobe contributions to episodic memory retrieval. Trends Cogn Sci 9(9):445–453

Wellman HM, Woolley JD (1990) From simple desires to ordinary beliefs: the early development of everyday psychology. Cognition 35(3):245–275

Wellman HM, Cross D, Watson J (2001) Meta-analysis of theory-of-mind development: The truth about false belief. Child Dev 72(3):655–684

Wimmer H, Perner J (1983) Beliefs about beliefs: representation and constraining function of wrong beliefs in young children's understanding of deception. Cognition 13(1):103–128

Wolpert DM, Flanagan JR (2001) Motor prediction. Curr Biol 11(18):R729–R732

Wolpert DM, Doya K, Kawato M (2003) A unifying computational framework for motor control and social interaction. Philos Trans R Soc B 358(1431):593–602

Wynn T, Coolidge FL (2011) The implications of the working memory model for the evolution of modern cognition. Int J Evol Biol. https://doi.org/10.4061/2011/741357

Chapter 11
The Acheulean Origins of Normativity

Ceri Shipton, Mark Nielsen, and Fabio Di Vincenzo

Abstract 'Normativity' refers to the human conformity to the behavioral modes of a society, which underpins diverse aspects of our behavior, including symbolism, cooperation, and morality. It has its developmental basis in overimitation, the uniquely human bias towards replicating the intentional actions of a demonstrator, regardless of their causal relevance. Using evidence from stone tool technology, we suggest that both overimitation and normativity have their evolutionary origins in the Acheulean cultural tradition. Overimitation can be seen in arbitrary biases in Acheulean toolmaking; while normativity is evident in geographically and temporally restricted sub-types of Acheulean tools, which do not appear to be functional specializations. We argue that normativity would have conferred particular advantages for the Acheulean niche of cooperative hunting of mega-herbivores and living in large groups, through coordinating hunting strategies, promoting equitable sharing, and enforcing the punishment of free-riders.

Keywords Overimitation · Handaxes · Cleavers · Cooperation · Large groups · Food-sharing

C. Shipton (✉)
Centre of Excellence for Australian Biodiversity and Heritage, College of Asia and the Pacific, Australian National University, Acton, ACT, Australia

Institute of Archaeology, University College London, London, UK
e-mail: ceri.shipton@anu.edu.au

M. Nielsen
School of Psychology, University of Queensland, Brisbane, QLD, Australia

F. Di Vincenzo
Dipartimento di Biologia Ambientale, Sapienza Università di Roma, Rome, Italy

© Springer Nature Switzerland AG 2021
A. Killin, S. Allen-Hermanson (eds.), *Explorations in Archaeology and Philosophy*, Synthese Library 433, https://doi.org/10.1007/978-3-030-61052-4_11

11.1   Introduction

Humans have a tendency to conform to the established behavioral modes of a social group that exist independently of dyadic relationships (Claidière and Whiten 2012; Sripada and Stich 2006). This conformity is often enhanced by social and moral judgements of what are considered appropriate or desirable behaviors, with specific variants that may be described as the norms of a society. The power of such norms are intuitively underappreciated (Anderson and Dunning 2014; Cialdini 2007), yet they are a key foundational trait for many important aspects of our behavior (Roughley and Bayertz 2019). Normativity is essential for symbolism as it is how we come to the collective, rather than dyadic, agreement of the meanings of particular arbitrary vocalizations or images (Itkonen 2008). It allows us to make assumptions about the intentions of others and thereby more accurately interpret the meaning of their communications (Tomasello 2008). Normative ways of doing cooperative tasks help children to understand role differentiation (Tomasello and Hamann 2012); and normativity fosters the expectation of similar behavior that is essential to the initiation of delayed reciprocity relationships (McElreath et al. 2003). Normativity is also critical to human morality: enabling rapid moral judgement and decision making (Conway and Gawronski 2013); promoting the application of reciprocal empathy to all group members regardless of prior dyadic relationships (Tomasello and Vaish 2013); fostering a sense of group affiliation and an expectation of group loyalty (Rossano 2012; Lakin et al. 2003); and moving the burden of punishing transgressors from victims or their relatives to the group (Henrich et al. 2006; Boyd et al. 2003).

Understanding the context of the evolutionary emergence of normativity is therefore critical to understanding the evolution of language, ultra-sociality, and morality. Existing evolutionary accounts of normativity are few and focus on its more elaborate manifestations in formal ritual behaviors in *Homo sapiens* (Sterelny 2014, 2019), and penecontemporaneous elaborate stone tool making sequences (Högberg and Lombard 2019). In this chapter we explore recent hypotheses that normativity (and ritual) has deeper roots in the Acheulean cultural tradition that predates our species (Finkel and Barkai 2018; Shipton 2019b).

As might be expected for a trait that is so pervasive across different spheres of human behavior, there are several different types of normativity and parameters according to which it can vary (Anderson and Dunning 2014). Three broad categories of normative behavior have been defined: conventions that are widespread in a society, but for which there is little consequence for flouting (e.g., in Western cultures holding forks in the left or right hand); social norms where the collective judgment of society enforces certain behaviors (such as chewing with your mouth closed or not burping during meals); and moral norms where internalized social motivations influence behaviors through emotional responses (such as not consuming certain animals) (Anderson and Dunning 2014; Heywood 1996; Bicchieri 2005). Norms may be either prescriptive for a particular behavior (we do it like this), or proscriptive against behavioral variants (we do not do that). Norms may correspond to one or more of these types depending on the society and will also vary between individuals. Norms can transition between types, as well as be lost or created, during the history of a society. Some norms are highly variable between societies, but there are general norm categories that are universal, such as against incest, the unsanctioned killing of group members, for cooperation and sharing, and the punishment of norm violations (Sripada and Stich 2006; Cashdan 1989; Boehm 2012; Joyce 2006). While most norms do not have an innate basis (incest being a possible exception), the regular acquisition of norms in ontogeny suggests that normativity itself is innate, perhaps including norms for certain realms of behavior. In human behavior studies, identifying and differentiating norms often depends upon people explicitly stating their desires and motivations, data that is not archaeologically knowable. However, normativity is also materially instantiated in artifacts (Sinha 2009), so the archaeological record has potential to inform about its origins.

11.2   The Developmental Emergence of Normativity

Regardless of where, we are born into environments that are heavily socially structured. The burden is thus on the developing child and his/her immediate community to identify and subsequently encourage engagement in normative behavior. Although indoctrination into a socially adherent life likely occurs from birth, clear evidence of its expression in behavior is experimentally evident in the third year. For example, in one experiment an adult showed children aged 2–3 years a wooden block being relocated from one side of a board to another (Rakoczy et al. 2008). In one instance the block was pushed using a small tool and labelled with the novel verb 'daxing' ('This is daxing. Now I am daxing!'). On the other, the board was lifted such that the block slid, an action labelled as a mistake ('oops! This is not how daxing goes!'). A puppet was then introduced, proceeding to undertake the second action, stating that it was 'daxing'. The majority of children responded by scolding the puppet that this was not 'daxing', demonstrating what the appropriate action was. This suggests the children understood what was normatively appropriate and were motivated to correct another's errant behavior, even though they themselves were not at risk of being directly impacted (for a review, see Rakoczy and Schmidt 2013). Similarly, in an action-copying task with large samples of children and bonobos (Pan paniscus), the former replicated visibly causally irrelevant actions whereas none of the latter did (Clay and Tennie 2018). This confirms experimental results from chimpanzees (Pan troglodytes) and orangutans (Pongo pygmaeus) suggesting what has come to be known as overimitation is a uniquely human trait (Nielsen and Susianto 2010; Horner and Whiten 2005).

As alluded to, the notion of overimitation refers to the engagement in normatively-guided social-learning behavior whereby experimentally induced responses to the demonstration of new behavior and subsequent copying incorporates each step of the modelled sequence, including actions that are obviously causally unnecessary (for a recent literature review, see Hoehl et al. 2019). For example, in one study an adult experimenter showed children aged 2–13 years living in a large, industrialized Western city, and in remote Bushman communities of the Kalahari how to retrieve a toy from a closed box (Nielsen and Tomaselli 2010). Although the box could easily be opened by hand, the experimenter complicated the demonstration by swiping a miscellaneous object across the top of the box in a causally irrelevant manner, then using the same object to open the box. Regardless of their cultural background (i.e., the phenomenon is not culturally specific), the children replicated the model's object use and incorporated the causally irrelevant actions into their response. Older children and adults are more inclined towards this behavior, indicating overimitation does not result from a lack of causal understanding in younger children (e.g., McGuigan et al. 2011).

Highlighting how overimitation can be seen as a normative act, Kenward and colleagues (2011) modelled a sequence of relevant and irrelevant actions to 3- and 5-year-old children, then asked them what they intended to do. The majority of children reported that they were going to perform all of the actions, and then, when asked to explain why, commonly responded by noting the causally relevant actions were necessary while expressing uncertainty for the value of unnecessary actions (see also Kenward 2012; Keupp et al. 2013). It is argued that the children thought they should perform the irrelevant actions because, in the course of the demonstration, they incorporated these into the arbitrary norms about the appropriate way to use the new artifact.

11.3 Overimitation in the Acheulean

As the forgoing section outlined, the developmental foundation of normativity is to be found in the uniquely human trait of overimitation. Previously we have laid out the case for the emergence of overimitation in the Acheulean (Nielsen 2012; Shipton and Nielsen 2015; Shipton 2019a): the longest lasting and most widespread of all prehistoric cultures. The Acheulean is defined by the presence of two distinctive large (5–35 cm long) symmetrical stone tool types called handaxes and cleavers. Cleavers have a large unworked bit as their principal cutting edge, similar to a modern adze or chisel, and are typically made on large flakes of stone and finished with minimal bifacial retouch. Handaxes are bifacially worked to a typically tear-drop shape with their sinuous durable cutting edge extending around much of their perimeter. The Acheulean first emerges in East Africa around 1.75 million years ago (Beyene et al. 2013) and is the dominant culture in Africa, the Middle East, India, and Europe until at least 0.3 million years ago (Deino et al. 2018).

Considering the correspondence between conceptually different manufacturing methods and the creation of closely related but distinct tool types (handaxes and cleavers), it was suggested there was a propensity for imitation among Acheulean hominins (Shipton 2010). Such a propensity would explain the reliable maintenance of manufacturing methods at individual sites over the course of tens of thousands of years (Sharon et al. 2011). The unparalleled ubiquity and longevity of the Acheulean tradition (Shipton 2020), suggests that even overimitation may have been a feature of hominin behavior among those species who manufactured it, which included *Homo erectus* and *Homo heidelbergensis* (Nielsen 2012). In particular, the causal opacity (Lycett and Eren 2019) and hierarchical complexity (Muller et al. 2017) of the steps involved in producing many of these tools may have required overimitation for their successful reproduction across generations of knappers (Shipton and Nielsen 2015). Notably, stone knapping has a discernable causal outcome that is linked to a functional purpose, unlike the actions demonstrated in experimental scenarios of overimitation (e.g., waving a stick in the air). However, in moving from a focus on the purely functional to expressions of tool construction that are socially constrained, we believe the emergence of overimitation can be found.

Some evidence for this comes from more complex cleaver manufacturing sequences that have arbitrary sidedness bias (Shipton 2019a), even though the finished products are symmetrical (Shipton et al. 2018b). The site of Chirki in India, estimated to be over 780,000 years old (Sangode et al. 2007), has a cleaver manufacturing sequence with four distinct stages (Corvinus 1983). Here there are predominant but arbitrary and opposing orientations to both the large preparatory bit strike and that of the large flake blank for the tool itself (Fig. 11.1). As hominins were able to remove large flakes from both sides of the core, there is no technological reason why the direction of the bit creation and blank strikes could not be reversed.

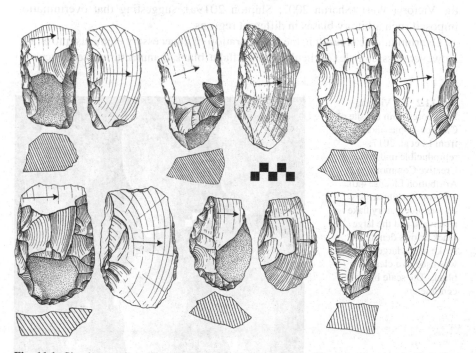

Fig. 11.1 Six cleavers from Chirki, oriented with their bits (working edges) upwards. Note that the preparatory scar to create the bit and the large flake on which these tools were made are consistently struck from the same (arbitrary) directions (indicated by the black arrow). The scale bar is in centimeters

In southern Africa, the Victoria West technique is one of the most complex cleaver production strategies, foreshadowing some of the characteristics of later Middle Stone Age Levallois technology. At the site of Canteen Kopje this technology dates to around 1 million years ago (Beaumont and Vogel 2006). Here large flake cleaver blanks were exclusively struck from the same arbitrary direction on Victoria West cores (Fig. 11.2) (Li et al. 2017). The Victoria West at Canteen Kopje is found throughout a 1.8 m thick layer, that may represent ~300,000 years of occupation (Li et al. 2017), indicating the robust reproduction of this arbitrary sidedness bias.

We do not think this sidedness bias can be explained by handedness for three reasons. Firstly, replicative knapping indicates that producing large flake blanks for Acheulean bifaces typically requires a large hammerstone held in two hands with the core stabilized by another means (Shipton et al. 2009; Madsen and Goren-Inbar 2004). Secondly, prior to the large flake blank being struck, these cores were symmetrical about their long axis so they could have been struck from the bottom right corner instead of the bottom left. Thirdly, following on from this there is a counter archaeological example where similar cores were in fact struck from the bottom right corner. In north Africa, the Tabelbala-Tachengit is a method of cleaver production similar to Victoria West, with hierarchical bifacial cores, tapering in plan view, from which individual cleaver blanks were struck. Crucially, the Tabelbala-Tachengit cleaver blanks were consistently struck from the other side of the core to the Victoria West (Sharon 2007; Shipton 2019a), suggesting that overimitation imposed such arbitrary biases in different regions.

Overimitation can result in the incorporation of unnecessary elements into a task. However, less complex Acheulean large flake blank manufacturing sequences do

Fig. 11.2 Six Victoria West cores from the site of Canteen Kopje (modified from Li et al. 2017; reproducible under the Creative Commons Attribution License 4.0). The small white arrows denote preparatory flake removals, and the large black arrows denote the strike of the large flake that was used as a cleaver blank. The scale bar is in centimeters

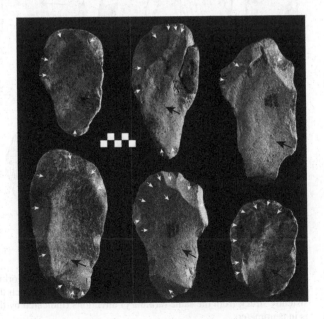

not seem to have the same sidedness biases as those of Victoria West and Tabelbala-Tachengit (Sharon 2007; Shipton 2019a), suggesting less reliance on social structuring and more toleration of variation in simpler sequences where causal understanding could be more rapidly achieved when learning to knap. In fact, overimitation likely filtered out the severe errors of inexperienced knappers while allowing neutral variations to be reproduced. These variations could then have been used as a substrate for adapting knapping sequences to new material types and clast forms (Di Paolo and Di Vincenzo 2018). This combination of reproduction of essential elements and toleration of neutral variation allowed Acheulean hominins to produce the same tool types on a broad variety of rocks and even on elephant bone (Costa 2010).

11.4 Handaxe Sub-Types

If overimitation was an emerging feature of Acheulean hominin behavior, then there was the developmental potential for normativity in Acheulean hominin groups. In this section we discuss handaxe sub-types as possible evidence for normativity. All Acheulean handaxe and cleaver assemblages may be described as varying around a modal shape (Schick and Clark 2003; Shipton 2013), but in some cases there appear to be distinct sub-types within an assemblage. Examples of this include the 'zoomorphic' bifaces from Bentadjine, north Africa (Berlant and Wynn 2018), the 'mandolin' ficron-cleavers from the Zebra Valley, southern Africa (Hardaker 2020), or the large ficron handaxes from Lewa Swamp, east Africa (Fig. 11.3). Some of the clearest expressions of handaxe sub-types are to be found in the late Acheulean of southern Britain (White et al. 2019; Wenban-Smith 2004; Bridgland and White 2014). Here multiple distinct types occur within the same assemblages such as the tranchet cordates and ovates from Boxgrove, or the plano-convex and twisted profile handaxes from Hitchin (Fig. 11.4). It has long been thought that the diversity of

Fig. 11.3 Three large ficrons from Lewa Swamp, east Africa

Fig. 11.4 Pairs of handaxes from the British late Acheulean sites of Boxgrove (**a** and **b**, **e** and **f**) and Hitchin (**c** and **d**, **g** and **h**). **a** and **b** are ovate tranchet, **c** and **d** are plano-convex, **e** and **f** are cordate tranchet, **g** and **h** are twisted profile. Tranchet scars are shown with arrows. The handaxes are shown at a standardized length to facilitate comparisons of shape. Note that *C* and *D* are elongate as well as having asymmetric profiles, while *G* and *H* are more rounded as well as having twisted profiles

sub-types goes beyond what functional needs would warrant (Bordaz 1970; Breuil 1932), and indeed experimental evidence suggests that handaxe shape variation does not confer appreciable functional differences in cutting efficiency (Key and Lycett 2017a). How then are such sub-types to be explained?

Conformity to the most common model but with a relatively wide range of variation around it, might mean handaxes perceived by archaeologists as different sub-types are actually at different ends of a unimodal continuum of variation. However, we think the diversity of some forms in British Acheulean sites belies this proposition. The plano-convex and twisted profile from Hitchin for example, are doubly distinctive in their planform *and* edge position and shape (Fig. 11.4), making it unlikely if not impossible for them to lie on the same continuum of form (Shipton and White 2020). The final Acheulean site of Broom displays a wide variety of different handaxe types in an assemblage that does not appear to be a palimpsest of multiple occupations (Hosfield and Chambers 2009; Shipton and White 2020). To arrive at such unusual handaxe shapes requires a high degree of knapping skill and a plan to achieve that particular shape from the outset of the knapping sequence.

Another possibility is that overimitation of the production process led to the exaggeration of minor variations that then became distinctive features. However, the small nodules of flint on which the British handaxes were made are irregular in shape and have unpredictable internal variations, so that knapping sequences need to be not merely accurately reproduced, but dynamically adapted to arrive at the same final form (García-Medrano et al. 2019). Experimental flint knapping shows that different canalized flaking routines do not result in the distinctive shape types seen in the British Acheulean (Shipton and Clarkson 2015). In fact, the plano-convex handaxes from Hitchin were made on both flake and cobble blanks which involve conceptually and procedurally contrasting production sequences.

A third possibility is that the idiosyncratic handaxe shapes favored by a handful of experts were preferentially replicated due to prestige bias. However, transmission chain experiments in which handaxe forms are reproduced in other media, indicate rapid departures from the starting forms within a handful of handaxe generations (Schillinger et al. 2016; Shipton et al. 2018b). Maintaining these forms over tens and even hundreds of thousands of years, as seems to have occurred in different temperate periods during the last 0.5 million years in Britain (White et al. 2018), must have involved a stronger motivation for particular sub-types than prestige bias.

The forms evident in the British late Acheulean are difficult to replicate and rare in the wider Acheulean world. One of the few other instances of these types are twisted profile pieces from Gombore in the Melka Kunture Formation in Ethiopia, where they occur in a particular horizon dated to around 0.9 ma, but not before this (Gallotti et al. 2010; Gallotti and Mussi 2017). The distance in space and time between twisted pieces from Gombore and a series of ~0.4 ma sites in Britain (White et al. 2019) indicates they are convergent forms. But in both cases the twisted profile pieces can constitute over a third of the handaxe assemblages, so that they were a repeatedly targeted form. We suggest that in the last million years, normative conventions to produce certain handaxe shapes were a feature of hominin behavior.

Many of the handaxes from the British site of Boxgrove (0.5 ma) have been finished with tranchet flakes, where the tip of the piece is removed in a single large blow oblique to the long axis of the tool (Fig. 11.4). While producing a particularly sharp tip, this is a highly risky strategy liable to break the tool in two. Making ficrons, evident at both Lewa Swamp (Fig. 11.3) as well as British sites such as Swanscombe is also risky as the narrow tips are easy to break in manufacture or use. Hominins seem to have been highly motivated to reproduce the forms evident in some Acheulean sites of the last 0.5 million years, perhaps as prescriptive social norms.

11.5   Why Normativity?

Conforming to norms can be costly and even maladaptive in strict survivalist terms (Anderson and Dunning 2014; Prentice and Miller 1996), so the evolution of normativity must have conferred an important advantage on the hominin lineages in which it appeared. We suggest that its selection may be related to the mega-herbivore hunting niche of Acheulean hominins (Shipton 2019b). The Acheulean emerged at a time when mega-herbivores, in particular elephants, were far more common than they have been in the recent past (Faith et al. 2019) and Acheulean artifacts are often associated with the butchered remains of these mega-herbivores (Barkai 2019b; Ben-Dor et al. 2011). One of the key advantages of the handaxe over simple stone flakes is the ergonomic shape and long durable bifacial cutting edge, which experiments indicate make it particularly efficient in lengthy cutting tasks such as elephant butchery (Gingerich and Stanford 2016; Toth and Schick 2019; Galán and Domínguez-Rodrigo 2014; Key and Lycett 2017b; Jones 1980). And indeed, there is direct residue evidence that Acheulean tools were used in elephant butchery (Solodenko et al. 2015).

From the outset Acheulean hominins had primary access to elephant carcasses, suggesting they were often hunted rather than scavenged (Domínguez-Rodrigo et al. 2014; Domíngucz-Rodrigo and Pickering 2017; Mosquera et al. 2015). Unsurprisingly, given the size discrepancy, hunting these animals appears to have been a group activity with instances of multiple handaxes associated with the remains of elephants and other mega-herbivores (Goren-Inbar et al. 1994; Piperno and Tagliacozzo 2001; Roberts and Parfitt 1999). Cooperative hunting may have given Acheulean hominins a tactical advantage over animals that normally rely on their size for predator defense. The transport of elephant heads to Acheulean sites suggests cooperation of more than two individuals as they can weigh over 400 kg (Agam and Barkai 2016). Elephants provide an extremely large amount of meat and therefore maximizing the fitness obtained from their carcasses required the transport and sharing of carcass elements with group members not involved in the hunt (Kaplan et al. 2000; Barkai 2019a; Isaac 1978). In comparison to other stone artifact types, handaxes appear to have been preferentially curated and deposited at repeatedly used Acheulean occupation sites (Shipton et al. 2018a; Pope and Roberts 2005), perhaps because of their role in the secondary butchery and sharing of transported carcass elements.

We hypothesize that the prominence of cooperative hunting and food sharing in the niche of Acheulean hominins would have created the selective conditions for the evolution of normativity. Some Acheulean sites are very large (e.g., Foley and Lahr 2015; Jennings et al. 2015), much larger than those of the preceding Oldowan period in the same landscapes (Domínguez-Rodrigo and Pickering 2017; de la Torre 2011; Rogers et al. 1994), suggesting hominin group sizes were correspondingly larger. In larger groups it becomes more difficult to develop and maintain close dyadic relations with all individuals (Roberts et al. 2009), so that deciding whether someone is a reliable cooperative partner is more difficult.

Reliable cooperation is particularly salient in elephant hunting, undoubtedly a dangerous activity in which you would want to be sure that all individuals involved are engaged in the common effort and not attempting to shirk or use their own maverick strategy. A social hunting norm would be a more efficient way to ensure commitment of all participants to the strategy than individual dyadic relationships.

Likewise, when transporting and apportioning elephant carcass elements, a social norm for sharing would be more efficient than determining individual contributions to the hunt or relying on agonistic dominance interactions. Food sharing enforced by social norms is a universal behavior in recent hunter-gatherer societies (Gurven 2004). Indeed, social norms have been suggested to have evolved in the context of multi-individual food-sharing, where dyadic and kinship relations would otherwise result in stochastic food shortages for group members who ultimately enhance cooperative endeavors (Kaplan et al. 2005). Furthermore, formal modeling suggests that for cooperation to become a stable strategy in hunter-gatherer groups of around 100 individuals or more, norms of group punishment for free-riders are an essential strategy (Boyd et al. 2003).

11.6 Conclusion

Normativity permeates and underpins diverse aspects of our behavior. It is what allows us to describe societies in a Durkheimian sense as having a separate existence from the individuals that constitute them. Determining when and why it emerged in our evolutionary history is thus a question worthy of some attention. Our argument is summarized in the following.

The early ontogenetic expression of normativity is overimitation (Nielsen et al. 2015; Rossano 2012; Keupp et al. 2013). It is possible that children overimitate because they follow norms, or equally that they follow norms as an outcome of overimitating. We do not think they can be developmentally separated, however. Rather, both are expressions of the *Homo*-specific inclination to satisfy social motivations, to do things not because they serve an immediate functional purpose but because they facilitate social connectivity. This trait is apparent in Acheulean hominin behavior (Nielsen 2012; Shipton and Nielsen 2015), in particular the arbitrary conformity in complex manufacturing sequences from around 1 million years ago (Shipton 2019a). Arbitrary conformity is also evident in the end products of Acheulean knapping sequences from around the same time, which we think is indicative of normative conventions in these societies. By the late Acheulean in the last 0.5 million years, handaxe sub-types are more prominent—including ones that are difficult to make and vulnerable to breakage. This suggests an enhanced motivation to produce them, such that they may be regarded as social norms with active pressure to obtain these sub-types.

We speculate that normativity may have evolved as a solution to three related problems in the Acheulean niche of mega-herbivore hunting in large groups. Firstly, in coordinating a common strategy in a highly dangerous activity. Secondly, in promoting the transport and equitable sharing of carcass elements even among group members who did not participate in the hunt. And thirdly in ensuring the group responsibility for the punishment of cowards, mavericks, and ungenerous individuals.

Around a million years ago our ancestors moved away from a tool industry characterized by a priority of causal-functional construction, to one incorporating socially important elements. This shift is likely to be the bedrock of normative behavior, from which then sprung human autapomorphies like symbolism and ritual (see Nielsen 2018, 2019; Nielsen et al. in press). Critical testing of this hypothesis promises to yield new insights into what it took to become human.

References

Agam A, Barkai R (2016) Not the brain alone: the nutritional potential of elephant heads in Paleolithic sites. Quat Int 406:218–226

Anderson JE, Dunning D (2014) Behavioral norms: variants and their identification. Soc Personal Psychol Compass 8(12):721–738

Barkai R (2019a) An elephant to share: rethinking the origins of meat and fat sharing in Palaeolithic societies. In: Lavi N, Friesem DE (eds) Towards a broader view of hunter-gatherer sharing. McDonald Institute for Archaeological Research, Cambridge, pp 153–167

Barkai R (2019b) When elephants roamed Asia: the significance of proboscideans in diet, culture and cosmology in Paleolithic Asia. In: Kowner R, Bar-Oz G, Biran M, Shahar M, Shelach-Lavi G (eds) Animals and human society in Asia. Springer, Cham, pp 33–62

Beaumont PB, Vogel JC (2006) On a timescale for the past million years of human history in central South Africa. S Afr J Sci 102:217–228

Ben-Dor M, Gopher A, Hershkovitz I, Barkai R (2011) Man the fat hunter: the demise of Homo erectus and the emergence of a new hominin lineage in the Middle Pleistocene (ca. 400 kyr) Levant. PLoS One 6(12):e28689

Berlant T, Wynn T (2018) First sculpture: handaxe to figure stone. Nasher Scupture Center, Dallas

Beyene Y, Katoh S, WoldeGabriel G, Hart WK, Uto K, Sudo M et al (2013) The characteristics and chronology of the earliest Acheulean at Konso, Ethiopia. Proc Natl Acad Sci 110(5):1584–1591

Bicchieri C (2005) The grammar of society: the nature and dynamics of social norms. Cambridge University Press, Cambridge

Boehm C (2012) Moral origins: the evolution of virtue, altruism, and shame. Basic Books, New York

Bordaz J (1970) Tools of the Old and New Stone Age. Natural History Press, New York

Boyd R, Gintis H, Bowles S, Richerson PJ (2003) The evolution of altruistic punishment. Proc Natl Acad Sci 100(6):3531–3535

Breuil H (1932) Les industries à éclats du Paléolithique ancien: le clactonien. Libraire Ernest Leroux, Paris

Bridgland DR, White MJ (2014) Fluvial archives as a framework for the Lower and Middle Palaeolithic: patterns of British artefact distribution and potential chronological implications. Boreas 43(2):543–555

Cashdan E (1989) Hunters and gatherers: economic behavior in bands. Econ Anthropol 1989:21–48

Cialdini RB (2007) Descriptive social norms as underappreciated sources of social control. Psychometrika 72(2):263

Claidière N, Whiten A (2012) Integrating the study of conformity and culture in humans and non-human animals. Psychol Bull 138(1):126

Clay Z, Tennie C (2018) Is overimitation a uniquely human phenomenon? Insights from human children as compared to bonobos. Child Dev 89(5):1535–1544

Conway P, Gawronski B (2013) Deontological and utilitarian inclinations in moral decision making: a process dissociation approach. J Pers Soc Psychol 104(2):216

Corvinus G (1983) A survey of the Pravara Rivers system in Western Maharashtra, India: the excavations of the Acheulian site of Chirki-on-Pravara. Tübinger Monographien zur Urgeschichte, Tübingen

Costa AG (2010) A geometric morphometric assessment of plan shape in bone and stone Acheulean bifaces from the Middle Pleistocene site of Castel di Guido, Latium, Italy. In: Lycett S, Chauhan P (eds) New perspectives on Old Stones. Springer, Cham, pp 23–41

de la Torre I (2011) The Early Stone Age lithic assemblages of Gadeb (Ethiopia) and the Developed Oldowan/early Acheulean in east Africa. J Hum Evol 60(6):768–812

Deino AL, Behrensmeyer AK, Brooks AS, Yellen JE, Sharp WD, Potts R (2018) Chronology of the Acheulean to Middle Stone Age transition in eastern Africa. Science 360(6384):95–98

Di Paolo LD, Di Vincenzo F (2018) Emulation, (over)imitation and social creation of cultural information. In: Di Paolo LD, Di Vincenzo F, De Petrillo F (eds) Evolution of primate social cognition. Springer, Cham, pp 267–282

Domínguez-Rodrigo M, Pickering TR (2017) The meat of the matter: an evolutionary perspective on human carnivory. Azania 52(1):4–32

Domínguez-Rodrigo M, Bunn H, Mabulla A, Baquedano E, Uribelarrea D, Pérez-González A et al (2014) On meat eating and human evolution: a taphonomic analysis of BK4b (Upper Bed II, Olduvai Gorge, Tanzania), and its bearing on hominin megafaunal consumption. Quat Int 322:129–152

Faith JT, Rowan J, Du A (2019) Early hominins evolved within non-analog ecosystems. Proc Natl Acad Sci 116(43):21478–21483

Finkel M, Barkai R (2018) The Acheulean handaxe technological persistence: a case of preferred cultural conservatism? Proc Prehistor Soc 84:1–19

Foley RA, Lahr MM (2015) Lithic landscapes: early human impact from stone tool production on the central Saharan environment. PLoS One 10(3):e0116482

Galán A, Domínguez-Rodrigo M (2014) Testing the efficiency of simple flakes, retouched flakes and small handaxes during butchery. Archaeometry 56(6):1054–1074

Gallotti R, Mussi M (2017) Two Acheuleans, two humankinds: from 1.5 to 0.85 Ma at Melka Kunture (Upper Awash, Ethiopian highlands). J Anthropol Sci 95:1–46

Gallotti R, Collina C, Raynal J-P, Kieffer G, Geraads D, Piperno M (2010) The early Middle Pleistocene site of Gombore II (Melka Kunture, Upper Awash, Ethiopia) and the issue of Acheulean bifacial shaping strategies. Afr Archaeol Rev 27(4):291–322

García-Medrano P, Ollé A, Ashton N, Roberts MB (2019) The mental template in handaxe manufacture: new insights into Acheulean lithic technological behavior at Boxgrove, Sussex, UK. J Archaeol Method Theory 26:396–422

Gingerich JA, Stanford DJ (2016) Lessons from Ginsberg: an analysis of elephant butchery tools. Quat Int 466:269–283

Goren-Inbar N, Lister A, Werker E, Chech M (1994) A butchered elephant skull and associated artifacts from the Acheulian site of Gesher Benot Ya'aqov, Israel. Paléorient 20:99–112

Gurven M (2004) To give and to give not: the behavioral ecology of human food transfers. Behav Brain Sci 27(4):543–559

Hardaker T (2020) A geological explanation for occupation patterns of ESA and early MSA humans in southwestern Namibia? An interdisciplinary study. Proc Geol Assoc 131(1):8–18

Henrich J, McElreath R, Barr A, Ensminger J, Barrett C, Bolyanatz A et al (2006) Costly punishment across human societies. Science 312(5781):1767–1770

Heywood JL (1996) Conventions, emerging norms, and norms in outdoor recreation. Leis Sci 18(4):355–363

Hoehl S, Keupp S, Schleihauf H, McGuigan N, Buttelmann D, Whiten A (2019) 'Over-imitation': a review and appraisal of a decade of research. Dev Rev 51:90–108

Högberg A, Lombard M (2019) 'I can do it' becomes 'we do it': Kimberley (Australia) and Still Bay (South Africa) points through a socio-technical framework lens. J Paleolithic Archaeol. https://doi.org/10.1007/s41982-019-00042-4

Horner V, Whiten A (2005) Causal knowledge and imitation/emulation switching in chimpanzees (Pan troglodytes) and children (Homo sapiens). Anim Cogn 8(3):164–181

Hosfield R, Chambers J (2009) Genuine diversity? The Broom biface assemblage. Proc Prehistor Soc 75:65–100

Isaac GL (1978) The Harvey lecture series, 1977–1978: food sharing and human evolution: archaeological evidence from the Plio-Pleistocene of east Africa. J Anthropol Res 34(3):311–325

Itkonen E (2008) The central role of normativity in language and linguistics. In: Zlatev J (ed) The shared mind: perspectives on intersubjectivity. John Benjamins Publishing Company, Philadelphia, pp 279–305

Jennings RP, Shipton C, Breeze P, Cuthbertson P, Bernal MA, Wedage WO et al (2015) Multi-scale Acheulean landscape survey in the Arabian Desert. Quat Int 382:58–81

Jones PR (1980) Experimental butchery with modern stone tools and its relevance for Palaeolithic archaeology. World Archaeol 12(2):153–165

Joyce R (2006) The evolution of morality. MIT Press, Cambridge, MA

Kaplan H, Hill K, Lancaster J, Hurtado AM (2000) A theory of human life history evolution: diet, intelligence, and longevity. Evol Anthropol Issues News Rev 9(4):156–185

Kaplan H, Gurven M, Hill K, Hurtado AM (2005) The natural history of human food sharing and cooperation: a review and a new multi-individual approach to the negotiation of norms. In: Gintins H, Bowles S, Boyd R, Fehr E (eds) Moral sentiments and material interests: the foundations of cooperation in economic life. MIT Press, Cambridge, MA, pp 75–113

Kenward B (2012) Over-imitating preschoolers believe unnecessary actions are normative and enforce their performance by a third party. J Exp Child Psychol 112(2):195–207

Kenward B, Karlsson M, Persson J (2011) Over-imitation is better explained by norm learning than by distorted causal learning. Proc R Soc B Biol Sci 278(1709):1239–1246

Keupp S, Behne T, Rakoczy H (2013) Why do children overimitate? Normativity is crucial. J Exp Child Psychol 116(2):392–406

Key AJ, Lycett SJ (2017a) Influence of handaxe size and shape on cutting efficiency: a large-scale experiment and morphometric analysis. J Archaeol Method Theory 24(2):514–541

Key AJ, Lycett SJ (2017b) Reassessing the production of handaxes versus flakes from a functional perspective. Archaeol Anthropol Sci 9:737–753

Lakin JL, Jefferis VE, Cheng CM, Chartrand TL (2003) The chameleon effect as social glue: evidence for the evolutionary significance of nonconscious mimicry. J Nonverbal Behav 27(3):145–162

Li H, Kuman K, Lotter MG, Leader GM, Gibbon RJ (2017) The Victoria West: earliest prepared core technology in the Acheulean at Canteen Kopje and implications for the cognitive evolution of early hominids. R Soc Open Sci 4(6):170288

Lycett SJ, Eren MI (2019) Built-in misdirection: on the difficulties of learning to knap. Lithic Technol 44(1):8–21

Madsen B, Goren-Inbar N (2004) Acheulian giant core technology and beyond: an archaeological and experimental case study. Eurasian Prehist 2(1):3–52

McElreath R, Boyd R, Richerson P (2003) Shared norms and the evolution of ethnic markers. Curr Anthropol 44(1):122–130

McGuigan N, Makinson J, Whiten A (2011) From over-imitation to super-copying: adults imitate causally irrelevant aspects of tool use with higher fidelity than young children. Br J Psychol 102(1):1–18

Mosquera M, Saladié P, Ollé A, Cáceres I, Huguet R, Villalaín J et al (2015) Barranc de la Boella (Catalonia, Spain): an Acheulean elephant butchering site from the European late Early Pleistocene. J Quat Sci 30(7):651–666

Muller A, Clarkson C, Shipton C (2017) Measuring behavioural and cognitive complexity in lithic technology throughout human evolution. J Anthropol Archaeol 48:166–180

Nielsen M (2012) Imitation, pretend play, and childhood: essential elements in the evolution of human culture? J Comp Psychol 126(2):170–181

Nielsen M (2018) The social glue of cumulative culture and ritual behavior. Child Dev Perspect 12(4):264–268

Nielsen M (2019) The human social mind and the inextricability of science and religion. In: Henley TB, Rossano MJ, Kardas EP (eds) Handbook of cognitive archaeology: psychology in prehistory. Routledge, New York, pp 296–310

Nielsen M, Susianto EW (2010) Failure to find over-imitation in captive orangutans (Pongo pygmaeus): implications for our understanding of cross-generation information transfer. Nova Science Publishers, New York

Nielsen M, Tomaselli K (2010) Overimitation in Kalahari Bushman children and the origins of human cultural cognition. Psychol Sci. https://doi.org/10.1177/0956797610368808

Nielsen M, Kapitány R, Elkins R (2015) The perpetuation of ritualistic actions as revealed by young children's transmission of normative behavior. Evol Hum Behav 36(3):191–198

Nielsen M, Langley MC, Shipton C, Kapitany R (in press) *Homo neanderthalensis* and the evolutionary origins of ritual in *Homo sapiens*. Philo Trans R Soc B 375:20190424

Piperno M, Tagliacozzo A (2001) The elephant butchery area at the Middle Pleistocene site of Notarchirico (Venosa, Basilicata, Italy). In: Cavarretta G, Gioia P, Mussi M, Palombo M (eds) La Terra degli Elefanti. Universita degli Studi di Roma 'La Sapienza', Rome, pp 230–236

Pope M, Roberts M (2005) Observations on the relationship between Palaeolithic individuals and artefact scatters at the Middle Pleistocene site of Boxgrove, UK. In: Gamble C, Porr M (eds) The hominid individual in context. Routledge, London, pp 81–97

Prentice DA, Miller DT (1996) Pluralistic ignorance and the perpetuation of social norms by unwitting actors. Adv Exp Soc Psychol 28:161–209

Rakoczy H, Schmidt MF (2013) The early ontogeny of social norms. Child Dev Perspect 7(1):17–21

Rakoczy H, Warneken F, Tomasello M (2008) The sources of normativity: young children's awareness of the normative structure of games. Dev Psychol 44(3):875

Roberts MB, Parfitt SA (1999) Boxgrove: a Middle Pleistocene hominid site at Eartham Quarry, Boxgrove. English Heritage, West Sussex

Roberts SG, Dunbar R, Pollet TV, Kuppens T (2009) Exploring variation in active network size: constraints and ego characteristics. Soc Netw 31(2):138–146

Rogers MJ, Harris JW, Feibel CS (1994) Changing patterns of land use by Plio-Pleistocene hominids in the Lake Turkana Basin. J Hum Evol 27(1/3):139–158

Rossano MJ (2012) The essential role of ritual in the transmission and reinforcement of social norms. Psychol Bull 138(3):529

Roughley N, Bayertz K (eds) (2019) The normative animal? On the anthropological significance of social, moral, and linguistic norms. Oxford University Press, Oxford

Sangode S, Mishra S, Naik S, Deo S (2007) Magnetostratigraphy of the Quaternary sediments associated with some Toba tephra and Acheulian artefact bearing localities in the western and central India. Gondwana Mag 10:111–121

Schick K, Clark JD (2003) Biface technological development and variability in the Acheulean industrial complex in the Middle Awash region of the Afar Rift, Ethiopia. In: Soressi M, Dibble HL (eds) Multiple approaches to the study of bifacial technologies. University of Pennsylvania Museum of Archaeology and Anthropology, Philadelphia, pp 1–30

Schillinger K, Mesoudi A, Lycett SJ (2016) Copying error, evolution, and phylogenetic signal in artifactual traditions: an experimental approach using "model artifacts". J Archaeol Sci 70:23–34

Sharon G (2007) Acheulian large flake industries: technology, chronology, and significance, BAR international series. Archaeopress, Oxford

Sharon G, Alperson-Afil N, Goren-Inbar N (2011) Cultural conservatism and variability in the Acheulian sequence of Gesher Benot Ya'aqov. J Hum Evol 60(4):387–397

Shipton C (2010) Imitation and shared intentionality in the Acheulean. Camb Archaeol J 20(2):197–210

Shipton C (2013) A million years of hominin sociality and cognition: Acheulean bifaces in the Hunsgi-Baichbal Valley, India, BAR international series. Archaeopress, Oxford

Shipton C (2019a) The evolution of social transmission in the Acheulean. In: Overmann K, Coolidge FL (eds) Squeezing minds from stones. Oxford University Press, Oxford, pp 332–354

Shipton C (2019b) Three stages in the evolution of human cognition: normativity, abstraction, and recursion. In: Henley TB, Rossano MJ, Kardas EP (eds) Handbook of cognitive archaeology: psychology in prehistory. Routledge, New York, pp 153–173

Shipton C (2020) The unity of Acheulean culture. In: Groucutt H (ed) Culture history and convergent evolution. Springer, Cham, pp 13–27

Shipton C, Clarkson C (2015) Handaxe reduction and its influence on shape: an experimental test and archaeological case study. J Archaeol Sci Rep 3:408–419

Shipton C, Nielsen M (2015) Before cumulative culture. Hum Nat 26(3):331–345

Shipton C, White MJ (2020) Handaxe types, colonization waves, and social norms in the British Acheulean. J Archaeol Sci Rep. https://doi.org/10.1016/j.jasrep.2020.102352

Shipton C, Petraglia M, Paddayya K (2009) Stone tool experiments and reduction methods at the Acheulean site of Isampur Quarry, India. Antiquity 83(321):769–785

Shipton C, Blinkhorn J, Breeze PS, Cuthbertson P, Drake N, Groucutt HS et al (2018a) Acheulean technology and landscape use at Dawadmi, central Arabia. PLoS One 13(7):e0200497

Shipton C, Clarkson C, Cobden R (2018b) Were Acheulean bifaces deliberately made symmetrical? Archaeological and experimental evidence. Camb Archaeol J 29(1):65–79

Sinha C (2009) Objects in a storied world: materiality, normativity, narrativity. J Conscious Stud 16(6/7):167–190

Solodenko N, Zupancich A, Cesaro SN, Marder O, Lemorini C, Barkai R (2015) Fat residue and use-wear found on Acheulian biface and scraper associated with butchered elephant remains at the site of Revadim, Israel. PLoS One 10(3):e0118572

Sripada CS, Stich S (2006) A framework for the psychology of norms. In: Carruthers P, Laurence S, Stich S (eds) The innate mind. Oxford University Press, Oxford, pp 280–301

Sterelny K (2014) A Paleolithic reciprocation crisis: symbols, signals, and norms. Biol Theory 9(1):65–77

Sterelny K (2019) Norms and their evolution. In: Henley TB, Rossano MJ, Kardas EP (eds) Handbook of cognitive archaeology: psychology in prehistory. Routledge, New York, pp 375–397

Tomasello M (2008) Origins of human communication. MIT Press, Cambridge, MA

Tomasello M, Hamann K (2012) Collaboration in young children. Q J Exp Psychol 65(1):1–12

Tomasello M, Vaish A (2013) Origins of human cooperation and morality. Annu Rev Psychol 64:231–255

Toth N, Schick K (2019) Why did the Acheulean happen? Experimental studies into the manufacture and function of Acheulean artifacts. Anthropologie 123(4/5):724–268

Wenban-Smith F (2004) Handaxe typology and Lower Palaeolithic cultural development: ficrons, cleavers and two giant handaxes from Cuxton. Lithics 25:11–21

White MJ, Bridgland DR, Schreve DC, White TS, Penkman KE (2018) Well-dated fluvial sequences as templates for patterns of handaxe distribution: understanding the record of Acheulean activity in the Thames and its correlatives. Quat Int 480:118–131

White MJ, Ashton N, Bridgland D (2019) Twisted handaxes in Middle Pleistocene Britain and their implications for regional-scale cultural variation and the deep history of Acheulean hominin groups. Proc Prehistor Soc. https://doi.org/10.1017/ppr.2019.1

Part IV
Ethical Issues

Chapter 12
Social Archaeology as the Study of Ethical Life: Agency, Intentionality, and Responsibility

Artur Ribeiro

> *"Action springs not from thought, but from a readiness for responsibility."*
>
> *G.M. Trevelyan*

Abstract Supporters of agency research in archaeology are currently divided into two groups: those who remain somewhat faithful to classical social theories that define agency as a manifestation of individual and/or collective intentions, and those who have embraced object agency and perceive it as a dynamic force distributed among human and non-human objects. This chapter argues that agency according to the view of these two groups is too labile, contradictory, and counterproductive, and suggests a notion of agency based on G.W.F. Hegel's practical philosophy—a notion of agency based on ethical normativism. For Hegel, agency concerns the normative conditions that allow agents to experience and claim actions as their own. Ownership over actions imply assuming responsibility for those same actions. From these ideas, it becomes clear that agency is not a property of human individuals nor of social collectives, as suggested by theorists such as Roy Bhaskar and Anthony Giddens, but rather a historical product enacted by social institutions. Rather than the causal descriptions in which most of archaeological science is embedded, social archaeology concerns the understanding (*Verstehen*) of human action under intentional descriptions. Social archaeology is the *de facto* study of agency in past societies, an archaeology built around historical content that contextualizes the intentional/ethical actions of past peoples.

Keywords Responsibility · Norm · Agent · Intention · Ethics

A. Ribeiro (✉)
Institute of Prehistoric and Protohistoric Archaeology, University of Kiel, Kiel, Germany
e-mail: aribeiro@gshdl.uni-kiel.de

© Springer Nature Switzerland AG 2021
A. Killin, S. Allen-Hermanson (eds.), *Explorations in Archaeology and Philosophy*, Synthese Library 433, https://doi.org/10.1007/978-3-030-61052-4_12

12.1 Introduction

Conceptions of agency have undergone several transformations in the past decades, some of which signal a radical departure of agency as a discussion of the human condition towards a discussion of the metaphysical nature of action. When it first hit the archaeological scene in the 1980s, agency had connections to Marxist theory, namely Marx's *The Eighteenth Brumaire of Louis Bonaparte* (2015 [1883]), and the social theories developed by Anthony Giddens (1979, 1984, 1993 [1976]), and to a lesser extent, Margaret Archer (1982, 2004) and Roy Bhaskar (1991). More recently, the supporters of object agency have challenged the classical notion of agency and turned to thinkers such as Alfred Gell (1998), Jane Bennett (2010a, b), and Bruno Latour (1999, 2005) for a less anthropocentric and a more materialist notion of agency. Those who followed and, to a certain extent, still follow the classical notion of agency (e.g., Barrett 2001; Gardner 2007) perceive agency as involving individual freedom, power, social constraints, and ultimately, what it means to be human. Those who follow more recent iterations of the agency concept perceive it more as a vital dynamic distributed among several human and non-human bodies (Žižek 2014). While this latter perception still claims links to social theory, it also departs completely from it by radically altering the metaphysical status of agency and of society in general.

As pointed out by Nigel Pleasants, 'agency' as understood by Giddens and Bhaskar also has metaphysical grounding (1997, 1999): in these social theorists we can recognize what Wittgenstein described as 'philosophical pictures'—theoretical concepts aimed at describing states of affairs of the world, but that have ended up becoming actual portrayals of the essence of some phenomenon (Pleasants 1997, p. 3). This essentialist notion of agency, however, has never been fully accepted in archaeology because most archaeological theorists reject the idea of inherent individual or group free-will/volition, that is to say, it is understood that people can never choose the conditions in which they act (Barrett 2000; Dobres and Robb 2000, p. 5). This remains one of the most important claims concerning agency in archaeology: the actions of an agent are wholly dependent on the social and historical conditions available to the agent, or in other words, agency is contextual (Barrett 1988, 2000, 2006; Robb 2004, p. 105). Nevertheless, agency and its dependence on context remains wrapped up in a certain degree of confusion, partly because agency is expressed in a dual way: as manifesting in individuals (Bell 1992; Gravina 2007; Hodder 2000; Johnson 1989; Morris 2007) but also reproduced by social structures. This idea of agents vs. social structure is most famously known through the work of Giddens (Johnson 1989, p. 191), but is also known as the methodological individualism-holism duality that has been debated among social theorists (Ritzer and Gindoff 1994; Zahle and Collin 2014). As we shall see below, the agent-structure duality leads to some problems and Hegel's conception of agency manages to overcome these problems.

The notion of object agency adopted by some New Materialists is not free from its own problems, some of which derive from the fact that the notion is largely based on a newly established metaphysical orthodoxy (Alberti 2016). This is not the time or the place to discuss what metaphysics can offer archaeology (see Ribeiro 2019)—but it is the time and the place to discuss whether our understanding of agency should have a metaphysical underpinning. The most glaring problem with traditional analytic metaphysics is that it divests agency from any empirical scrutiny. Through metaphysics, agency becomes part of a universal ontology that is applicable to all times and places and a property of matter itself. Under this perspective, it becomes impossible to locate agency. As Tim Flohr Sørensen contends, this new conception of agency remains, nonetheless, contextual (2016), however, this just diverts attention from 'agency' to 'context' and it seems that anything and everything can be qualified as a context (Ribeiro 2016a, p. 230). In the materialist sense, agency seems to share more similarities to vitalist concepts such as Hans Driesch's *Entelechy* (Bennett 2010b; Sander 1993) and Henri Bergson's *Élan Vital* (Bergson 1944 [1911]) than it does to agency as it is conventionally understood among social theorists. In and of itself, distancing agency from social theory in favor of metaphysics is not automatically wrong but it does carry some consequences—agency becomes divested of its links to the sociological notions of freedom, individuality, and power. It is precisely on research concerning freedom and power that the advocates of object agency have been facing most criticism (Appadurai 2015, p. 22; Choat 2016; Ribeiro 2016a).

These are only some of the problems that afflict agency as it is currently understood in archaeology, and it seems sensible to say that these problems might also be the reason as to why agency has never left its status as a 'platitude' (Dobres and Robb 2000). As a Wittgensteinian 'picture', it might be desirable to keep the agency concept as labile as possible. This is tacitly suggested by Marcia-Anne Dobres and John Robb (2000, p. 10) and explicitly suggested by Henrietta Moore (2000, 2004). For this latter author, it seems obvious that archaeological practice automatically implies submitting oneself to some pre-theoretical assumptions, namely that individuals in other places and times were endowed with some degree of social competency (2000, p. 260). However, it is also dangerous to perceive past social actors in light of our own experiences—that would be, in Moore's own word—anachronistic (2000, p. 261). Thus, it makes sense to employ 'agency' more as an umbrella term, a 'concept-metaphor' (Moore 2004), which comprises understandings of subjectivities, the self, and personhood. While a seemingly sensible solution, this leaves archaeologists in a rather undesirable position: a lack of an anchor. Recognizing agency as a concept-metaphor does have its advantages but some effort should be employed in order to pin down a clearer understanding of agency as a working concept—and not as an essentialist property or object.

Given the problems faced by both classical notions of agency and its materialist variant, I suggest a third notion, which has yet to find a home in archaeology. This alternative comes from the work of the idealist philosopher G.W.F. Hegel, and it is best understood through the Hegelian philosophy of Robert Pippin (1989, 1997, 2008), Vincent Descombes (2004, 2007, 2014), Slavoj Žižek (2006, 2013, 2014), Charles Taylor (1975, 1989, 2015) and Allen Wood (1990). Although the Hegelian conception of agency does share some similarities with more classical notions, such as that of Giddens or Bhaskar, it also distances itself from them by conceiving social action as *ethical behavior*. While the addition of ethics might seem like a minor difference in relation to classical options, thinking in terms of ethics marks a radical change in our understanding of intentions, freedom, and individuality, and consequently, how we should conduct research under the rubric of social archaeology.

Thus in Sect. 12.2, I outline the core conceptual framework. This involves discussions of intentionality, meaningful action, and the relation between individuals and social collectives. Section 12.3 offers an analysis of objective and subjective agency, and Sect. 12.4, normativity as simultaneously individual and social. Section 12.5 contains my positive proposal for the study of agency and ethical life, and finally, Sect. 12.6 concludes. Throughout I argue for a reconceptualization of agency, not as an object to be identified in the archaeological record, but as a concept-metaphor for how freedom is constituted in social contexts. I outline, motivate and defend a revived and revised 'social archaeology' research program investigating how and why human societies created institutions throughout history that have provided individuals with more freedom of action, and consequently, given humans more responsibility.

12.2   Intentionality and Responsibility

12.2.1   Hegel's Ethical Life and Intention

Central to the Hegelian concept of agency[1] is the ethical notion of responsibility.[2] This notion is one of the core themes of the book *Outlines of the Philosophy of Right* (2008 [1820]) in which Hegel explains how responsibility can be recognized. First, to avoid confusion, it is necessary to distinguish morality (*Moralität*) from ethical life (*Sittlichkeit*). From the perspective of morality, actions are usually bound to a system of moral beliefs; for instance, actions can be 'good' or 'evil'. Recognizing whether an action is 'good' or 'evil' implies a tacit agreement to accept the historical notions of 'good' and 'evil' and their association to moral systems such as that of Christianity and Manicheism (Asad 2003). While social action can and should be studied under moral notions, this remains far from providing a widespread and useful framework for understanding agency. My aim is to understand what Hegel meant by 'ethical life' or *Sittlichkeit*.

In German, *Sittlichkeit* involves the notion of what is 'customary' and what is 'ethical', which is what provides *Sittlichkeit* its normativist basis (Wood 1990, p. 195). As argued by Pippin, normativism is a central element of Hegel's

---

[1] It should be noted that Hegel did not discuss agency *per se* as this concept does not exist in the German language; nevertheless, his ideas on philosophy of action are equivalent to what has been discussed in archaeology under the rubric 'agency studies'. In German, the closest translation of 'agency' is '*Handlung*' which translates roughly to 'social action'.

[2] Barry Barnes (2000) and Jeanette Kennett (2001), among others, have also explored the link between agency and responsibility, although not under a Hegelian framework.

understanding of agency (2008). From the perspective of ethical life (*Sittlichkeit*), the performance of an action relies not on whether it is morally right or wrong but on whether it is *conceivable*. As Descombes explains, under a normativist conception of ethical life, the actor is placed in a position in which he or she questions whether an action *can be* performed (Descombes 2007). Thinking of what actions can be performed, in our case, refers not to those actions that are either logically or physically impossible, i.e., count through cardinal numbers or swim across the Atlantic Ocean (Wittgenstein 1960, p. 54) but those actions that are sustained by social institutions (norms). These last types of actions are those we would typically assign an intent and consequently justify with *reasons* (Anscombe 2000; Pippin 2008, p. 3).

While it is possible to identify remarkable similarities between these premises and those that support Giddens' theory of structuration, there are also critical differences that consequently affect the methodologies associated to agency in archaeology. Let us parse some of these differences in detail. Giddens tried to distance himself from intentionality, as it was understood in hermeneutical voluntarism, by employing the notion of 'reflexivity' (Dornan 2002, p. 308; Giddens 1984, p. 3ff.). Unlike intentionality, reflexivity has a metaphysical grounding—it is a process of continuous monitoring of actions, embedded in space-time (Giddens 1984, p. 4). For Giddens, the notion of reflexivity is more adequate than intentionality alone because, as he points out, "the reasons actors offer discursively for what they do may diverge from the rationalization of action as actually involved in the stream of conduct of those actors" (Giddens 1984, p. 4). This has led Giddens to disregard the intentions of actions in favor of 'reflexive monitoring'. Giddens' conceptualization of intentionality, however, might not be fully accurate. Giddens provides the following example of intentionality: a submarine officer wants to change the course of the submarine but unintentionally pulls the wrong lever and ends up shooting and sinking the Bismarck (1984, p. 8). This example presupposes that intentionality is closely linked to outcomes of actions, and how some of these outcomes are unintentional (Bell 1992; Dornan 2002, p. 319ff.; McCall 1999; Vanpool and Vanpool 2003). The central problem is that intentionality is understood as 'goal-orientation' according to Gidden's Bismarck example, which is not necessarily correct, as has been pointed out by Timothy Pauketat (2001, pp. 79–80).

Intentionality was conceived in medieval Scholastic philosophy but was ultimately rejected by Cartesianists during late seventeenth century. However, in the late nineteenth century, it was reintroduced into philosophy by Franz Brentano (2015 [1874]). According to Brentano, intentionality is a property of psychical or mental phenomena, as opposed to physical or natural phenomena. Unlike physical phenomena, mental phenomena are characterized by reference to a content, i.e., a direction towards an object (Brentano 2015 [1874], p. 92). The meaning of 'direction towards an object' is not goal-orientation, but rather having a *direction*. It is entirely possible to have intentions without actively performing actions that have an expected outcome. For instance, acts of loving and hating are intentional: they must be directed towards something, given that it is impossible to love or to hate without there being something that is being loved or hated. In other words, the love or hate is about something. At the same time, to love or to hate are not 'goal-oriented' in the sense that a subject does not love an object with the hope of achieving an outcome.

The property of intentionality is understood under philosophy of language as *transitive*. Following Brentano, the intention of 'wanting' to change the course of a submarine is purely intentional whereas pulling a lever is not. Descombes clarifies this issue by distinguishing intransitive verbs from transitive intentional and unintentional[3] verbs (2014, p. 26ff.). Very briefly, an intransitive verb refers to those action-verbs that do not have a direct object, like *to sleep* or *to sneeze* whereas transitive verbs refer to action-verbs that require an object be the recipient of the action, like *to love* or *to push*. In this latter group, we can further distinguish intentionality and unintentionality in the following way: if someone has kicked a ball, the statement only makes sense if the kicker's foot actually connected physically to the ball. This case can be described as unintentional simply because it focuses on the mechanics by which a foot physically hits a ball. Analogously, this is what happens in the case of the submarine officer who pushes the wrong lever. But let us look at intentional transitivity—a detective who is searching for a criminal. In this case, the verb 'to search' refers to an intention but not in a physical or mechanical sense. When someone 'searches' there is no biological process occurring that corresponds to the act of searching whereas the act of kicking does actually involve the articulation of the muscles of a leg. What we mean when we say that 'a detective is searching for a criminal', is *providing meaning or giving a purpose* to the detective when trying to find or identify a criminal (e.g., perusing a criminal database, looking for clues at the crime scene, interrogating suspects, etc.).

12.2.2   Action as Meaningful

It was with post-processual archaeology, especially the work produced by Ian Hodder, that the basis for understanding 'meaning' in past societies became popularized. However, some crucial distinctions between Hodder's understanding of 'meaning' and the same concept employed here must be clarified. Whereas Hodder conceived 'meaning' as ideas and concepts that could be embedded into material culture (1992, p. 12), this chapter enforces the notion that meaning and intention are one and the same. Once again, we must remember that an intentional action does not entail goal-orientation, but it does entail direction, following Brentano (2015 [1874]). Thus, when we say that an action has a 'meaning' or is 'meaningful', what is implied is that the action is directed, that is to say, it *follows a norm* (Williams 2015).

As argued by Paul Boghossian (1989) and Stephen Turner (1994, 2010), in and of themselves, norms can be somewhat mysterious in that norms cannot have causal powers—they cannot force people to act in certain ways. This is why, as argued by

---

[3] Descombes does not describe transitivity as 'intentional' and 'unintentional' although his ideas remain clearer in this context if we use these words.

Hegel, normative behavior has to be ethical. Normative behavior involves *responsibility*, and this is what makes it *meaningful* at the same time. Pulling a lever, in and of itself, is not meaningful unless the action is also viewed as a responsibility. This is the key difference that separates Hegel from Giddens because for Hegel, agency only exists when responsibility is recognized over the action. Without responsibility, an action cannot be intentional—It remains merely physical. This needs some further elucidation: in a poignant critique of the computational theory of mind, John Searle asks us to imagine a room in which a monolingual English speaker converts one set of Chinese symbols into another set of Chinese symbols through instructions explained to the speaker in English (1980). In short, this thought-experiment was to illustrate that knowing how to convert symbols does not necessarily imply understanding what those symbols *mean*. I find this incredibly relevant to Giddens' Bismarck example because pulling levers is not necessarily an intentional act—a machine can be programmed to do the same thing, just like in Searle's Chinese room where the English speaker acts as if he or she were a machine. Given that the English speaker in the Chinese room is only transcribing Chinese text but not understanding (*Verstehen*) it, we must realize that the content of the transcribed text is not of the English speaker's responsibility. In order for the English speaker to become a responsible agent, it is required that he or she want something and express it of his or her *own* volition. Therefore, it is crucial that we differentiate the English speaker's transcription of Chinese script, which is a physical/mechanical act, and those expressions that entail the speaker's intentions—those which the agent can claim ownership over, in the sense of being responsible for them.

Hegel states that the freedom of action that underlines agency requires that the agent recognize an action as his or her *own* (Pippin 2008, p. 7ff.). In Giddens' example, shooting the Bismarck is a direct consequence of pulling the wrong lever—but not of wanting to change the course of the submarine and thus, the submarine officer is responsible for making a mistake (wrong lever) but not of making the wrong decision (wanting to change course). From a legal point-of-view, having shot the Bismarck would have qualified as *involuntary*: literally an action without volition. This brings up a very important question: should social theory, and consequently, agency in archaeology, be the study of the involuntary mistakes made by people in the past? Not only would this be rather vapid, it would be nigh impossible in practice. The aim of agency research should not be the study of the consequences of actions but rather normative and ethical bases underlying human action. This means that our attention should be on the *socio-historical* contexts in which these actions occurred.

12.2.3   Individuals and Collectives

The notion of 'individual' has a number of definitions ranging from designating a basic 'unit' to designating a 'social agent'. In this latter sense, Hegelianism recognizes individuals as those who are subjects of an intentional action. As stated in the previous section, in order for this to be possible, an agent can only be considered an agent if his or her deeds and actions are perceived as his or her own. This is in the literal sense: an agent can only be considered an agent when carrying out action that can be considered *privately owned by the agent* (Pippin 2008, p. 5).

The literature on agency in archaeology remains quite confusing on this issue because discussions regarding individuals remains quite heavily reliant on the traditional distinction between individual agents (methodological individualism) and collective structure (collectivism). For instance, Gardner states in the opening of *Agency Uncovered* that agency "...concerns the nature of individual freedom in the face of social constraints" (2007, p. 1). This is a very common view in archaeology: individual agency as self-determination that is constrained by social conditions. This dialectic forces us to recognize a separation of the agent and the social structure, both of which shape how agency manifests.

Under the Hegelian paradigm, this duality cannot be accepted because it presupposes an essentialist notion of agency, that is to say, as a universal property of all humans at all times and places. This leads to the contradictory premise found in agency theories, such as that of Anthony Giddens and Roy Bhaskar (Bhaskar 1979, p. 18; Giddens 1984, p. 3), which claims that agency is a universal feature of being human yet at the same time dependent on social constraints (Giddens 1984, p. 169). I realize the attraction of accepting such a premise, as it provides agency with the best of methodological individualism and collectivism by conceiving agents as subjects who enable and create new norms, but also as objects who are constrained by previously established structural norms. However, a detailed analysis of some case-studies demonstrate that it is impossible to differentiate what counts as an act of a self-determined individual and what counts as constraints which are structurally imposed by the society.

Matthew Johnson (2000) has presented a case-study that addresses the rise of an individual in the Renaissance, a period in which individuality was constructed with reference to new economic practices and notions of nation-state. Johnson's case-study, in specific, focuses on Lord Cromwell, a fifteenth century 'self-made man' from a Lincolnshire family, who rose to become one of England's most influential and wealthy men. In order to highlight Cromwell as a knowledgeable social agent, Johnson focuses on his bold reconstruction of Tattersall Castle and proceeds to describe the decisions made by Cromwell with regards to the reconstruction. As Johnson emphasizes himself—looking at Cromwell in search of individual intention can draw attention towards specific goals and strategies—which is not necessarily wrong but is shallow in terms of research (2000, p. 226).

There is, nevertheless, something in Johnson's case-study that requires emphasizing: first, in rebuilding Tattersall Castle, Cromwell built a tower in order to "mark a certain form of lordship, [which] is a recurrent motif of the period, part of an *established symbolic vocabulary*" (2000, p. 220; emphasis mine); second, the symbolic vocabulary is further pursued by Cromwell through marks of 'elite identity', which are reinforced by passages through gatehouses that contain heraldic marks and artificial moats—all of which serve to impress the visitor (2000, p. 229). When it comes to Cromwell's reconstruction program, it is impossible to differentiate between those elements that can be claimed to be his, as a social actor, and those that belong to the social structure and this is because Cromwell's intentional actions need to be framed within an established 'language' that his contemporaries could *recognize*. Cromwell could have only intended to construct a tower and other elements denoting his lordship if he and his contemporaries recognized the 'symbolic vocabulary' associated to those elements. Without that recognition, there would have been no reason for Cromwell to want to reconstruct Tattersall Castle in the way he did. Thus, the established symbolic vocabulary that Johnson refers to cannot be understood as a social *constraint* upon the individual, much on the contrary, it would have to be understood as something that *allowed* Cromwell the possibility to denote his lordship.

This is the inherent paradox of freedom: in order to be free to act, one must commit oneself to norms, and in doing so, place responsibilities upon oneself (Appiah 2005, p. 6; Brandom 2004, p. 248; Pippin 2008, p. 65ff.). With this in mind, the individualist logic implied in Giddens' theory of structuration needs to be turned on its head. This requires us to accept that social norms are not the outcome of individually held beliefs and objectives—much on the contrary, social norms are the elements that *enable* individual beliefs and objectives (Pippin 2008, pp. 4–5).

12.3   Objective Mind

In order to fully apprehend this idea requires understanding what Hegel meant by the Objective and Subjective Mind.[4] The logic underlying the Objective and Subjective Mind is remarkably simple: for every subjective action—and here this must be understood as a *personal action* performed by a subject—that same action requires an impersonal objective meaning. What this translates to is that the actions of an agent can only be deemed as truly his or hers (subjective) if they are framed through norms that are acknowledged by others (objective). This is distinct from Giddens' agent-structure dialectic in the following way: whereas in Giddens' theory, an agent manifests his or her agency by challenging social and historical structures, from a Hegelian viewpoint, the agent and the social and historical structures are one and the same—every agent is a representative of the *Zeitgeist* he or she belongs to. It is in this sense that an agent is always a historical product.

This seems to imply that an agent is always somewhat of a 'cultural dupe' (Giddens 1979, p. 71ff.; Johnson 1989); that an agent never actually has absolute

---

[4] These concepts are based on the German word '*Geist*' which has no direct translation in English. The closest equivalents are 'mind' and 'spirit'. Both these English terms highlight some of the properties associated to the Hegelian notion of '*Geist*': 'mind' emphasizes the mental property of intentional action (as argued by Franz Brentano (2015 [1874]) whereas the English meaning of 'spirit' emphasizes prevailing mood or attitude of a person or group of people (*Zeitgeist* = the mood or attitude of a time). I have opted for using the English term 'mind' following the convention of Vincent Descombes' translator (Descombes 2001, ch 10, fn 1).

freedom from cultural contexts. In a way, this is true: an agent is always a 'cultural dupe' because what is recognized as 'social freedom' always requires a cultural grammar in which it can be understood. It is perhaps according to this logic that we can best understand the argument inspired by Wittgenstein's later philosophy (1967, 2009 [1953]) that a 'private language', that is, a language that only one person understands, is a logical impossibility. This topic was addressed by Michael Shanks and Cristopher Tilley in *Re-constructing Archaeology* (1987, p. 122ff.) and in a similar vein to Hegelianism, they also saw the agent-structure dualism as unnecessary. Shanks and Tilley's ideas on agency were indirectly inspired by Hegel (through Marx) but their ideas did not translate well into case-studies (Johnson 1989, p. 193). This is primarily because they were, at the time, missing some essential notions, namely that of Objective and Subjective Mind. According to Hegel, an agent can never be ignorant of his or her status as an agent, otherwise the agent would be nothing more than a set of aleatoric physical behaviors. For example, if an ant walks across a patch of sand and traces lines that resemble Winston Churchill, ought it be said that the ant chose to draw Winston Churchill in the sand (Putnam 1981, p. 1)? No ant could make such a choice, because in order to choose to draw Winston Churchill it is necessary to know of him and to know that it is conceivable to draw him.

It is in this sense that agency follows the Objective and Subjective Mind. It is first objective because an agent's actions need to be recognized normatively, that is to say, by yourself and others, and it is subjective because it requires ownership by an agent. So in a way, an agent is always simultaneously following rules while being free. For example, in order to intentionally steal something, one must necessarily share the background normative knowledge about private property, i.e., what is mine and what is thine (Taylor 2004, p. 205). Without this knowledge, it would be impossible to steal. This is patent when observing the behavior of very young children who have not yet taken as their own the normative knowledge of private property—they tend to pick up things and displace them because they have yet to understand that things can be owned by specific people. So for example if a toddler enters an office, picks up a pen, and leaves with it, it seems safe to say that the child is not *intentionally* stealing a pen.

A pertinent question that can be asked is whether it is possible to challenge norms at all? Of course it is, but by challenging one norm, there are other norms that must be followed. I am overemphasizing the role played by the normative knowledge because it is an important element in understanding how the tension between individualism and social structure can be overcome. Under the notions of Objective and Subjective Mind, the agent is neither a self-determinate individual who can initiate actions (e.g., Gell 1998, p. 16) nor simply part of the social structure.

12.4   Normativism and Triadic Relations

To recognize why individualism is inadequate requires understanding how normativism operates at a social level. Vincent Descombes has provided excellent examples as to what it means to follow a norm. For instance, in a defense of semantic atomism, Jerry Fodor and Ernest Lepore argue that in order to become a cat-owner, there needs to be a relational property between a person and a cat (1992, p. 3). This is technically correct, however, there needs to be something more than just a person and a cat: there also needs to be at least another entity who is *not the owner of that same cat* (Descombes 2014, pp. 116–18). Thus, being a cat owner requires more than just a dyadic relation between a human and a cat, it requires a triadic relation between a cat, its owner, and those who are not its owners (Descombes 2014, based on Peirce 1960, paragraph 471). This relation between cat owners and non-owners is necessarily normative because it behooves the objective acceptance of the binding notion of property, or in other words, in order to act as a subject in relation to a cat requires that other people recognize that same freedom. Moreover, with that freedom comes responsibility because being a cat owner implies responsibility over a cat, responsibility which is acknowledged by others by being non-owners of the same cat.

This manner of reasoning abolishes the individual vs. collective social structure dualism in archaeology and the misunderstandings it generates. It might seem that I am advocating a theory of agency that ignores both individuals and social structure, but what I am providing is the exact opposite: through Hegel's work we begin to understand that intentional action is dialectical by being *simultaneously individual and social*. However, in order to act intentionally, like wanting to become a cat owner, requires being in a historical and social context in which such a relation is recognized by others.

A further clarification is necessary: behaving according to social norms does not necessarily mean that one is acting collectively. For example, it is entirely possible for two or more people to act intentionally towards the same object, such as when two people decide to push a car together (Searle 1992, p. 22). This type of cooperative action, however, is very different than cases in which football teams win football matches or when armies invade foreign countries, and that is because soccer teams and armies refer not to a multitude of individuals, strictly speaking, but rather to social institutions. When we refer to two people pushing a car we are referring specifically to the number of people involved in pushing a car, but that is not the case when referring to a soccer team or an army whose numbers can vary greatly. So when we say a soccer team has won a game, we are not referring exclusively to the 11 players that are on the field, we are also referring to the coaches and the players who are on the bench. The same goes for an army: when it is stated that an army has invaded a country, this means much more than the fact that a large number of soldiers has entered foreign territory given that an army also contains generals and technical staff who stay behind to coordinate strategies.

Now, the actions we accord to individuals are dependent on the institutional norms in which humans are embedded and given this dependency, what we mean by 'agent' cannot be anything more than a social status, not unlike being a teacher or a taxpayer. This is clearly distinct from the self-positing dogma of Johann Gottlieb Fichte or the self-legislative principle of Immanuel Kant (Pippin 2008, p. 65) given that these two ideas require thinking of a single human in dualistic terms: as a psychological *Ego* that legislates and as the *Body* that follows the law (Descombes 2004, p. 320; 2014, p. xxvi). Under the Hegelian paradigm, acting as an individual within a soccer team or within an army should be understood as simultaneously subjective and objective because the actions of a soccer team member or of a soldier are both his or her own and of the team or army, at the same time.

This has implications on the traditional models based either on individuals or collectives because under the Hegelian conception of agency, any attempt at the strict study of humans as social members require recognizing their social statuses (brother, citizen, tribesman, soldier, craftsman, priest, unemployed, married, single, student, retired, etc.). As explained earlier, intentionality is transitive, which means that all intentional actions have a direction towards an object. Thus, if agency is intentional it means that agency can never be a property inherent to humans, but rather an aspect of how humans have directed their actions in different times and places. Following Sartre (2003 [1943]), one could argue that humans are always in a position to act intentionally, even in extreme cases such as imprisonment. This is true, which means that the study of agency should not be the study of who has or does not have agency, since if we conceive research on agency in such simple terms it can be assumed that everyone has some 'degree' of agency; what seems a much more fruitful avenue of research is understanding how and why human societies created institutions throughout history that have provided more freedom of action, and consequently, given humans more responsibility.

12.5   What Does It Mean to Study Agency?

As stated earlier in the chapter, agency serves a better purpose as a concept-metaphor than it does as some sort of essentialist property of people (and/or things or the universe), which can be studied empirically. Therefore, I do not recommend an archaeology of agency *strictu sensu*. Similarly, it also makes little sense to have 'archaeologies of...' some of the concepts highlighted in this chapter: archaeology of intentionality, of responsibility, or of subjectivity. This is because an archaeology of agency would imply that agency is a domain that is opposed to other domains, or in other words, that there can be an archaeology of agency as opposed to an archaeology of non-agency. A much more flexible way of applying the ideas outlined in this chapter is to revive the nigh-defunct notion of a *social archaeology*.[5]

At face value, many of the ideas contained in this chapter might be difficult to accept because archaeological practice is assumed to be subservient to two main modes of analysis: systems-centered and actor-centered: the former mode focuses on top-down processes in which human actors are framed in terms of how they respond to external stimuli, whereas the latter mode are bottom-up and focus on historical processes as enacted by the decisions of human actors (Stanton 2004, p. 30). The reason why research is conducted in this manner is partly due to how processual archaeologists have defined archaeological methodologies as primarily top-down, and how in response, post-processual archaeologists have defined bottom up methodologies. Even though many years have passed, and the theoretical implications of both processual and post-processual archaeology have been somewhat mitigated, or just plainly ignored (Ribeiro 2016b, pp. 149–50), the *Zeitgeist* that gave rise to both these movements is still around. This *Zeitgeist* is based on the outdated view of the human being who acts according to either external influences or internal choices, but never both at the same time.

For example, in a review of archaeological practice, Elizabeth Arkush framed the work presented in 2010 as either perceiving human societies as controlled by 'external' influences or as shaped by the humans themselves (2011). This is a particularly crude view of human reality because it sees humans in a dualistic fashion. At no point, in a person's daily life, does one act either from arbitrary internal choices or simply as a response to externalities. A more accurate viewpoint would be to subsume human action as either under *causal* or *intentional descriptions* (Ribeiro 2018).

As described earlier, to understand humans (or animals and objects) as agents requires seeing their intentions in light of the norms that provide their freedom of action. Opposed to this is the view that human action is (merely) mechanical, as a causally determined response or effect of prior physical states. In Hegel, agency is an outcome of Mind (*Geist*) but Mind always presupposes Nature (*Natur*). What this means is that agency is a natural feature of the world, in fact, for Hegel agency is the underlying element which sustains the universe (Pippin 2008, p. 9). As an anti-dualist, Hegel never believed Mind to be of a different substance from Nature; he did however believe that natural organisms can evolve to the point in which they could not be *explained through the boundaries of nature* (Pippin 2008, p. 47). This aligns with the philosophy of the later Wittgenstein and his supporters who viewed the social sciences as requiring methods that were irreducible to those of the natural sciences (Von Wright 1971; Winch 2008 [1953]). What these philosophers defend is a *methodological pluralist* (perhaps even a methodological anarchist) approach that includes both causal and intentional descriptions. Therefore, it is not that humans either act causally or internally, but rather, that they can be described under causal or intentional terms. So for instance, while the migration of Dust Bowl farmers can

---

[5] I assume many people would disagree with the idea that social archaeology is defunct. What is meant by this is that archaeology of recent years seems to have lost supporters of social theory, in favor of post-humanist theory and archaeological science (Kristiansen 2014).

be described as an effect of ecological pressure (Hempel 1942, pp. 40–41), one should also take into account the intention of the farmers wanting to go to California and not somewhere else. In the causal process that led the Dust Bowl farmers to migrate, intentional actions were performed that were necessary for the migration to occur (Dray 1957).

In light of these ideas, it seems clear that a method applied exclusively to agency is unnecessary. Nevertheless, there are disciplines and methods that favor either causal or intentional descriptions. History holds a distinct advantage in producing intentional descriptions. I refer to history here not in the disciplinary sense, i.e., as research on the past based on historical documents, but as what Roger Chartier terms 'intellectual history' (1982), Ian Morris calls 'cultural history' (2000), and Carlo Ginzburg calls 'history of mentalities' (1993).

There are two distinct advantages in thinking about agency from a historical viewpoint. First, once agency is recognized not as a universal capacity that can be assigned to all humans (and/or objects) for all places and times, it becomes clear that to recognize the specific ways in which agency manifests also requires histori-cally tracing how normative meanings came into existence. As explained earlier in the chapter, intentionality requires an action directed towards an object, and as such these objects have to be located within a given period and place (if they are locatable at all). It would be impossible for an agent to have wanted to change the course of a submarine during the Bronze Age, or for a Neanderthal to have wanted to go to the bank. The conditions under which we can understand why agents act in the way they do can be subsumed under the principle of 'narrative intelligibility' (Descombes 2001, p. 182ff.; Ribeiro 2018)—if an actor wants to change the course of a subma-rine or go to the bank, our understanding of these intentions presuppose a determi-nate past that has provided the conditions in which these intentions can be considered meaningful to their respective agents. Denying this past is to assume universality, that humans are agents regardless of time and place, that a Palaeolithic hunter-gather in prehistoric France was just as much of an agent as a billionaire in twenty-first century North America. Secondly, as some archaeologists have pointed out, humans and non-human objects should not be regarded as ontologically distinct and that archaeology should follow the Latourian explanatory principle of 'symmetry' (Webmoor 2007; Webmoor and Witmore 2008; Witmore 2007, 2014). History, however, does not recognize this distinction as has been pointed out by Latour him-self (1993, p. 92). This is probably the reason why the discipline of history, unlike anthropology and archaeology, is currently not burdened by the choice between an ontological/materialist view and an epistemological view.

In addition to all this, one must also consider the role of responsibility when conceiving how agency can be studied. The current social frameworks in archaeol-ogy that are based on normativist principles (e.g., theory of practice, communities of practice) are excellent at providing a contextualization of how practices are cre-ated within societies, but they lack the explanatory capacity provided by the Hegelian conception of agency. Normativism, understood under the concept of 'practice', can explain the process through which people acquire knowledge as to how to brush their teeth, express dissent, build a church, smoke a pipe, ring a bell, make a pot, or hunt a rabbit, but practice theories are remarkably limited in explaining *why* people brush their teeth, express dissent, build a church, or smoke a pipe. This is because a norm, in and of itself, cannot be the cause or the reason provided for why people act in the way they do.

This is because practices need to be framed within a specific and historical context that provide the agent with the ethical reasons as to why they act in the way they do (Hacking 1995). So for instance, people who brush their teeth regularly can provide several justifications as to why they brush their teeth, from wanting to have good breath to wanting to please their dentist. These justifications all depend on a historically specific social context—one in which it makes sense to want to have good breath and wanting to impress your dentist is something desirable. Thus, even when freedom to not brush teeth is provided, many people in modern western society see several reasons as to why they should brush their teeth regularly nonetheless—they see it as their ethical duty to do so. Whereas practice theories can still provide great research frameworks to understand how patterned behavior is established, the Hegelian conception of agency can provide explanations as to why those behaviors became dominant in the first place.

12.6 Concluding Remark

As stated by Timothy Webmoor and Christopher Witmore, it seems that anything and everything can be considered 'social' nowadays (Webmoor and Witmore 2008). The issue pointed out by Webmoor and Witmore, however, concerns not what is social or not, but in what ways sociality can be uncovered and understood. This has to take into account that agency is not an object to be identified in the material record but a general term to describe how freedom is constituted in social contexts. By adopting this Hegelian conception of agency, research in social archaeology would progress on a more stable ground, one that is not limited by the contradictions of classical social theories nor the limitations of the new materialist conception of agency.

References

Alberti B (2016) Archaeologies of ontology. Annu Rev Anthropol 45(1):163–179
Anscombe GEM (2000) Intention, 2nd edn. Harvard University Press, Cambridge, MA
Appadurai A (2015) Mediants, materiality, normativity. Publ Cult 27:221–237
Appiah A (2005) The ethics of identity. Princeton University Press, Princeton
Archer MS (1982) Morphogenesis versus structuration: on combining structure and action. Br J Sociol 33(4):455–483
Archer MS (2004) Being human: the problem of agency. Cambridge University Press, Cambridge/New York
Arkush E (2011) Explaining the past in 2010. Am Anthropol 113(2):200–212

Asad T (2003) Formations of the secular: Christianity, Islam, modernity. Stanford University Press, Stanford

Barnes B (2000) Understanding agency: social theory and responsible action. Sage, London

Barrett JC (1988) Fields of discourse: reconstituting a social archaeology. Crit Anthropol 7(5):249

Barrett JC (2000) A thesis on agency. In: Dobres M-A, Robb J (eds) Agency in archaeology. Routledge, New York, pp 61–68

Barrett JC (2001) Agency, the duality of structure, and the problem of the archaeological record. In: Hodder I (ed) Archaeological theory today. Polity, Cambridge, pp 141–164

Barrett JC (2006) Archaeology as the investigation of the contexts of humanity. In: Papaconstatinou D (ed) Deconstructing context: a critical approach to archaeological practice. Oxbow, Oxford, pp 194–211

Bell J (1992) On capturing agency in theories about prehistory. In: Gardin J-C, Peebles CS (eds) Representations in archaeology. Indiana University Press, Bloomington, pp 30–55

Bennett J (2010a) Vibrant matter: a political ecology of things. Duke University Press, London

Bennett J (2010b) A vitalist stopover on the way to a new materialism. In: Coole D, Frost S (eds) New materialisms: ontology, agency, and politics. Duke University Press, London, pp 47–69

Bergson H (1944 [1911]) Creative evolution. Modern Library, New York

Bhaskar R (1979) The possibility of naturalism: a philosophical critique of the contemporary human sciences. Routledge, London/New York

Bhaskar R (1991) Philosophy and the idea of freedom. Blackwell, Oxford

Boghossian PA (1989) The rule-following considerations. Mind 98(392):507–549

Brandom R (2004) From a critique of cognitive internalism to a conception of objective spirit: reflections on Descombes' anthropological holism. Inquiry 47(3):236–253

Brentano F (2015 [1874]) Psychology from an empirical standpoint. Routledge, London/New York

Chartier R (1982) Intellectual history or sociocultural history? The French trajectories. In: LaCapra D, Kaplan SL (eds) Modern European intellectual history: reappraisals and new perspectives. Cornell University Press, Ithaca, pp 13–46

Choat S (2016) Review. Glob Discourse 6(1/2):136–139

Descombes V (2001) The mind's provisions: a critique of cognitivism. Princeton University Press, Princeton

Descombes V (2004) Le Complement de Sujet: Enquête sur le fait d'agir de soi-même. Gallimard, Paris

Descombes V (2007) Le raisonnement de l'ours: et autres essais de philosophie pratique. Seuil, Paris

Descombes V (2014) The institutions of meaning: a defense of anthropological holism. Harvard University Press, Cambridge, MA

Dobres M-A, Robb J (2000) Agency in archaeology: paradigm or platitude? In: Dobres M-A, Robb J (eds) Agency in archaeology. Routledge, London/New York, pp 3–17

Dornan JL (2002) Agency and archaeology: past, present, and future directions. J Archaeol Method Theory 9(4):303–329

Dray WH (1957) Laws and explanation in history. Oxford University Press, Oxford

Fodor JA, LePore E (1992) Holism: a shopper's guide. Blackwell, Oxford

Gardner A (2007) Introduction: social agency, power, and being human. In: Gardner A (ed) Agency uncovered: archaeological perspectives on social agency, power, and being human. Left Coast Press, Walnut Creek, pp 1–15

Gell A (1998) Art and agency: an anthropological theory. Clarendon, Oxford

Giddens A (1979) Central problems in social theory: action, structure and contradiction in social analysis. Macmillan, Houndmills

Giddens A (1984) The constitution of society: outline of the theory of structuration. Polity Press, Cambridge

Giddens A (1993 [1976]) New rules of sociological method: a positive critique of interpretative sociologies, 2nd edn. Stanford University Press, Stanford

Ginzburg C (1993) Microhistory: two or three things that I know about it. Crit Inq 20(1):10–35

Gravina B (2007) Agency, technology, and the 'muddle in the middle': the case of the Middle Palaeolithic. In: Gardner A (ed) Agency uncovered: archaeological perspectives on social agency, power, and being human. Left Coast Press, Walnut Creek, pp 65–78

Hacking I (1995) Rewriting the soul: multiple personality and the sciences of memory. Princeton University Press, Princeton

Hegel GWF (2008 [1820]) Outlines of the philosophy of right. Oxford University Press, Oxford

Hempel CG (1942) The function of general laws in history. J Philos 39(2):35–48

Hodder I (1992) Theory and practice in archaeology. Routledge, London/New York

Hodder I (2000) Agency and individuals in long-term processes. In: Dobres M-A, Robb J (eds) Agency in archaeology. Routledge, London/New York, pp 21–33

Johnson M (1989) Conceptions of agency in archaeological interpretation. J Anthropol Archaeol 8(2):189–211

Johnson M (2000) Self-made men and the staging of agency. In: Dobres M-A, Robb J (eds) Agency in archaeology. Routledge, London/New York, pp 213–231

Kennett J (2001) Agency and responsibility: a common-sense moral psychology. Clarendon Press, Oxford

Kristiansen K (2014) Towards a new paradigm: the third science revolution and its possible consequences in archaeology. Curr Swed Archaeol 22:11–34

Latour B (1993) We have never been modern. Harvester Wheatsheaf, London/New York

Latour B (1999) Pandora's hope: essays on the reality of science studies. Harvard University Press, Cambridge, MA

Latour B (2005) Reassembling the social: an introduction to actor-network-theory. Oxford University Press, Oxford

Marx K (2015 [1883]) The eighteenth brumaire of Louis Bonaparte. Wallachia Publishers, New York

McCall JC (1999) Structure, agency, and the locus of the social: why poststructural theory is good for archaeology. In: Robb J (ed) Material symbols: culture and economy in prehistory. Center for Archaeological Investigations, Carbondale, pp 16–20

Moore HL (2000) Ethics and ontology: why agent and agency matter. In: Dobres M-A, Robb J (eds) Agency in archaeology. Routledge, London/New York, pp 259–263

Moore HL (2004) Global anxieties: concept-metaphors and pre-theoretical commitments in anthropology. Anthropol Theory 4(1):71–88

Morris I (2000) Archaeology as cultural history: words and things in Iron Age Greece. Blackwell, Oxford

Morris J (2007) 'Agency' theory applied: a study of later prehistoric lithic assemblages from northwest Pakistan. In: Gardner A (ed) Agency uncovered: archaeological perspectives on social agency, power, and being human. Left Coast Press, Walnut Creek, pp 51–63

Pauketat TR (2001) Practice and history in archaeology: an emerging paradigm. Anthropol Theory 1(1):73–98

Peirce CS (1960) Collected papers of Charles Sander Peirce. Belknap Press, Cambridge, MA

Pippin R (1989) Hegel's idealism: the satisfactions of self-consciousness. Cambridge University Press, Cambridge

Pippin R (1997) Idealism as modernism: Hegelian variations. Cambridge University Press, Cambridge

Pippin R (2008) Hegel's practical philosophy: rational agency as ethical life. Cambridge University Press, Cambridge

Pleasants N (1997) Free to act otherwise? A Wittgensteinian deconstruction of the concept of agency in contemporary social and political theory. Hist Hum Sci 10(4):1–28

Pleasants N (1999) Wittgenstein and the idea of a critical social theory: a critique of Giddens, Habermas and Bhaskar. Routledge, London/New York

Putnam H (1981) Reason, truth and history. Cambridge University Press, Cambridge

Ribeiro A (2016a) Against object agency. A counterreaction to Sørensen's 'Hammers and nails'. Archaeol Dialogues 23(2):229–235

Ribeiro A (2016b) Archaeology will be just fine. Archaeol Dialogues 23(2):146–151
Ribeiro A (2018) Death of the passive subject: intentional action and narrative explanation in archaeological studies. Hist Hum Sci 31(3):105–121
Ribeiro A (2019) Archaeology and the new metaphysical dogmas: comments on ontologies and reality. Forum Kritische Archäologie 8:25–38
Ritzer G, Gindoff P (1994) Agency-structure, micro-macro, individualism-holism-relationism: a metatheoretical explanation of theoretical convergence between the United States and Europe. In: Sztompka P (ed) Agency and structure: reorienting social theory. Gordon and Breach, Yverdon, pp 3–23
Robb J (2004) Agency: a personal view. Archaeol Dialogues 11(2):103–107
Sander K (1993) Entelechy and the ontogenetic machine—work and views of Hans Driesch from 1895 to 1910. Dev Genes Evol 202(2):67–69
Sartre J-P (2003 [1943]) Being and nothingness: an essay on phenomenological ontology. Routledge, London/New York
Searle J (1980) Minds, brains, and programs. Behav Brain Sci 3:417–424
Searle J (1992) Conversation. In: Parret H, Verschueren J (eds) (On) Searle on conversation. John Benjamins Publishing Company, Philadelphia, pp 7–29
Shanks M, Tilley C (1987) Re-constructing archaeology: theory and practice. Routledge, London/New York
Sørensen TF (2016) Hammers and nails: a response to Lindstrøm and to Olsen and Witmore. Archaeol Dialogues 23(01):115–127
Stanton TW (2004) Concepts of determinism and free will in archaeology. Anal Antropol 38(1):29–83
Taylor C (1975) Hegel. Cambridge University Press, Cambridge
Taylor C (1989) Sources of the self: the making of the modern identity. Harvard University Press, Cambridge, MA
Taylor C (2004) Descombes' critique of cognitivism: symposium: Vincent Descombes, the mind's provisions. Inquiry 47(3):203–218
Taylor C (2015) Hegel and modern society. Cambridge University Press, Cambridge
Turner SP (1994) The social theory of practices: tradition, tacit knowledge, and presuppositions. University of Chicago Press, Chicago
Turner SP (2010) Explaining the normative. Polity Press, Cambridge
VanPool TL, VanPool CS (2003) Agency and evolution: the role of intended and unintended consequences of action. In: Vanpool TL, VanPool CS (eds) Essential tensions in archaeological method and theory. University of Utah Press, Salt Lake City, pp 89–113
Von Wright GH (1971) Explanation and understanding. Routledge and Kegan Paul, London
Webmoor T (2007) What about 'one more turn after the social' in archaeological reasoning? Taking things seriously. World Archaeol 39(4):563–578
Webmoor T, Witmore CL (2008) Things are us! A commentary on human/things relations under the banner of a 'social' archaeology. Nor Archaeol Rev 41(1):53–70
Williams M (2015) Blind obedience: the structure and content of Wittgenstein's later philosophy. Routledge, London/New York
Winch P (2008 [1953]) The idea of a social science and its relation to philosophy. Routledge, London/New York
Witmore CL (2007) Symmetrical archaeology: excerpts of a manifesto. World Archaeol 39(4):546–562
Witmore CL (2014) Archaeology and the new materialisms. J Contemp Archaeol 1(2):203–246
Wittgenstein L (1960) The blue and brown books, 2nd edn. Harper Perennial, New York
Wittgenstein L (1967) Remarks on the foundations of mathematics. MIT Press, Cambridge, MA
Wittgenstein L (2009 [1953]) Philosophical investigations. Wiley-Blackwell, Chichester

Wood AW (1990) Hegel's ethical thought. Cambridge University Press, Cambridge

Zahle J, Collin F (2014) Rethinking the individualism-holism debate: essays in the philosophy of social science. Springer, Cham

ŽiŽek S (2006) The parallax view. MIT Press, Cambridge, MA

ŽiŽek S (2013) Less than nothing: Hegel and the shadow of dialectical materialism. Verso, London/New York

Žižek S (2014) Absolute recoil: towards a new foundation of dialectical materialism. Verso, London/New York

Chapter 13
Are Archaeological Parks the New Amusement Parks? UNESCO World Heritage Status and Tourism

Elizabeth Scarbrough

Abstract In this chapter I address the concern that UNESCO World Heritage designation leads to unregulated tourism. I argue that heritage tourism not only has a negative impact on the site but may adversely impact local populations and descendant communities. I detail two related worries, UNESCO-cide and the Disneyfication of cultural heritage. The term 'UNESCO-cide' was coined by Marco d'Eramo to describe the role overtourism has played in the death of cities listed on UNESCO's World Heritage list. Disneyfication is the process of sanitizing potentially controversial or seemingly negative narratives from the tourist site to make the experience more palatable. I focus my analysis on two UNESCO World Heritage sites: Angkor Archaeological Complex in Cambodia and George Town in Malaysia. After a discussion about the negative impacts World Heritage designation has had on these sites, I suggest some mitigating strategies for tourism.

Keywords World Heritage · Tourism · UNESCO-cide · Disneyfication · Angkor Archaeological Complex · George Town

13.1 Introduction

The rise of cultural heritage tourism in recent years has spurred discussion about the relationship between tourism, damage to sites, and harm to local communities. The term 'UNESCO-cide' was coined by Marco d'Eramo to describe the role overtourism has played in the death of cities listed on UNESCO's World Heritage list. In this chapter I discuss some of the worries associated with UNESCO World Heritage

E. Scarbrough (✉)
Department of Philosophy, Florida International University, Miami, FL, USA
e-mail: escarbro@fiu.edu

© Springer Nature Switzerland AG 2021   235
A. Killin, S. Allen-Hermanson (eds.), *Explorations in Archaeology and Philosophy*, Synthese Library 433, https://doi.org/10.1007/978-3-030-61052-4_13

status. Specifically, I address the worry that World Heritage designation leads to unregulated tourism which has a negative impact on the site as well as the local and descendant communities. Does World Heritage designation turn archaeological parks into 'cultural heritage Disneylands'? What would that even mean?

In this chapter I focus on two examples from Southeast Asia: Angkor Archaeological Complex in Cambodia and George Town in Malaysia. Both of these sites have World Heritage designation. Angkor was voted the world's topmost site by Trip Advisor in 2018 (Sotheary 2018), while George Town is a more recent addition to UNESCO World Heritage designation. In what follows I will first provide a bit of background on UNESCO World Heritage designation and how it has increased tourism, followed by a brief discussion on how the term 'UNESCO-cide' is applied (and its relationship to the Disneyfication of culture). I will then outline my two examples and detail the negative impacts UNESCO World Heritage designation has caused. The penultimate section of the chapter will address these negative impacts by suggesting some mitigating strategies for tourism, before I present some general conclusions.

13.2 UNESCO World Heritage Designation and Its Relationship to Tourism

UNESCO began their World Heritage designation in 1972 to protect places of 'outstanding universal value'. UNESCO World Heritage (henceforth WH) sites can receive the UNESCO designation for their universal cultural and/or natural value. With this badge comes a recognition that the site is valuable for humanity at large and is worthy of protection. However, there is a tension in UNESCO's prescriptions for the protection of cultural property with UNESCO stating both:

> Being convinced that damage to cultural property belonging to any people whatsoever means damage to the *cultural heritage of all mankind*, since each people makes its contribution to the culture of the world. (UNESCO, *Convention for the Protection of Cultural Property in the Event of Armed Conflict*, May 14, 1954; emphasis added)[1]

And:

> ...cultural property constitutes one of the basic elements of civilization and *national culture, and that its true value can be appreciated only in relation to the fullest possible information regarding its origin, history and traditional setting*. (UNESCO, *General Conference in Paris*, October 12 to November 14, 1970; emphasis added)[2]

---

[1] See UNESCO: Convention for the Protection of Cultural Property in the Event of Armed Conflict. http://www.unesco.org/new/en/culture/themes/armed-conflict-and-heritage/convention-and-protocols/1954-hague-convention/. Accessed March 14, 2020.

[2] See UNESCO: General Conference in Paris. http://www.unesco.org/new/en/culture/themes/armed-conflict-and-heritage/convention-and-protocols/1954-hague-convention/. Accessed March 14, 2020.

The tension identified here is between seeing cultural heritage as valuable to all mankind and seeing it as importantly embedded in a specific cultural community. UNESCO's generally universalist perspective (Benhamou 2010) or cosmopolitan perspective (Appiah 2006) as evidenced in the first quote is in contrast with the more regionalist language in the second. In the second quote we see evidence that UNESCO acknowledges that cultural heritage's 'true value' can only be appreciated with regard to its origin, history and its traditional setting.[3]

UNESCO recognizes this tension and has taken steps in recent years to ramp up the participation and collaboration between the international body and local peoples. They recognize that only with local buy-in do these sites avoid danger. Nevertheless, the worry still remains that there exists an inherent tension between the interests of the UN and the interests of local and descendant populations. I point to this tension now and will return to it when discussing the potential negative impacts UNESCO WH designation might actuate.

UNESCO WH designation is also used as a marketing tool for heritage tourism and has recently been criticized for being nothing more than a tool for Destination Marketing Organizations (see Morgan et al. 2003; Ryan and Silvanto 2009). 2018 was the European Year of Cultural Heritage; unsurprisingly, tourism to heritage sites increased dramatically. Francesco Bandarin, former world heritage director at UNESCO, said that overtourism was an inevitable consequence of the WH designation. "[T]he very reasons why a property is chosen for inscription on the world heritage list are also the reasons why millions of tourists flock to those sites year after year" (Barron 2017). And while WH designation has spurred increased tourism, it does not seem as if that was the original purpose of the designation. As evidence, 'tourism' is mentioned only once in the 38 articles of the 1972 "Convention Concerning the Protection of the World Cultural and Natural Heritage".[4] Now it is hard to imagine World Heritage sites without tourists taking photographs, souvenir stands, and tour busses. In recent years UNESCO has collaborated with the tourism industry. For example, in 2008 UNESCO worked with a popular promotional guide *1001 Historic Sites You Must See Before You Die* (Cavendish 2008). Becoming a WH site not only lends a sort of 'trans-national' importance which yields an economic premium but promises the tourist a certain type of experience (Bourdeau et al. 2015).

By any measure, tourism is booming. It isn't just that there are more people on the planet, but there are more people with money. Cheaper flights, cheaper accommodations, higher volume cruise ships and tour busses, all encourage volume tourism. The United Nations World Tourism Organization reports that international tourism has grown 40-fold since commercial jet traffic began six decades ago (Tourtellot 2017). From 1950 to 2008 the number of international tourists increased from 25 million to 924 million (González Tirados 2011). Many of the sites these

---

[3] For an overview of recent work on the ethics of ownership of cultural heritage, spelling out 'regionalist', 'nationalist' and 'universalist' debates, see Matthes (2016).

[4] For more on the original role of tourism and the UNESCO WH program, see Bourdeau et al. (2015).

tourists visit—including archaeological ruins, natural attractions, and historic cities—have not changed in size. Nor have their tourism models adequately addressed the influx of new tourists.

Tourism is an $8 trillion (USD) industry and is one of the largest employers on earth: one in 11 people work in the tourism or travel industry (Becker 2017). Besides an increasingly wealthy population, technology has had a serious impact on tourism and travel. Smartphone mapping, home sharing apps, car sharing apps, and travel review websites have not only changed the way we travel but have increased tourism. Perhaps the biggest influence on travel in the past few years has been social media. Justin Francis, the chief executive of Responsible Travel says, "You can't talk about overtourism without mentioning Instagram and Facebook… Seventy-five years ago, tourism was about experience seeking. Now it's about using photography and social media to build a personal brand. In a sense, for a lot of people, the photos you take on a trip become more important than the experience" (quoted in Manjoo 2018). The 'Instagenic' or 'Instagrammability' of the site is a factor WH sites market towards. The need many tourists have to take photos will be discussed as a significant factor in the Disneyfication of cultural heritage tourism later in this chapter.

There are 759 cultural heritage sites on the UNESCO WH list (out of 1092 as of April 2019). 254 of these are cities, and the vast majority of these cities (138) are in Europe. Currently there are 48 sites on the UNESCO 'danger list' for being threatened by humans or nature. Only two sites have ever been delisted for gross violations. Many sites, especially those which have been on the list for many years, have yet to present paperwork detailing their required conservation and management plans (Gray 2016).

The process for receiving a WH designation is analogous to hosting the Olympics. The local community (government) must submit a proposal and vie for the designation in hopes to inject cultural heritage tourism dollars into their economy. Nations have obligations as well once their sites get listed. They are expected to protect the cultural heritage and work with UNESCO for strategic plans for sustainable tourism. However, as stated earlier, many countries have failed to submit reports and very few countries have been censured for non-compliance.

13.3   Value and Harm

How do we value World Heritage sites, and how might UNESCO WH status negatively impact sites and peoples?

13.3.1 Value

How do we value cultural heritage? In "The Values of the Past", James O. Young provides a useful framework for thinking about the different valuing practices with cultural heritage. He asserts that there are four core values when thinking about cultural heritage. There is cognitive value, economic value, cultural value, and cosmopolitan value. Cognitive value encompasses the archaeological value the archaeological data provides but is broader than just archaeological data. Young divides cognitive value into two types: intrinsic and extrinsic (Young 2013, p. 28). Extrinsic cognitive value encompasses archaeological data broadly construed. Archaeological data are not only sources of knowledge about past civilizations but can also be sources of medical knowledge. Young also believes that cognitive value may be intrinsic—a source of knowledge that is valuable for its own sake.

Economic value is the value the site might yield in tourism dollars but also the money individual artifacts might be sold for (either legally or illegally). Unlike cognitive value, economic value is "almost universal" (Young 2013, p. 32). While economic value often reigns supreme, some have argued that certain objects have so much intrinsic value they ought not be considered fungible. This is evidenced by many prescriptions against the buying and selling of human remains, including by the United States' NAGPRA (the Native American Graves Protection and Repatriation Act). By contrast, cultural value, according to Young, comes from a culture valuing something in virtue of its contribution to the culture's well-being (Young 2013, p. 34). Those who view a site as bound-up with their own heritage are likely to engage with the site differently and value the site differently than those who view themselves as outsiders (Salazar 2010, p. 195). This cultural value corresponds to the regionalist language found in UNESCO. This picture is complicated by the value such sites have in national dialogues. 'Cultural nationalism', a term coined by John Marryman, expresses the role nation-states have in this debate. Cultural nationalism does not squarely fit within Young's framework (Merryman 1986).

Cosmopolitan value states that cultural property belongs, in a real sense, to all of mankind. Young uses philosopher Kwame Anthony Appiah's work on cosmopolitanism to elucidate this type of value. In *Cosmopolitanism: Ethics in a World of Strangers*, Appiah states, "The connection people feel to cultural objects that are symbolically theirs, because they were produced from within a world of meaning created by their ancestors—the connection to art through identity—is powerful. It should be acknowledged. The cosmopolitan, though, wants to remind us of other connections" (Appiah 2006, p. 135). Appiah reminds us that the cosmopolitan citizen appreciates "the connection not through identity but despite difference" (in Cuno 2012, p. 85). He asserts that we respond powerfully to artworks that are not from our own cultural heritage because they spur our imaginative powers in important ways. These 'imaginary connections' to other peoples help engender feelings of solidarity with human beings different from one's own cultural traditions. In a particularly powerful quote Appiah states, "My people—human beings—made the

Great Wall of China, the Chrysler Building, the Sistine Chapel: these things were made by creatures like me, through the exercise of skill and imagination" (in Cuno 2012, p. 85). It is in this spirit that UNESCO positions themselves as the guardian of culture and natural heritage, with roughly 190 member stations ratifying the Convention. The convergence of UNESCO WH and tourism essentially asks us to weigh these four different values.

Tourism commodifies World Heritage and the vast tourism literature discusses this commodification. In Marilena Vecco's "Value and values of cultural heritage", she attempts to provide a formula for the "total economic value of the heritage asset" (Vecco 2019). Total economic value (TEV) is equivalent to the use value (UV) and non-use (NUV) of the site. More specifically it is the sum of the direct use value (DUV), indirect use value (IUV), option value (OV), bequest value (BV) and the existence value (EV). Her equation is as follows:

$$TEV = UV + NUV = DUV + IDV + OV + BV + EV$$

Of course, she defines all of these terms, but not necessarily in ways that are helpful to employ. To quantify these various variables, she unhelpfully reminds us we could employ numerous techniques including the Contingent Valuation Method, the Hedonic Price Method, the Travel Cost Method, or the Petition Method. Strangely, 'bequest value' (the value the 'heritage asset' will have to future generations) is listed as a non-use value. In Young's framework, bequest value may be economic, cosmopolitan, cultural, or cognitive. Importantly, this formula, for merely adding up the various values, does not help us weigh the disparate values nor does it help us determine how to weigh the interests of future generations. The way we ought to weigh these various values guides my thinking in the remainder of this chapter.

13.3.2   UNESCO-Cide and Disneyfication

First, UNESCO-cide. UNESCO-cide occurs when a city is named a heritage site and consequently, "dies out, becoming the stuff of taxidermy, a mausoleum with dormitory suburbs attached" (Gray 2016). The worry is that the city becomes a 'dead site'—frozen in time, not allowed to thrive or change. As Marco d'Eramo writes: "A serial killer of cities is wandering about the planet. Its name is UNESCO and its lethal weapon is the label 'World Heritage', with which it drains the life-blood from glorious villages and ancient metropolises, embalming them in a brand-name time warp" (d'Eramo 2014). The term 'UNESCO-cide' can be applied to cities near archaeological parks, such as Siem Reap near Angkor Archaeological Park, just as it has been to 'Heritage Cities' such as George Town.

Second, Disneyfication. The Society of American Archivists have defined Disneyfication as "the process of increasing the marketability of a story by removing or downplaying distasteful or potentially controversial facts from the historical record to make the story as unobjectionable to as wide an audience as possible".[5]

In reaffirming the universal role of cultural heritage, UNESCO has been charged with imposing a meta-narrative on cultural sites (Gravari-Barbas et al. 2015, p. 4). This narrative, in line with Appiah's "cosmopolitan experience" is one of sameness through difference, or to put it in Disney's terms, 'it's a small world after all'. The worry is that focusing on the universal role of cultural heritage subordinates local and subaltern narratives even further.

Another related worry, as articulated by Sarah Ellen Shortliffe, this meta-narrative is often Eurocentric and male-oriented (Shortliffe 2015). Realizing this UNESCO formed the gender trainings, and publications such as the Passport to Equality which cites their Convention for the Elimination of All Forms of Discrimination Against Women (CEDAW). A study by Sophia Labadi determined that the sites being nominated for WH designations were associated with historical men. When women were mentioned they were described in a 'neutral way' while "famous men tend to be described in a more positive and praiseworthy manner" (Labadi 2007, p. 162). There has been a shift in trying to balance the 'voices' of heritage, but for too long this notion that cultural heritage is the heritage of all mankind has overlooked issues of gender (Vernooy 2006, p. 232). This is exacerbated by the fact that tourism is the "most sex-segregated industry or the world's most sex role stereotyped industry" (Aitchison 1999, p. 61). Tourism Studies has grappled with their history, acknowledging that the abstract notion of a tourist is male (Shortliffe 2015). Thus not only do sites memorialize famous men, the default tourist is also male.

Disneyland's motto, 'the happiest place on earth', is also relevant here. In an effort to attract and appease tourists, heritage destinations may refrain from highlighting the darker parts of their stories: colonialism, slavery, torture, etc. In doing so they are silencing local and descendant populations' stories and, as we will see, are contributing to UNESCO-cide.

These two phenomena, while usually discussed separately, are often related. The more we turn a site into a sort of happy-place 'Disneyland', the more we risk UNESCO-cide. Through the use of my two examples I show that Disneyfication is a main contributor to UNESCO-cide.

13.4   Discussion of Two Examples

In this section I discuss UNESCO-cide and Disneyfication via the case studies of George Town and Angkor Wat Archaeological Park.

---

[5] Society of American Archivists. A Glossary of Archival and Records Terminology. https://www2.archivists.org/glossary/terms/d/disneyfication. Accessed March 14, 2020.

13.4.1 George Town

George Town in Malaysia is often referred to as an example of UNESCO-cide. The city was awarded WH status in 2008. A base for the British East India Company in 1786, the area became a hub for traders between India and China, and a jumping off point for pilgrims who departed Malay to visit Mecca. The mix of disparate architectural styles resulting from the jumble of traders is one of the reasons the town was awarded WH status. A travel journal from 1811 describes the spirit of the historic city:

> Turning the eye southward, Georgetown and the harbor are seen. The various styles in the construction of habitations of this small town have a strange effect—the European house, the Hindoo [sic] bungalow, the Malay cottage, the Chinese dwelling and the Burman hut are mingled together with regularity and apparently without plan, the first settlers having each built his residence according to the customs of the country. (Clodd 1948, p. 120)

For a site such as George Town to become a World Heritage site it must first be nominated by a 'state party' (e.g., countries which have adhered to the World Heritage Convention). Two advisory boards, the International Council on Monuments and Sites (ICOMOS) and the International Union for Conservation of Nature (IUNC), independently evaluate the application, providing their recommendations to the World Heritage Committee. The World Heritage Committee then consults with a third advisory body, the International Centre for the Study of the Preservation and Restoration of Cultural Property (ICCROM). The World Heritage Committee has the final say on World Heritage status. In order to become a World Heritage site, the World Heritage Committee must believe that the site is of 'outstanding universal value' and has met one of ten criteria set out by UNESCO. George Town was deemed to have 'outstanding universal value' and while it only needed to satisfy one, the World Heritage Committee believed it satisfied three criteria, specifically (ii) (iii) and (iv):

> (ii) to exhibit an important interchange of human values, over a span of time or within a cultural area of the world, on developments in architecture or technology, monumental arts, town-planning or landscape design;
> (iii) to bear a unique or at least exceptional testimony to a cultural tradition or to a civilization which is living or which has disappeared;

And:

> (iv) to be an outstanding example of a type of building, architectural or technological ensemble or landscape which illustrates (a) significant stage(s) in human history.[6]

George Town (ii) is an exceptional example of a multi-cultural trading town, (iii) is a living site of intangible heritage expressing the multi-cultural heritage of Asia with European colonial influences, and (iv) has a unique mix of architecture, including the stilt houses on the Penang jetty. One of the reasons George Town has been awarded WH status is due to the clan jetties on Penang Island. Each family owned

---

[6] See UNESCO: Criteria for Selection. https://whc.unesco.org/en/criteria/. Accessed May 13, 2019.

Fig. 13.1 Clan Jetties, George Town Malaysia. Reproducible under the terms of the Creative Commons Attribution 2.0 Generic License. Attribution: David Johnson. (Source: https://ccsearch. creativecommons.org/photos/11a82fee-5952-44ec-a8cb-04ba96571677.   Accessed   September 8, 2019)

their own jetty and as their families grew, they added more stilt homes connected by wooden walkways (see Fig. 13.1). The jetties were named for the clans who owned them. Journalist Laignee Barron described the jetties as, "A ramshackle collection of stilts houses and sheds, stretching along a line of wooden piers each bearing the surname of its Chinese clan they are one of the last intact bastions of Malaysia's old Chinese settlements" (Barron 2017).

George Town lost their free port status in 1969, which resulted in the city's decline. In 2000 the 1966 Rent Control Act was repealed and many inhabitants were forced to leave. The stilt city turned into a hub for illegal gambling and was called by locals a 'squatters' slum'. Two of the jetties were torn down to erect modern housing but receiving WH status in 2008 saved the extant jetties. The families who owned the remaining seven jetties felt that UNESCO was their 'hail Mary' pass to save their ancestral homes as these clan jetties were the last bastion of the old Chinese settlements in this region. While the area had high unemployment for the 30 years prior to their WH designation, after receiving WH status George Town became a tourist hot-spot: the city's hotels now receive between six and seven million people each year.

Clement Liang, a member of the Penang Heritage Trust, who lobbied for the clan jetties' inclusion in George Town's world heritage zone, believes that while UNESCO named 2017 the year of 'sustainable tourism', the organization had no concrete plan: "UNESCO has no clear guidelines or effective methods to control the commercialization of world heritage sites, and its talk on sustainability is more a verbal exercise than enforceable" (Liang, quoted in Barron 2017). Worries that tourism has killed their community are widespread.

Do the benefits of WH status outweigh the various burdens? The locals admit that UNESCO saved the remaining seven jetties. They believed the 'heritage economy' would benefit all parties: the local area receives economic growth and revitalization of areas that might have fallen into disrepair. It allows for cultural heritage to be protected and preserved. However, local communities feel like they have struck a deal with the devil: they saved the site but at the expense of its soul. This is at the heart of UNESCO-cide.

The first symptom of UNESCO-cide I would like to discuss is that local populations feel that much of the resultant tourism involves morally problematic voyeurism. Tourists watch locals as if monkeys in a zoo; they walk into clan homes uninvited (this is in direct violation of UNESCO's Hoi An protocols). Although the area is festooned with 'no photo' signs on local houses, many locals have either moved and/or boarded up their windows to avoid intrusive stares.

Another symptom of UNESCO-cide is when locals can no longer afford their homes or local services. Fishermen and oyster farmers have been moved out to make way for snack bars and souvenir shops. Local Siew Oheng says the tourism "affected our privacy. Our jetty has become commercialized. People are moving. During the December holidays like Chinese New Year and Malay Raya, it's not even a place to live" (quoted in Barron 2017).

The third symptom of UNESCO-cide, and the symptom I will be focusing the majority of this discussion on, is the Disneyfication of culture. One hoped that George Town would be 'preserved' as a living-cultural heritage community. A multi-ethnic, multi-cultural port town with so many disparate cultural traditions would be ideal for a tourist. The Malaysian Domestic Promotion division attempted to package the area for short-stay (2–3 day) holidays, for the Malaysian tourist. Originally the plan was to advertise the area as 'sun-and-sea', but currently heritage tourism is the predominate tourism in the area. There was a tourist boom in the 1930s and 40s, offering small city hotels (with accompanying brothel services) to sailors. The 60s saw tourism as a 'hippy trail', and became a destination for British, American and Australian soldiers and general travelers during the Vietnam War. What tourists see now is very different. The Malaysian government promoted a 'Malaysia, My Second Home' campaign, aimed at retired expatriates, enticing non-Malaysian individuals to buy up beach-front property. Most of these expats come from the UK, Indonesia, Hong Kong, Taiwan, and Singapore. By 2004 the heritage aspect of the area had been decidedly left to rot. The tsunami of 2004 and Bali bombings of 2002 and 2005 also wreaked havoc upon the tourism industry. In creating a WH site, no time or effort was given to preserve the lifeblood of the town. Rather, expatriates and tourists were welcomed and the historic jetties were ignored until they were seen as profitable.

When WH sites sanitize the negative parts of the past and pander to cultural heritage tourists, they run the risk of alienating the very tourists they are trying to attract. With the influx of tourism, even tourists feel that locations have become 'too touristy' or a 'tourist trap'. There is a delicate balance between catering to tourists and keeping the regional flavor of a location. Evidence for this claim can be seen on the review for the clan jetties on Trip Advisor. One reviewer stated (March 2019), "Not really worth the visit, most of the tour groups visit here thus lots and lots of souvenir shops and food stalls." Another reviewer states, "Very busy with cruise ship hoards when we visited Chew Jetty is very commercialized. Not really our thing." Finally: "This may be because we visited during the Chinese new year, but we found that we were missing the history of the jetties. Each jetty is named after a clan that resides there. I found myself asking a lot of questions but not having anybody to answer them. The tourist shops do not have items that are unique to the jetties or the clans. We walked down to the end of the Chew jetty, looked at the Strait of Malacca, and turned back." As this last reviewer complains, there is little information upon arrival to the Penang Island about the unique history of the island. The Penang Tourism Action Council produced a brochure, Penang Heritage Trails, which details two self-guided tours: Historic George Town Trails and Traditional Trades. The first lays out the original routes of the American Express, and the second discusses the 'endangered trades'. However, these brochures are hard to find and placed in the airport upon arrival and in the ferry terminal. This is all evidence for the claim that very little has been done to contextualize the area or provide guidance for interpretation. Not only has the history of colonialism been erased from the site, much of the various religious traditions as well as the regular hustle and bustle has been erased as well.

In sum, WH designation has saved the physical 'tangible' culture but has done little for the local community and has done nothing to help preserve the various stories that make the island worthy of its WH designation. Local Siew Pheng runs a homestay in one of the stilt houses and worries that if the clans don't impose their own plan for sustainable tourism, "the identity, the history of this place will be gone" (quoted in Barron 2017).

13.4.2 Angor Wat

My second example will perhaps be more familiar to readers. Angkor Wat Archaeological Complex in Cambodia is one of the most famous archaeological parks in the world (see Fig. 13.2). It received its UNESCO World Heritage designation in 1992. It was built successively from the ninth to fifteenth century during the Angkor period of the Khmer civilization. It comprises over 400 square kilometers of temples, hydraulic structures (e.g., dykes, reservoirs, canals) and other monuments. Angkor was the center of the Khmer Kingdom for centuries, and the archaeological park shows off the beautiful Khmer architecture as well as Khmer engineering marvels. Its sprawling temples are connected by a very complex early water-management system that delivered water throughout the plateau, turning sandy land into land excellent for rice farming.

Angkor was one of the most powerful and advanced civilizations in Southeast Asia. At one point the kingdom comprised Vientaine (Laos) to the Malay Peninsula,

Fig. 13.2 Angkor Wat. Reproducible under the terms of the Creative Commons Attribution 2.0 Generic License. Attribution: Jean-Pierre Dalbéra. (Source: https://commons.wikimedia.org/wiki/File:Angkor_Vat_(6931599619).jpg. Accessed May 13, 2019)

from southern Vietnam to the border of Pagan in Burma. At its peak it may have housed over one million people, making it the largest city in the world until the Industrial Revolution. During the reign of Jayavarman VII (c. 1181–1200), the sandstone towers were a record achievement. Perhaps no king afterwards built temples on such a scale again. Jayavarman VII made Mahayana Buddhism the Khmer Empire's state religion, changing the iconography on some of the older temples from Hindu to Buddhist. It is rumored that it is the king's smiling face etched into images of the Buddhist divinity that adorn the various temples (Hammer 2016). In the fifteenth century Angkor was conquered by Siam who ceded it to the French in 1907. Angkor Park was established in 1925 as an open-air archaeological museum. It was under the direction of French archaeologists until 1972. During the American War of the 1970s the Angkor temples and monuments provided locals with literal shelter from the bombing of Siem Reap.[7] At the time the area was occupied by the Khmer Rouge, and during this time little conservation or restoration work was done. Pol Pot's anti-intellectualism resulted in the death of many local archaeologists and historians: "The greatest harm to Angkor during this time was the murder of all the Cambodian archaeological experts and the permanent loss of historical knowledge

---

[7] The United States originally denied the bombing. See https://www.nytimes.com/1976/03/05/archives/cambodia-shows-a-bombed-town-conducts-foreign-envoys-on-tour-of.html. Accessed May 13, 2019.

that should have been passed down to the next generation" (Candelaria 2005). At the end of the Pol Pot regime in 1979, locals started to taking care of the temples. Vietnamese troops left Cambodia in 1989.

Angkor Wat received WH designation on December 14, 1992 for satisfying criteria (ii), (iii) and (iv) (cited above) as well as (i) which states, "to represent a masterpiece of human creative genius". In 2015 Angkor was named 'The Most Famous Historic Site in the World' by Trip Advisor and was ranked the world's topmost site by them in 2018.

The site has three main areas: Roluous, Banteay Srei, and Angkor (which includes the temples of Angkor Wat and Angkor Thom). UNESCO created five zones:

1. Monumental Sites
2. Protected Archaeological Reserves
3. Protected Cultural Landscapes
4. Sites of Archaeological, Anthropological or Historic Interest
5. Socio-Economic and Cultural Development Zone of Siem Reap/Angkor Region

In the 90s the site was in disrepair and in major need of stabilization. It took quite a while for tourism to tick up due to the land mines which still riddled the site and the site still had occasional visits from Khmer Rouge guerillas (who had been exiled into the forest since the end of the war). The site was generally unsafe after dark. The Cambodian government paired with the private company, Sokha Hotel Company (aka Sokimex), who managed the site. Sokha Hotel received a 50/50 split on the first $3 million (USD) in ticket sales, after which it received a whopping 70% of ticket sales.

Following the bid for WH designation the ICC (International Coordinating Committee for the Safeguarding and Development of the Historic Site of Angkor) was founded. Japan and France co-chaired the committee which was mandated to set the priorities for conservation on the site. The two countries also held the purse strings. The national (Cambodian) body responsible for Angkor was the APSARA (Autorite pour in Protection du Site et l'Amenagement de la Region d'Angkor). Created in 1995, APSARA is still in charge of all the monuments, the ancient hydraulic system, cultural tourism, and the local community living on the site. In 1997 APSARA established the 'heritage police' (which I will discuss later).

Angkor Wat's numerous temples are still used today. It holds the record for the world's largest religious structure. According to a 2004 report there are perhaps 100,000 inhabitants over 112 historic settlements scattered throughout the site.[8] So not only is Angkor Wat an historical site of immense value, but due to the local residents living there, UNESCO has also designated it a so-called "living heritage site" (Miura 2010, p. 104).

From 2004 to 2014 Angkor Wat saw its tourism increase by more than 300% (Caust 2017). One of the poorest countries in Southeast Asia, the tickets for the

---

[8] See UNESCO: Angkor. https://whc.unesco.org/en/list/668. Accessed September 27, 2019.

Archaeological Park alone provide Cambodia with more than $85 million dollars a year in the first nine months of 2018 (Chan 2018). This is a 13% increase over 2017 numbers. In 2016 the government took over ticket sales to the archaeological complex from the Sokimex group. Shortly after it increase ticket prices from $20 to $37 for one day tickets, $40 to $62 for three-day tickets and week-long tickets from $60 to $72. Tourism accounts for 16% of Cambodia's GDP.[9]

Of course, tourism to the site has caused some damage. Normal wear and tear are exacerbated by the fact that many of the structures on the site have never been adequately stabilized. Uneven pathways are common, a result of the never-ending struggle between man's artifacts and the Banyan trees. There are very few guards (or 'heritage police') to stop tourists from climbing on the temples. There are other dangers in letting in so many tourists. For example, in 2014 a New Zealand woman destroyed a Buddha statue at the park reportedly because she thought it fake and that it did not belong there (Kitching 2014). Further, with a nearly 20% year-to-year increase on tourism, environmental pollution has become a problem, including rubbish heaps and waste control.

Angkor has also had to deal with disrespectful behavior on the site. Angkor Wat is still an active site of worship and as such has signs explaining respectful dress and behavior to visitors. Despite these signs the site has had problems with international tourists stripping down to take naked photos with the temples as a backdrop. To combat this in 2015 Angkor Wat released a video to inform visitors of their 'code of conduct' which includes: no smoking, modest dress, not giving money or candy to begging children, no taking photos of monks, and following all posted signs (no touching or climbing).

However, one of the biggest challenges for Angkor Wat is uncontrolled tourist development *around* the archaeological park. The nearby city of Siem Reap host tourists making their way to Angkor Wat. Massive hotels have sprung up around the park; their tapping the groundwater has impacted the water table beneath the temples of Angkor impacting their stability. The groundwater is only five feet below the surface, so thousands have made illegal private wells. Unsurprisingly, it is primarily the luxury hotel industry and resultant golf resort which have overtaxed the water. Among the list of international hotels include a Park Hyatt, Sofitel, Raffles, Belmond, Marriot (Le Meridien Angkor), and more. Development is ongoing and is increasing. In 2019 Macau Legend purchased 1200 hectares in Siem Reap Province to develop a resort/casino, investing $90 million (USD) in the price of land alone.

UNESCO has deemed the situation a 'water crisis', investigating the possibility of pumping water from the Tonlé Sap Biosphere Reserve in to help sustain management of the water table. Residents of Siem Reap no longer feel comfortable bathing in local water, stating that it has become too dirty. Only a decade before local communities used it as drinking water (Sok 2017).

---

[9] Moreover, the film industry has pumped some much-needed funds into the economy. For example, *Laura Croft: Tomb Raider,* reportedly cost Paramount Pictures $10,000 a day to film in the archaeological complex (McWhinnie 2015). By contrast, one third of the Cambodian population live on less than $1 a day.

Although there is a 20% year-to-year increase to tourism at the archaeological park, overall tourism in Siem Reap (the closest city to Angkor Wat) has declined. According to Ho Vandy, the secretary-general for the Cambodia National Tourism Alliance, many tourists choose to spend fewer days in Siem Reap and in Cambodia, so tour operators, hotels and other vendors have reported that business has declined (Neth 2011, p. 48). There does not seem to be a plan for the development of the town which has the potential to destroy the very temples its livelihood depends on.

Measures instituted on the Angkor complex have been more successful than relying on the local government to regulate the expansion of Siem Reap. One measure that is seen as generally successful is that tourists cannot visit the complex without hiring a local guide. Cambodia's Tourism Authority recently banned food inside the heritage complex to offset the amount of litter that otherwise would be left behind by visitors. In 2017, the APSARA Authority banned food vendors from setting up their stalls in front of the entrance to Angkor Wat. Angkor Wat was on UNESCO's Danger List from 1992 to 2004. While it was delisted in 2004, with the exogenous threat of Siem Reap's never-ending expansion as a tourist hub, the site might be relisted.

Thus, we have already discussed some of the problems tourism poses to the archaeological complex. The sheer number of tourists puts the site in danger, not just from footfall but from rubbish piles. Outside Angkor Wat, we see that the resultant tourism industry in Siem Reap has led to serious changes in the water table that not only put the temples in danger, but also pollute the water for local usage.

We have already seen some of the symptoms of UNESCO-cide. The destruction of the water table has impacted both the temples and the health of the local population making the area less habitable. We've also seen tourists being disrespectful at locals' sites of worship, again making the area less welcoming of everyday life. Additionally, Angkor Wat has become such an icon to worldwide travelers that third parties are profiting off of the religious iconography. Earlier in 2019 the Cambodian government complained to Amazon over an item described as 'Angkor Wat Toilet Products', which were toilet seat covers adorned with the outline of the famous temple (Sopheakpanha 2019). Angkor Wat is a national symbol of pride, and a religious site; understandably the Cambodian government and local worshipers were far from impressed.

Siem Reap has been deeply impacted by tourism at the Angkor complex. Shockingly, 50% of those living in Siem Reap are from other areas of Cambodia (Neth 2011; Chheang 2010). Public policy analyst Vannarith Chheang reports that 50% of those in Siem Reap work in the tourism industry and that the average length of stay in the city is about five years. People move from other parts of Cambodia for economic opportunity but do not put down roots. Hotel staff commented in a personal interview (January 9, 2008): "The salary of the local [hotel] staff is just enough to survive but not enough to have a good life. We just get only about $100 per month. With such rapidly [increasing] living cost[s], we find it more and more difficult to live within the current salary" (quoted in Chheang 2010, p. 95).

Sok An, the Minister in charge of the Council of Ministers and the Chairman of the National Tourism Authority of Cambodia, said that if the Khmer had no Angkor, nobody would recognize them (see Miura 2010, p. 111). The locals have never been invited to the ICC and have very few opportunities to voice their concerns. Prum Tevi (of the Royal Academy of Cambodia) states:

The people living around the Angkor Park in particular and Siem Reap province in general are still poor. They could not get much benefit from tourism, but on the other hand they are the victims of tourism, given they suffer from the rapid increase in living costs. In order to have a sustainable tourism, it requires a strong participation from the local people. Now we can't see it happening in Cambodia. The poor are becoming poorer while a small group of rich are becoming richer. The government never pays attention to improve the livelihood of the local people here. (quoted in Chheang 2010, p. 96)

In sum, while wealth is being generated for a few, locals are receiving inadequate benefits from the WH designation of the neighboring Archaeological park. Further the WH designation has caused environmental degradation and water management issues which, if not adequately addressed, will damage the very structures WH status was designed to protect.

Before 1992, the attitude of UNESCO toward WH sites was to remove all locals from the designated area. In the early 1990s UNESCO shifted their policy and learned their lesson from sites such as Ayutthaya (Thailand) and Borobudur (Indonesia). We see this shift in the criteria for enshrinement, which shows a shift from viewing these sites as monuments frozen in time to seeing the site as a 'living' site. As such, people were not resettled from the Angkor Wat archaeological site. Furthermore, there are no reliable numbers on how many people live on the site. In 1992 there were approximately 22,000 people living in Zones 1 and 2. By 2005 the number reached 100,000 (in a study done by APSARA; see Miura 2010, p. 112). Most of these locals are rice farmers, but they consider themselves 'forest people'; people who rely on the forest for subsistence. Their families have traditionally owned land which is now considered part of the temple complex, and have harvested rice on this land. There are also trees that are owned by certain locals, who have the rights to harvest their fruit or resin. Lands are used for cattle to graze and for water buffalo to bathe. During French colonialization, the French government tried to ban local inhabitants from the land and outlawed logging and harvesting of resin. In 2000 the Minister of Interior, without consultation from the local government or APSARA, outlawed many cultural practices within the archaeological park including: cultivating of rice, resin collection, grazing cattle, killing birds, entering the forest, brining cutting instruments inside Angkor Tom. No compensation was offered to the locals, nor job training. However, because of a 2004 decision from the government (APSARA), local inhabitants cannot be physically removed from the site (see Miura 2010, p. 119). Further, it is well known that the 'Heritage Police' extort money from Cambodians working in the complex. This includes souvenir vendors, caretakers of religious statues, beggars, and rice cultivators. Thus, not only are locals harmed by the inclusion of Angkor as a WH site, the part of the site that involves 'living heritage' is in danger of being eliminated, pushing out locals and denying them the ability to earn a living.

D'Eramo's UNESCO-cide worries remain. Has Angkor 'died out' becoming a 'mausoleum with dormitory suburbs attached'? The forest people have lost much of their livelihood and other locals have lost access to clean water. It appears that the protection of the monuments of Angkor has taken precedence over 'living heritage', although even these monuments are in danger with the expansion of Siem Reap. Siem Reap has not been turned into a 'suburb', but has become a bifurcated city, home for many poor Cambodians on the one hand, and rich tourists, able to afford golf and luxury hotels, on the other.

Has Angkor fallen prey to the Disneyfication of Cambodian culture? Has the site papered over 'distasteful or potentially controversial facts from the historical record to make the story as unobjectionable to as wide an audience as possible'? As seen by recent tourist data, travel to Angkor has increased, but travel around Cambodia has not. Many bemoan the fact that tourists don't go beyond Angkor. In fact, many tourists don't go beyond the main temple (Angkor Wat) in the temple complex. Tourists fly into Siem Reap's international airport, stay at a luxury hotel outside of the main thoroughfare of Siem Reap, and visit the archaeological park for a few hours over 1–3 days. During their visit, they must be accompanied by a local guide, who will provide them with as little or as much information as requested. The guides know the spots most tourists like to take photos and do not discuss any historical details unless specifically requested to do so. Based upon my own experience during research trips, most of the guides are men in their late 40s–60s. This means most were alive during the time of the Khmer Rouge; the genocide of their people is recent history to them, within their living memory. Yet very little about the Khmer Rouge is discussed at the site. Most of the larger signs in the archaeological complex are dedicated to explaining the conservation efforts of the temples, detailing which countries contributed what funds. Perhaps the lack of information about the recent genocide is understandable. The official Ministry of Tourism Cambodia ('Kingdom of Wonder') website, which lists guides to all the provinces in Cambodia and major tourist attractions, does not list either the Tuol Sleng Genocide Museum (S-21 Prison) or the Choeung Ek Genocidal Center (Killing Fields).

According to UNESCO's former Scientific Advisor for Angkor, Azedine Beschaouch:

> The last 25 years have been a period of intense philosophical reflection for the Cambodians. They have asked themselves whether the decline of the Angkor Empire … signaled the end of a Cambodia that was peaceful and highly cultured, and whether they themselves were part of another Cambodia, one which is destructive and barbaric. Angkor above all allows them to be reconciled with their own history: aligning themselves to this great civilization allows them to draw a line under the barbaric times. (Boukari 2002)

Perhaps Angkor is just taking cues from UNESCO. As seen in this quote, Cambodia is portrayed as either 'highly cultured' or 'barbaric'. If this is how the international community discusses Cambodia it is no wonder that tourism practices offer only the picture of the beauty of the Khmer Empire. Also seen in this quote is how heritage tourism becomes a tool for identification: is Cambodia the Khmer Empire, or is it the Khmer Rouge? Is Cambodia just Ta Prohm, the 'Tomb Raider' temple? What bothers me most about this is that Beschaouch glosses over 500 years of the history of this region and ignores the rich cultural traditions of peoples who were subjected to genocide. Cambodia is neither the Khmer Empire or Khmer Rouge, but both and more. The 'and more' part is missing from most of heritage tourism.

This idealization of the past is not uncommon in heritage tourism. Bomb craters are left as pox marks on the landscape, yet no sign contextualizes them. Beggars walk around the complex, many missing limbs yet there is no attempt to discuss the generational impact of war. There is no discussion of the American bombings or Vietnamese occupation. A nearby landmine museum however does a good job of informing tourists about the harm of landmines, and the work that still needs to be done to demine the area.

13.4.3   Summary of Symptoms of UNESCO-Cide

UNESCO World Heritage sites are the site of battle between the global and the local. As alluded at the outset of this chapter, organizations like UNESCO, World Monuments Fund, International Council of Museums, the International Council on Monuments and Sites (ICOMS) stress the 'universal' aspect of the value of cultural heritage. As such concepts of 'tradition' and 'authenticity' are brutally used to force the 'unchangeableness' of culture. UNESCO (and ICOMOS and others) define the authenticity of material culture often by the continuity of original materials. According to Article 15 of the Venice Charter for the Conservation and Restoration of Monuments, "All reconstruction work should however be ruled out a priori. Only anastylosis, that is to say, the reassembling of existing but dismembered parts can be permitted. The material used for integration should always be recognizable and this use should be the least that will ensure the conservation of a monument and the reinstatement of its form".[10] Another approach to the authenticity of built heritage can be found in ICOMOS's 1994 NARA Document on Authenticity, which defines authenticity of a heritage building via continuity of use rather than continuity of materials.[11] This document recognizes that a respect for cultures also mean a respect for different standards of identity and authenticity.

Heritage tourism is in the 'history-telling' business and as such carries a heavy burden. While locals offering sand-and-sea vacations have to worry about the environmental damage tourism does to their site, they rarely have to worry about neo-colonialization and the Disneyfication of culture. Dallen Timothy, a cultural tourism scholar at Arizona State University has argued that WH designation has turned sites into commodities for outsiders, "rather than remaining in control of the people whose cultural heritage it really is. It's a matter of powerful versus the powerless" (quoted in Gray 2016). This retelling of history in service of the international heritage tourist can silence certain narratives, leading to worries of 'cultural

---

[10] See UNESCO-ICOMOS: The Venice Charter. https://www.icomos.org/centre_documentation/bib/2012_charte%20de%20venise.pdf. Accessed March 14, 2020.

[11] See UNESCO-ICOMOS: NARA Document on Authenticity. https://whc.unesco.org/archive/nara94.htm. Accessed March 14, 2020.

colonization' and tourism as 'neo-colonialism'. We've seen this in both of our cases, and now I turn to possible solutions to these worries.

13.5 Possible Solutions

In this section I will address some possible solutions to the problems UNESCO World Heritage designation imposes on the site. As we have seen, the problems are myriad: from wear and tear on the physical site, to the exploitation of the locals, and the silencing of narratives unfriendly to heritage tourism. Solutions should foreground the ways we can prevent these sites from turning into mausoleums, reproducing 'culture' from a stagnant period of time. In un-Disneyfying the sites, I believe we can mitigate worries of UNESCO-cide.

13.5.1 Privatization

If we return to our four values (i.e., cognitive, economic, cultural, and cosmopolitan) you will see that the story I have told about Angkor Wat and George Town rests on economic value. In both cases a major problem is that the economic value is not benefiting the people who live in or near the heritage site. The local population isn't in a fight over cognitive, cultural, or cosmopolitan values but rather is advocating for sustainable tourism that would economically benefit them.

One might think that many of these problems could be solved with an injection of additional funds. We could fix the wear and tear on the site, and perhaps with more money an interpretive center could be built on the site to help contextualize the monuments and give voice to local and descendant populations. Perhaps unsurprisingly, I will suggest that this is not a good solution and, unfortunately, this is becoming an ever-popular path. In a desperate attempt to receive more funds, more and more World Heritage sites are seeking out private over public funding. In April 2018, Delhi's Red Fort, a seventeenth century fort, became 'sponsored' by a corporation. An open bidding process took place and Dalmia Bharat Corporation was the highest bidder. It will pay the Indian government Rs 25 crore ($3.56 million USD) for the sponsorship under India's 'Adopt A Heritage' scheme. While the Archaeological Society of India is still in charge of all conservation efforts, Dalmia are expected to provide the following: façade illumination, installation and maintenance of water fountains and bathroom facilities, installing tourist check-point turnstiles, a 1000 square foot visitor center with 'thematic' cafeteria, and fencing (Goyal 2018). The company will also be hosting the annual Independence Day celebrations. The Minister of Tourism admits that this scheme is more about providing tourist amenities than archaeological conservation. And of course, the corporate sponsors will be able to use the sites for advertising.

The response to this adoption has not been univocal. The Ministry of Tourism believes that the adoption of the Red Fort, "is an important first step and that there are many more monuments and heritage places that need to be managed by private entities as they come on board as Monument Mitras (friends)" (Goyal 2018). Objectors believe the government has handed over the symbol of India's independence to private companies. India is blaming the slow start to this program on the Archaeological Society of India, who they argue is holding up the process (Sharma 2018).

The Red Fort is only one piece of immovable cultural heritage available for adoption. Next is the Taj Mahal; GMR Sports (consumer sports product company) and ITC Ltd. (a cigarette company) are believed to be the finalists in the race to adopt it. The Indian government argues that without corporate sponsorship the seventeenth century marble mausoleum will succumb to discoloration, insect infestation, and the spread of algae.

While India does not currently allow corporate entities to directly fund their archaeological restoration efforts, Italy does. The Italian luxury goods brand Tod's funded a 2014 restoration of the Colosseum in Rome (Pianigiani and Yardley 2014). The fashion brand Fendi kicked in some money for the Trevi Fountain restoration, and Bulgari gave $2 million (USD) to help clean the Spanish steps. Salvatore Ferragamo fashion house donated just under a million dollars to help restore a wing of the Uffizi Gallery. Unlike the Red Fort, none of these Italian fashion brands indulged in large corporate branding on the sites. Fendi will receive a small plaque near the fountain (the size of a shoebox). For $33 million (USD) Tod's earned the right to put its corporate logo on entrance tickets to the Colosseum, although it will not be putting up any logos on site. For its donation to the Colosseum Tod's will also have the rights to associate its brand with the Colosseum for up to 15 years. Of the private sponsorship of Italy's cultural heritage Maria Luisa Catoni, Professor of Ancient Art History and Archaeology at IMT Lucca University, comments, "I am very worried that the Italian government doesn't have a line. This is a question of preservation and restoration, but also a question of taste" (quoted in Faiola 2014). In 2019 it was announced that the fashion house Bulgari will dedicate over a million dollars to the restoration of Largo di Torre Argentina, the site where Caesar was stabbed (Daly 2019). While the fashion house has promised locals not to de-home the feral cats that have taken over the site, we do not yet know if their corporate sponsorship will lead to advertising on the site. Besides the worry of corporate advertisements, we might worry that private businesses have no, or at least less, motivation than UNESCO to be mindful of both living culture and un-Disneyfied narratives. Thus, adding funds via private firms does not seem to mitigate our UNESCO-cide worries.

13.5.2 Digital Technologies

Another possible mitigation is the use of new technologies to memorialize built heritage before it is destroyed. For example, drones are helping conserve the Great Wall of China (Deng 2018). Due to a partnership between Intel and the China Foundation for Cultural Heritage Conservation, Intel's Falcon 8+ drones were used to map and make aerial photographs of the site. The next stage is for the AI data capture system to create a visual representation of the Jiankou section. (China has the second most number of UNESCO World Heritage sites.) However, without this partnership with a private corporation (Intel) this would not have happened. Similarly, Palmyra's monuments are being mapped by the Arc/k Project. Worries from our last section on privatization of heritage remain. The worry here concerns the ownership of this digital data (Olson 2018). Obviously, there are epistemic benefits/epistemic value in this project. And if such projects were free there would be no harm in following through with them. But the serious question is: who will benefit from this knowledge? And who will this digital heritage belong to? Will it belong to indigenous communities? Local or national governments? Private organizations? And who should it belong to? Further, if there are economic benefits, who will reap them? Also notice here that the projects' recording immovable cultural materials does nothing to help preserve 'living' culture. As such it does nothing to mitigate our UNESCO-cide worries and adds additional worries about the ownership of such digital heritage.

13.5.3 Revert to Local Control

When worries of UNESCO-cide (and the contributing factor of Disneyfication) are raised, a common solution offered is to return the cultural heritage management exclusively to the local population. According to cultural heritage tourism experts Caust and Vecco (2017), "an inherent danger in the awarding of UNESCO Status is that it takes the planning and control away from the local community so that the locale becomes a 'play-thing' for national and international interests" (p. 2). The proposed solution would say that we should invest in cultural value above and beyond cosmopolitan value. But this solution too is very complicated and, I believe, will not solve our UNESCO-cide worries. If we take both of our examples, the interests of the local governments, local populations, and national governments conflict. We still retain the worry that people in power with more money exploit those without. Malaysia is interested in George Town insofar as it can capitalize on tourism, which is much more profitable than supporting the various clans. Perhaps the solution is to show the Malaysian government that by allowing for run-away tourism, they are devaluing the very thing they hope to make money from. In Cambodia, the APSARA has used the UNESCO designation to extract money from locals and it is not clear that they would stop doing so if UNESCO was not involved. Further, the jobs available to the local populations usually require English or another second language. These jobs often go to the urban elite, those with the education to interact with tourists. This is evidenced by the influx of tourism workers to Siem Reap, who only stay in the city for five years. Simply stating that local populations should be the caretakers of their own cultural heritage passes the buck. In the two cases I detailed, it would be asking locals who are poor and have little political capital to do something near impossible without international help. Which leads us back where we started.

13.5.4 Work with UNESCO Against UNESCO-Cide

I think the most productive solution to the crisis of UNESCO-cide is to work with UNESCO to adopt a series of protocols to mitigate these worries. Some of these will be quite easy, and others quite difficult. Some will be near impossible without buy-in from their respective local and national governments. In what follows I will list a series of suggestions that enlist the help of UNESCO.

One plausible solution is to impose higher entrance fees and timed tickets for sites, and earmarking the fees collected for services for the local population. Siew Pheng, the local woman mentioned earlier who runs a homestay in George Town, thinks that the clans should impose an entrance fee. She also believes that this fee shouldn't be used exclusively to update services for tourism but should also funnel into improvements for the sake of residents (Barron 2017). Many sites have chosen to impose higher entrance fees and have dedicated funds back to local residences. Sites such as Angkor have differing fees for local, national, and international tourists. UNESCO could make this mandatory for sites to retain WH designation. I suspect the reason that they don't is because they want to retain local and national input into the management of the site. But UNESCO could do a better job of enforcing the code breaking that already occurs and demand that sites have a fully fleshed out sustainable tourism plan that they adhere to or lose their WH designation.

The hardest, and perhaps most effective, solution to the UNESCO-cide problem is to change the culture of heritage tourism. UNESCO's focus on intangible culture and diversity of cultural expression goes a long way to mitigating worries of turning sites into mausoleums. UNESCO has invested in protecting more and more intangible culture, with two more conventions written in 2003 and 2005. Specifically, UNESCO has three conventions: Tangible Heritage, Intangible Heritage and Diversity of Cultural Expression. Intangible heritage includes (but is not limited to) oral traditions, customs, performing arts, ways of life, social practices, rituals, festive events, knowledge, and skill to produce traditional crafts.[12] Article 2.1 of the 2003 Convention for Safeguarding Intangible Cultural Heritage defines intangible

---

[12] See UNESCO: Intangible Cultural Heritage. https://ich.unesco.org/. Accessed September 27, 2019.

culture as "practices, artifacts and cultural spaces associated therewith—that communities, groups and in some cases individuals recognize as part of their heritage".[13] Focusing on how tangible and intangible heritage intersect would help mitigate our UNESCO-cide and Disneyfication worries. The importance of intangible culture and minorities cannot be understated. "Immigrants and other minorities often have a close relationship with a heritage that is vulnerable in the face of social and economic pressures; preserving their intangible heritage is a way of reproducing their cultural identity. Intangible heritage can be a source of ontological security for minorities in a society where other citizens have distant physical characteristics, cultural beliefs, informal practices, religions and languages" (Seglow 2019, p. 13).

The focus on the history of the site, and on a single (or few) narratives present history as stagnant and a nostalgic version of a difficult past. D'Eramo calls this 'temporal fundamentalism' where tourists and others believe "What dates from an earlier time is worthier of merit" (d'Eramo 2014). Combating this will be difficult but presenting tourists with more information, from different sources, about the site could mitigate this worry. Tourists are eager for the information as well. Tourists to the George Town clan jetties bemoaned the lack of contextualizing information. And tourists to the Angkor archaeological complex are curious about landmines and the recent turbulent past of Cambodia, as well as the current lifestyles of the local populations.

'Temporal fundamentalism' is a contributing factor in Disneyfication. In showcasing one narrative of a town or heritage site, you turn the site into a mausoleum (UNESCO-cide) and importantly erase subaltern voices. One solution to Disneyfication and temporal fundamentalism is to train local and descendant populations to be tour guides and encourage them to share their life stories, making the tour more about how the heritage site is still used and impacts the local population today. As stated earlier, hiring local tour guides is common in many World Heritage sites. In 2005 the Asia and Pacific region of UNESCO (Bangkok, Thailand) worked with the Asian Academy of Heritage Management to offer regional-based programs for heritage tour guide training. The power of a personal tour would go a long way towards mitigating our Disneyfication worries and could easily be made a prerequisite for entrance at many WH sites.

Several researchers have noted the connection between cultural heritage, identity formation, and testimonial injustice (e.g., Matthes 2016; Pantazatos 2017). Miranda Fricker defined testimonial injustice, a form of epistemic injustice, as "Testimonial injustice occurs when the testimony of an individual or collective is perceived as less credible owing to an identity prejudice" (Fricker 2007, p. 28). Heritage is partially valued as a source of knowledge, one that is relevant to our identity formation in the present. As seen in processes of Disneyfication, WH designation has been linked to silencing contributions from disparate peoples, from local populations at large to underrepresented groups within. When people (local and descendant

---

[13] See UNESCO: Convention for the Safeguarding of the Intangible Cultural Heritage. http://portal. unesco.org/en/ev.php-URL_ID=17716&URL_DO=DO_TOPIC&URL_SECTION=201.html. Accessed March 14, 2020.

populations) are left out of the narratives of cultural heritage, this is a type of testimonial injustice. As Andreas Pantazatos states, "Non-expert stakeholders are usually victims of epistemic injustice because they are charged by experts with lack of the knowledge or conceptual resources to provide adequate interpretations of what is in transit... Stakeholders who are not experts might have developed associations and thus different kinds of knowledge from experts but their information is likewise highly relevant to heritage" (Holtorf et al. 2019, p. 134). Article 4 of the UNESCO Faro 2005 Convention on the Value of Cultural Heritage to Society states that:

(a) Everyone, alone or collectively, has the right to benefit from the cultural heritage and contribute toward its enrichment;
(b) Everyone, alone or collectively, has the responsibility to respect the cultural heritage of others as much as their own heritage, and consequently the common heritage of Europe;
(c) Exercise of the right to cultural heritage may be subject only to those restrictions which are necessary in a democratic society for the protection of the public interest and the rights and freedoms of others.

In demanding that site interpretation has input for local and descendant populations we may help reduce the amount of epistemic injustice done to populations, thus increasing the site's cognitive value. With UNESCO's help we can put local communities at the center of heritage management.

We need to ask for whom do we preserve this monument or site? Currently these sites are used to acquire tourists' dollars, at the expense of the local populations. Hard questions need to be answered, and quickly. How many people are too many people? Once figuring out the numbers the site can sustainably hold, a limit needs to be set and adhered to. How long do tourists stay and how much money do they spend on other goods to local community members? In the case of Angkor and George Town the locals are not seeing the benefit of tourist dollars. Finally, how does their presence impact local culture and the local environment? Countries, cities, and archaeological sites need to think about their local to tourist ratio. The more tourists a place welcomes, the more likely the site will be impacted by those tourists. A country or site with a high 'tourism density index' will adversely affect local culture and environment, to the ultimate detriment of tourist and local alike. More tourism does not mean better tourism. As we've seen, unregulated tourism leads to UNESCO-cide and Disneyfication of culture.

13.6 Conclusions

All of us interested in the intersection of archaeology, culture, and tourism are stakeholders in this ethical debate. UNESCO World Heritage designation has caused damage to the sites I discussed in this chapter. Although branded as a WH site, George Town does little to avail its unique heritage to tourists. Nor does the site offer many benefits to the local and descendant populations. With no or little contextualization of the site, and a bevy of tourist-only food and souvenir shops, tourisms will wane. The locals' worries are well founded; without a plan in place the identity and culture of the site will be lost. Angkor Wat Archaeological Complex fairs no better. Angkor's locals are losing their livelihood. And although tourism is increasing at the site, without better water management strategy, the archaeological complex will suffer great losses.

I've argued that the problem cannot be solved with money alone. Privatizing cultural heritage may lead to disastrous results. Digital Cultural Heritage raises its own worries of 'digital colonialism' among other concerns. I've also argued that throwing the problem back to local control (as many critics of UNESCO have done) is nothing more than a buck-passing solution. The most plausible solutions to the worries of UNESCO-cide and the Disneyfication of culture are perhaps the hardest to advocate for: common sense restrictions on unregulated numbers of tourists, adopting more sites of intangible culture, providing tourists with a deeper and richer sense of the site by telling various stories from disparate populations.

Overtourism is in no danger of ceasing. If we want to preserve monuments for the future while allowing cultures to thrive and change, we will have to discourage the sort of tourism which got us into this mess in the first place. In a world of selfie-stick tourism, UNESCO World Heritage sites are in a unique position to offer tourists a deep and rich experience of the world.

References

Aitchison C (1999) Heritage and nationalism: gender and the performance of power. In: Crouch D (ed) Leisure/tourism geographies: practices and geographical knowledge. Routledge, London/New York, pp 59–73

Appiah KA (2006) Cosmopolitanism: ethics in a world of strangers. Norton, London/New York

Barron L (2017) 'Unesco-cide': does world heritage status do cities more harm than good? The Kathmandu Post, September 2, 2017

Becker E (2017) Only governments can stem the tide of tourism sweeping the globe. The Guardian, August 5, 2017

Benhamou F (2010) L'inscription au patrimoine mondial de l'humanité. Rev Tiers Monde 202(2):113–130

Boukari S (2002) Angkor's role in the search for a lost unity. The UNESCO New Courier, May 2002, pp 13–15

Bourdeau L, Gravari-Barbas M, Robinson M (eds) (2015) World heritage, tourism and identity: inscription and co-production. Routledge, London/New York

Candelaria MAF (2005) The Angkor sites of Cambodia: the conflicting values of sustainable tourism and state sovereignty. Brooklyn J Int Law 31(1):253–288

Caust J (2017) Is UNESCO World Heritage status for cultural sites killing the things it loves? The Conversation. http://theconversation.com/is-unesco-world-heritage-status-for-cultural-sites-killing-the-things-it-loves-98546. Accessed 13 May 2019

Caust J, Vecco M (2017) Is UNESCO World Heritage recognition a blessing or burden? Evidence from developing Asian countries. J Cult Herit 27:1–9

Cavendish R (ed) (2008) 1001 historic sites you must see before you die. BES Publishing, Hauppauge

Chan S (2018) Ticket revenue at Angkor soars to more than $85 million. Khmer Times, October 2, 2018

Chheang V (2010) Tourism and local community development in Siem Reap. Ritsumeikan J Asia Pac Stud 27:85–101

Clodd HP (1948) Malaya's first British pioneer: the life of Francis light. Luzac and Company, London

Cuno J (2012) Whose culture? The promise of museums and the debate over antiquities. Princeton University Press, Princeton

D'Eramo M (2014) Urbanicide in all good faith. Domus, August 20, 2014

Daly J (2019) Site where Julius Caesar was stabbed will finally open to the public. Smithsonian Magazine, March 5, 2019

Deng I (2018) China's crumbling Great Wall is getting some hi-tech conservation help from drones. South China Morning Post, May 2, 2018

Faiola A (2014) Fears that corporate cash will lead to Disneyfication of Italy's cultural heritage? News Network Archaeology, September 7, 2014

Fricker M (2007) Epistemic injustice: power and the ethics of knowing. Oxford University Press, Oxford

González Tirados RM (2011) Half a century of mass tourism: evolution and expectations. Serv Ind J 31(10):1589–1601

Goyal S (2018) Blog: the Red Fort sponsorship controversy. India Campaign, April 30, 2018. https://www.campaignindia.in/article/blog-the-red-fort-sponsorship-controversy/444264. Accessed 13 May 2019

Gravari-Barbas M, Bourdeau L, Robinson M (2015) World heritage and tourism: from opposition to co-production. In: Bourdeau L, Gravari-Barbas M, Robinson M (eds) World heritage, tourism and identity: inscription and co-production. Routledge, London/New York, pp 1–24

Gray DD (2016) UNESCO World Heritage sites and the downside of cultural tourism. Skift, January 28, 2016. https://skift.com/2016/01/28/unesco-world-heritage-sites-and-the-downside-of-cultural-tourism/. Accessed 13 May 2019

Hammer J (2016) The lost city of Cambodia. Smithsonian Magazine. https://www.smithsonian-mag.com/history/lost-city-cambodia-180958508/. Accessed 13 May 2019

Holtorf C, Pantazatos A, Scarre G (eds) (2019) Cultural heritage, ethics and contemporary migrations. Routledge, London/New York

Kitching C (2014) Tourist who smashed Buddha statue while 'hearing voices' inside Cambodia's Angkor Wat says it was fake and didn't belong there. Daily Mail Australia, October 14, 2014

Labadi S (2007) Representations of the nation and cultural diversity in discourses on world heritage. J Soc Archaeol 7(2):147–170

Manjoo F (2018) 'Overtourism' worries Europe: how much did technology help get us there? The New York Times, August 29, 2018

Matthes EH (2016) The ethics of historic preservation. Philos Compass 11(12):786–794. https://doi.org/10.1111/phc3.12379

McWhinnie L (2015) 8 Angkor Wat facts that will blow your mind. The Collective, December 23, 2015. http://blog.topdeck.travel/8-angkor-wat-facts-will-blow-mind/. Accessed 13 May 2019

Merryman JH (1986) Two ways of thinking about cultural property. Am J Int Law 80(4):831–853

Miura K (2010) World heritage sites in Southeast Asia: Angkor and beyond. In: Hitchcock M, King VT, Parnwell M (eds) Heritage tourism in Southeast Asia. University of Hawaii Press, Honolulu, pp 103–129

Morgan NJ, Pritchard A, Piggott R (2003) Destination branding and the role of the stakeholder: the case of New Zealand. J Vacat Mark 9(3):285–299

Neth B (2011) World heritage Angkor and beyond: circumstances and implications of UNESCO listings in Cambodia. Göttingen University Press, Göttingen

Olson E (2018) How heritage became sexy (and what that means for the future of conservation). Global Heritage Fund, July 24, 2018. https://globalheritagefund.org/2018/07/24/how-heritage-became-sexy-future-conservation/. Accessed 13 May 2019

Pantazatos A (2017) Epistemic injustice and cultural heritage. In: Kidd IJ, Medina J, Pohlhaus G (eds) Routledge handbook of epistemic injustice. Routledge, New York, pp 370–386

Pianigiani G, Yardley J (2014) Corporate medicis to the rescue. The New York Times, July 15, 2014

Ryan J, Silvanto S (2009) The world heritage list: the making and management of a brand. Place Brand Public Dipl 5(4):290–300

Salazar NB (2010) From local to global (and back): towards local ethnographies of cultural tourism. In: Richards G, Musters W (eds) Cultural tourism research methods. CABI, Oxfordshire, pp 188–198

Seglow J (2019) Cultural heritage, minorities and self-respect. In: Holtorf C, Pantazatos A, Scarre G (eds) Cultural heritage, ethics and contemporary migrations. Routledge, New York, pp 13–26

Sharma JP (2018) Alphons writes to culture ministry to expedite heritage adoption scheme. Hindustan Times, June 22, 2018. https://www.hindustantimes.com/india-news/alphons-writes-to-culture-ministry-to-expedite-heritage-adoption-scheme/story-HbgtYs4oxkpx7oAJb0mgGN.html. Accessed 13 May 2019

Shortliffe SE (2015) Gender and (world) heritage: the myth of a gender neutral heritage. In: Bourdeau L, Gravari-Barbas M, Robinson M (eds) World heritage, tourism and identity: inscription and co-production. Routledge, London/New York, pp 107–120

Sok C (2017) Angkor water crisis. The UNESCO Courier: Many Voices, One World. https://en.unesco.org/courier/2017-april-june/angkor-water-crisis. Accessed 13 May 2019

Sopheakpanha N (2019) Cambodian gov't complains to Amazon over 'Angkor Wat toilet products'. VOA Khmer, January 21, 2019. https://www.voacambodia.com/a/cambodian-government-complains-to-amazon-over-ankor-wat-toilet-products/4752123.html. Accessed 13 May 2019

Sotheary P (2018) Angkor Temple voted world's topmost site. Khmer Times, August 6, 2018. https://www.khmertimeskh.com/519338/angkor-temple-voted-worlds-topmost-site/. Accessed 13 May 2019

Tourtellot J (2017) 'Overtourism' plagues great destinations; here's why. National Geographic, October 29, 2017

Vecco M (2019) Value and values of cultural heritage. In: Campelo A, Reynolds L, Lindgreen A, Beverland M (eds) Cultural heritage. Routledge, New York, pp 23–38

Vernooy R (ed) (2006) Social and gender analysis in natural resource management: learning studies and lessons from Asia. Sage, London

Young JO (2013) The values of the past. In: Scarre G, Coningham R (eds) Appropriating the past: philosophical perspectives on the practice of archaeology. Cambridge University Press, Cambridge, pp 25–41

Printed in the United States
by Baker & Taylor Publisher Services